Lecture Notes in Computer Science 16073

Founding Editors

Gerhard Goos
Juris Hartmanis

AF167585

The series Lecture Notes in Computer Science (LNCS), including its subseries Lecture Notes in Artificial Intelligence (LNAI) and Lecture Notes in Bioinformatics (LNBI), has established itself as a medium for the publication of new developments in computer science and information technology research, teaching, and education.

LNCS enjoys close cooperation with the computer science R & D community, the series counts many renowned academics among its volume editors and paper authors, and collaborates with prestigious societies. Its mission is to serve this international community by providing an invaluable service, mainly focused on the publication of conference and workshop proceedings and postproceedings. LNCS commenced publication in 1973.

Golnaz Badkobeh · Jakub Radoszewski ·
Nicola Tonellotto · Ricardo Baeza-Yates
Editors

String Processing and Information Retrieval

32nd International Symposium, SPIRE 2025
London, UK, September 8–11, 2025
Proceedings

 Springer

Editors

Golnaz Badkobeh ⓘ
University of London
London, UK

Nicola Tonellotto ⓘ
University of Pisa
Pisa, Italy

Jakub Radoszewski ⓘ
University of Warsaw
Warsaw, Poland

Ricardo Baeza-Yates ⓘ
KTH, Royal Institute of Technology
Stockholm, Sweden

Universitat Pompeu Fabra
Barcelona, Spain

University of Chile
Santiago, Chile

ISSN 0302-9743   ISSN 1611-3349 (electronic)
Lecture Notes in Computer Science
ISBN 978-3-032-05227-8   ISBN 978-3-032-05228-5 (eBook)
https://doi.org/10.1007/978-3-032-05228-5

This Springer imprint is published by the registered company Springer Nature Switzerland AG
The registered company address is: Gewerbestrasse 11, 6330 Cham, Switzerland

If disposing of this product, please recycle the paper.

Preface

The 32nd International Symposium on String Processing and Information Retrieval (SPIRE) was held on September 8–11, 2025, in London (UK), followed by the 19th Workshop on Compression, Text, and Algorithms (WCTA), held on September 11, 2025.

SPIRE started in 1993 as the South American Workshop on String Processing. It was held in Latin America until 2000. Then, SPIRE moved to Europe; overall, the conference has been held in Australia, Argentina, Bolivia, Brazil, Chile, Colombia, Finland, France, Israel, Italy, Japan, Mexico, Peru, Portugal, Spain, the UK, and the USA. SPIRE continues the long and well-established tradition of encouraging high-quality research at the broad nexus of string processing, information retrieval, and computational biology.

This volume contains the accepted papers presented at SPIRE 2025. SPIRE 2025 received a total of 42 submissions, 34 long papers and 8 short papers. Each submission received three single-blind reviews. After the discussion phase, the Scientific Program Committee accepted 17 long papers and 6 short papers. We thank all the authors for their valuable contributions and presentations at the conference and thank the Program Committee members and additional reviewers for their valuable work during the review and discussion phases. We also thank the members of the Local Organizing Committee for their support in organizing SPIRE.

We appreciate the high-quality talks included in the scientific program from three renowned researchers: Hideo Bannai (Institute of Science Tokyo, Japan), Camille Marchet (CNRS, Lille, France), and Rajeev Raman (University of Leicester, UK). This edition also had a Best Paper Award, sponsored by Springer. The award was announced during the conference.

We thank our sponsors: Springer, Web4Good, and Department of Computer Science at City St George's, University of London. Their generous support was instrumental in making this conference a reality, fostering academic excellence and enabling us to bring together a diverse group of researchers and students. Finally, we thank Springer for publishing the proceedings of SPIRE 2025 in the LNCS series.

September 2025

Golnaz Badkobeh
Jakub Radoszewski
Nicola Tonellotto
Ricardo Baeza-Yates

Organization

General Chairs

Golnaz Badkobeh   City St George's, University of London, UK
Ricardo Baeza-Yates   KTH, Royal Institute of Technology, Sweden, Universitat Pompeu Fabra, Spain, and University of Chile, Chile

Program Committee Chairs

Golnaz Badkobeh   City St George's, University of London, UK
Jakub Radoszewski   University of Warsaw, Poland
Nicola Tonellotto   University of Pisa, Italy

Steering Committee

Diego Arroyuelo   Pontificia Universidad Católica de Chile and Millennium Institute for Foundational Research on Data, Chile

Ricardo Baeza-Yates (Chair}   KTH, Royal Institute of Technology, Sweden, Universitat Pompeu Fabra, Spain, and University of Chile, Chile

Zsuzsanna Lipták   University of Verona, Italy
Franco Maria Nardini   ISTI-CNR, Italy
Nadia Pisanti   University of Pisa, Italy
Barbara Poblete   University of Chile, Chile and Amazon, USA
Berthier Ribeiro-Neto   Federal University of Minas Gerais, Brazil
Edleno Silva de Moura   Federal University of Amazonas and Jusbrasil, Brazil
Rossano Venturini   University of Pisa, Italy
Nivio Ziviani   Federal University of Minas Gerais, Brazil

Program Committee

Christina Boucher	University of Florida, USA
Laurent Bulteau	CNRS - Université Paris-Est Marne-la-Vallée, France
Panagiotis Charalampopoulos	Birkbeck, University of London, UK
Jonas Ellert	ENS Paris, France
Johannes Fischer	TU Dortmund, Germany
Paweł Gawrychowski	University of Wrocław, Poland
Daniel Gibney	Georgia Tech, USA
Meng He	Dalhousie University, Canada
Stepan Holub	Charles University, Czech Republic
Wing-Kai Hon	National Tsing Hua University, Taiwan
Dominik Kempa	Stony Brook University, USA
Tomasz Kociumaka	Max Planck Institute for Informatics, Germany
Dominik Köppl	University of Yamanashi, Japan
Dmitry Kosolobov	Ural Federal University, Russia
Susana Ladra	University of A Coruña, Spain
Thierry Lecroq	Université de Rouen Normandie, France
Moshe Lewenstein	Bar-Ilan University, Israel
Zsuzsanna Lipták	University of Verona, Italy
Felipe A. Louza	Universidade Federal de Uberlândia, Brazil
Takuya Mieno	University of Electro-Communications, Japan
Gonzalo Navarro	University of Chile, Chile
Giulio Ermanno Pibiri	Università Ca' Foscari Venezia, Italy
Nadia Pisanti	University of Pisa, Italy
Solon P. Pissis	CWI Amsterdam, The Netherlands
Nicola Prezza	Ca' Foscari University of Venice, Italy
Simon J. Puglisi	University of Helsinki, Finland
Arseny Shur	Bar-Ilan University, Israel
Blerina Sinaimeri	Università LUISS Guido Carli, Italy
Jouni Sirén	University of California, Santa Cruz, USA
Rossano Venturini	University of Pisa, Italy
Oren Weimann	University of Haifa, Israel

Additional Reviewers

Jarno Alanko	Adrián Gómez Brandón
Gabriel Bathie	Alessio Campanelli
Giulia Bernardini	Arnab Ganguly
Itai Boneh	Samah Ghazawi

Shay Golan
Jackson Huffstutler
Tomohiro I
Sung-Hwan Kim
Ragnar Groot Koerkamp
Manal Mohamed
Yuto Nakashima
Yakov Nekrich
Daniel Puttini
Narad Rampersad
Brian Riccardi

Guido Rocchietti
Daniel Saad
Rahul Shah
Jorma Tarhio
Guilherme Telles
Che-Wei Tsao
Cristian Urbina
Rahul Varki
Kaiyu Wu
Wiktor Zuba

Abstract of Invited Talks

New Perspectives on the Burrows–Wheeler Transform

Hideo Bannai ⓘ

M&D Data Science Center, Institute of Integrated Research, Institute of Science
Tokyo, Japan
hdbn.dsc@tmd.ac.jp

Abstract. The Burrows–Wheeler Transform (BWT) is a mapping from a given string to the string obtained by concatenating the last symbols of all rotations of the string, in the lexicographic order of the rotations. BWT is well known for its countless applications in data compression and text indexing. The "magic" of BWT stems from two important properties: (1) the reversibility of the transform (albeit only up to rotations of the string), and (2) the so-called "clustering effect" in which the transformed string is somehow "easier to compress".

It is important to note, however, that the BWT is not a bijection: Reversing a BWT image requires additional information, typically in the form of a special end-of-string symbol. Furthermore, there exist strings that are not BWT images in the first place. The bijective BWT (BBWT) addresses this issue, and can be described as the extended BWT (eBWT) applied to the Lyndon factors of the Lyndon factorization of the input string. BBWT can be considered as a generalization of BWT as they are equivalent for strings that are lexicographically smallest rotations. In this talk, we discuss recent results on the BWT from the perspectives of bijectivity (via BBWT) and compressiveness of the transformed string, showing that BBWT retains many properties of the BWT while introducing additional structure that enables novel compression schemes as well as new analyses on quantifying the clustering effect of the transform in terms of repetitiveness measures based on dictionary compression.

Scaling Genomic Reuse: Hypothesis and Algorithms for k-mer Collections

Camille Marchet [iD]

CRIStAL, University of Lille & CNRS
`camille.marchet@univ-lille.fr`

Abstract. The rapid growth of genomic sequencing has created an urgent demand for methods that not only store massive amounts of data efficiently but also support its effective reuse across analyses. At the heart of many such methods lies the concept of k-mer sets, compact representations of sequence content that enable fast querying, indexing, and comparison. This talk will explore recent advances in data structures and algorithms designed to manage large collections of k-mer sets, and will discuss scalability, dynamic updates, and query efficiency. We will highlight innovations that bridge theory and practice, from algorithmic breakthroughs to real-world applications in areas such as RNA-seq analysis in clinical research.

Succinct Dynamic Data Structures (25 Years on)

Rajeev Raman 🆔

School of Computing and Mathematical Sciences, University of Leicester, Leicester
LE1 7RH, UK
r.raman@leicester.ac.uk
https://le.ac.uk/people/rajeev-raman

Abstract. Succinct data structures (SDS) store data in a compact way, often using information-theoretically optimal space (to within lowerorder terms). In addition, they support queries on the data, often in time comparable to their classical (non-succinct) counterparts. This theoretical performance is often reflected in practice; indeed, due to their much smaller memory footprint, SDS can often considerably outperform classical data structures in terms of speed. This win-win situation has led to numerous real-world applications of SDS.

Unfortunately, the previous characterization applies only to *static* SDS. In other words, the entire data is given in advance, is pre-processed, and is then queried — if the data changes, the pre-processing must be performed anew. While this is acceptable for many applications, it is not the most typical use case, where updates to data are interspersed among the queries. To deal with the more general case, many authors have published papers on Dynamic SDS (DSDS). Unfortunately, DSDS run into time lower bounds that make *both* query and update operations slower than their classical counterparts, usually by a multiplicative logarithmic factor. This logarithmic factor is not only theoretically unpleasant, it is also a major factor in ensuring that the speed of DSDS in practice is not competitive with their classical counterparts (typically, DSDS are unable to fully utilize data locality). In this talk, I will survey the 25-year history of DSDS, and discuss approaches to ensuring that DSDS can indeed close the speed gap with their classical counterparts.

Contents

Testing Quasiperiodicity 1
  Christine Awofeso, Ben Bals, Oded Lachish, and Solon P. Pissis

KeBaB: k-mer Based Breaking for Finding Long MEMs 10
  Nathaniel K. Brown, Lore Depuydt, Mohsen Zakeri, Anas Alhadi,
  Nour Allam, Dove Begleiter, Nithin Bharathi Kabilan Karpagavalli,
  Suchith Sridhar Khajjayam, Hamza Wahed, Travis Gagie,
  and Ben Langmead

Analysing New Entropy Measures for Tries 18
  Lorenzo Carfagna and Carlo Tosoni

Depth First Representations of k^2-trees 28
  Gabriel Carmona and Giovanni Manzini

Dorst–Smeulders Coding for Arbitrary Binary Words 45
  Alessandro De Luca and Gabriele Fici

Prefix-Free Parsing for Merging Big BWTs 54
  Diego Díaz-Domínguez, Travis Gagie, Veronica Guerrini,
  Ben Langmead, Zsuzsanna Lipták, Giovanni Manzini,
  Francesco Masillo, and Vikram Shivakumar

RLZ-r and LZ-End-r: Enhancing Move-r 64
  Patrick Dinklage, Johannes Fischer, Lukas Nalbach, and Jan Zumbrink

Massively Parallel Computation of Matching Statistics 79
  Anastasia C. Diseth, Keijo Heljanko, and Simon J. Puglisi

Cache-Friendly Compressed Boolean Matrices 95
  Antonio Fariña, Adrián Gómez-Brandón, Asunción Gómez-Colomer,
  and Gonzalo Navarro

Tight Additive Sensitivity on LZ-Style Compressors and String Attractors 109
  Yuto Fujie, Hiroki Shibata, Yuto Nakashima, and Shunsuke Inenaga

On the Number of MUSs Crossing a Position 124
  Hiroto Fujimaru, Takuya Mieno, and Shunsuke Inenaga

String Consensus Problems with Swaps and Substitutions 133
   Estéban Gabory, Laurent Bulteau, Gabriele Fici, and Hilde Verbeek

Two-Player Communication Complexity of Pattern Matching 148
   Paweł Gawrychowski and Wojciech Janczewski

REINDEER2: Practical Abundance Index at Scale 156
   *Yohan Hernandez–Courbevoie, Mikaël Salson, Chloé Bessière,
   Haoliang Xue, Daniel Gautheret, Camille Marchet,
   and Antoine Limasset*

Efficient Computation of Closed Substrings 172
   Samkith K. Jain and Neerja Mhaskar

Nyldon Factorization of Thue-Morse Words and Fibonacci Words 188
   *Kaisei Kishi, Kazuki Kai, Yuto Nakashima, Shunsuke Inenaga,
   and Hideo Bannai*

String Matching with a Dynamic Pattern 202
   Bruno Monteiro and Vinicius dos Santos

Smallest Suffixient Sets as a Repetitiveness Measure 217
   Gonzalo Navarro, Giuseppe Romana, and Cristian Urbina

Longest Unbordered Factors on Run-Length Encoded Strings 233
   Shoma Sekizaki and Takuya Mieno

Longest Common Subsequence in K-Length Substrings for Run-Length
Encoded Strings 248
   B. Riva Shalom, Eitan Kondratovsky, and Ely Porat

Practical Algorithms for Hierarchical Overlap Graphs 265
   Saumya Talera, Parth Bansal, Shabnam Khan, and Shahbaz Khan

Counting Distinct (Non-)crossing Substrings 281
   *Haruki Umezaki, Hiroki Shibata, Dominik Köppl, Yuto Nakashima,
   Shunsuke Inenaga, and Hideo Bannai*

Faster Algorithm for Bounded Damerau–Levenshtein Distance 291
   Ryosuke Yamano and Tetsuo Shibuya

Author Index 305

Testing Quasiperiodicity

Christine Awofeso[1][iD], Ben Bals[2,3][iD], Oded Lachish[1][iD],
and Solon P. Pissis[2,3(✉)][iD]

[1] Birkbeck, University of London, London, UK
[2] CWI, Amsterdam, The Netherlands
[3] Vrije Universiteit, Amsterdam, The Netherlands
solon.pissis@cwi.nl

Abstract. A cover (or quasiperiod) of a string S is a shorter string C such that every position of S is contained in some occurrence of C as a substring. The notion of cover was introduced by Apostolico and Ehrenfeucht over 30 years ago [Theor. Comput. Sci. 1993] and it has received significant attention from the combinatorial pattern matching community. In this note, we show how to efficiently test whether S admits a cover. We design an algorithm that, given $n = |S|$, $q \in [n]$, $\epsilon \in \mathbb{R}^+$, and oracle access to S, uses $\mathcal{O}(q^3 \epsilon^{-1} \log q)$ letter queries to test whether S has a cover C of length at most q or is ϵ-far from having such a cover. Our insights also lead to a simple streaming algorithm for short covers.

Keywords: Property testing · Quasiperiodicity · Cover · Seed

1 Introduction

A cover (or quasiperiod) of a string S is a shorter string C such that every position of S is contained in some occurrence of C as a substring. The notion of cover generalizes the notion of period. It was introduced by Apostolico and Ehrenfeucht in 1993 [1], and since then it has received a lot of attention from the combinatorial pattern matching community. For example, the shortest cover of a string of length n can be computed in $\mathcal{O}(n)$ time [2,4]; see [5,15] for surveys.

In this note, we present a tester to determine whether a string S of length n has a cover of length at most q or the minimum Hamming distance of S and a string that has such a cover is at least ϵn, for some small $\epsilon \in \mathbb{R}^+$. The tester does not access S directly and instead uses queries to an oracle of the form: *what is the letter at position $i \in [n]$ of S?* Our algorithm uses $\mathcal{O}(q^3 \epsilon^{-1} \log q)$ such queries, which is independent of n. It is randomized and can provide false positives sometimes; see Sect. 3. Notably, our combinatorial insights yield a simple streaming algorithm for short covers; see Sect. 4. We start with Sect. 2, which provides the necessary notation, definitions, and tools. Our work proceeds along the lines of [14], where the authors provide testers for periodicity.

G. Badkobeh et al. (Eds.): SPIRE 2025, LNCS 16073, pp. 1–9, 2026.
https://doi.org/10.1007/978-3-032-05228-5_1

2 Preliminaries

We consider finite strings on an integer *alphabet* $\Sigma = [\sigma] = \{1, 2, \ldots, \sigma\}$. The elements of Σ are called *letters*. For a string $S = S[1] \cdots S[n]$ on Σ, its *length* is $|S| = n$. For any $1 \leq i \leq j \leq n$, the string $S[i] \cdots S[j]$ is called a *substring* of S. By $S[i \mathinner{.\,.} j]$, we denote its occurrence at the (starting) position i, and we call it a *fragment* of S. When $i = 1$, this fragment is called a *prefix*, and when $j = n$, it is called a *suffix*. The *Hamming distance* of two strings $S, S' \in \Sigma^n$ is $|\{i \in [n] : S[i] \neq S'[i]\}|$. An integer p, $1 \leq p \leq n$, is a *period* of a string S if $S[i] = S[i+p]$, for all $1 \leq i \leq |S| - p$. A string B is a *border* of S if it is a prefix and a suffix of S. The notion of cover generalizes the notion of period.

Definition 1 (Cover and Quasiperiod). *A string C is a* cover *of a string S if every position in S lies within an occurrence of C as a substring; that is, for all $i \in [|S|]$, there is an occurrence of C in S that starts at one of*

$$\max(i - |C| + 1, 1), \ldots, \min(i, |S| - |C| + 1).$$

If C is a cover of S, we say that S has a quasiperiod *$|C|$.*

Example 1. $C = \mathsf{aba}$ and $C' = \mathsf{abaaba}$ are covers of $S = \mathsf{abaababaababaaba}$.

For any $q \in \mathbb{N}$, by $\mathrm{QP}_\Sigma(q)$ we denote the set of all strings on Σ with a quasiperiod at most q. Since every cover of a string must be a border, any string $S \in \Sigma^n$ has $\mathcal{O}(n)$ covers, and, in fact, all covers of S can be computed in $\mathcal{O}(n)$ time [17,18]. The notion of seed [10] generalizes the notion of cover.

Definition 2 (Seed). *A string C is a* seed *of a string S, if $|C| \leq |S|$ and C is a cover of some string containing S as a substring.*

Example 2. $C = \mathsf{aba}$ and $C' = \mathsf{abaab}$ are seeds of $S = \mathsf{aabaababaababaabaa}$.

Unlike covers, the number of distinct seeds of $S \in \Sigma^n$ can be $\Theta(n^2)$ [13]. Theorem 1 allows checking whether any fragment of S is a seed efficiently.

Theorem 1 ([12,19]). *An $\mathcal{O}(n)$-size representation of all seeds of a string $S \in \Sigma^n$ can be computed in $\mathcal{O}(n \log n)$ time and $\mathcal{O}(n)$ space. If $\Sigma = [\sigma]$, with $\sigma = n^{\mathcal{O}(1)}$, the same representation can be computed in $\mathcal{O}(n)$ time.*

We use simple tools from Diophantine number theory in our analysis.

Definition 3 (Conical Combination). *A* conical combination *of the natural numbers $a_1, \ldots, a_k$ is a number $n = x_1 a_1 + \cdots + x_k a_k$, for some $x_1, \ldots, x_k \in \mathbb{N}$.*

This well-known result of Erdős and Graham underlies our algorithm.[1]

---

[1] In particular, the result for 6, 9, and 20 is famous as the Chicken McNugget theorem. It became a trend among Mathematics enthusiasts to order the largest number of Chicken McNuggets that could not be formed from the offered box sizes [9].

Theorem 2 (Frobenius Number Bound [6]). *Let $a_1 < \cdots < a_k \in \mathbb{N}^+$ be set-wise co-prime (i.e., $\gcd(a_1, \ldots, a_k) = 1$). Any number $n > 2a_{k-1}\lfloor a_k/k \rfloor - a_k$ can be written as $n = x_1 a_1 + \cdots + x_k a_k$, for some $x_1, \ldots, x_k \in \mathbb{N}$.*

Note that Theorem 2 does not require the numbers to be pairwise co-prime.

Corollary 1 ([6]). *Let $A \subseteq \mathbb{N}^+$ be bounded by $q \in \mathbb{N}$ and assume $\gcd(A) = 1$. Then any number $n \geq 2q^2$ can be written as a conical combination of A.*

Corollary 2. *Let $A \subseteq \mathbb{N}^+$ be bounded by $q \in \mathbb{N}$. Then any number $n \geq 2q^3$ such that $\gcd(A) | n$ can be written as a conical combination of A.*

Proof. Apply Corollary 1 to $A' := \{a/\gcd(A) \mid a \in A\}$ and $n' := n/\gcd(A)$. Multiply the resulting conical combination by $\gcd(A)$. □

Definition 4 (q-Cover Tester). *A q-cover tester is a randomized algorithm that receives as input $q, n \in \mathbb{N}$ and $\epsilon \in \mathbb{R}^+$ and has oracle access to a string $S \in \Sigma^n$. It returns YES if S has a quasiperiod at most q and NO with probability at least $3/4$ if S is ϵ-far from having a quasiperiod at most q; i.e., the minimum Hamming distance of S and a string S' that has such a cover is at least ϵn.*

A q-cover tester does not access the input string S directly; instead, it uses queries to the oracle. A *query* is an integer $i \in [n]$ provided to the oracle, on which the oracle returns the letter $S[i]$. The *query complexity* of a tester is the maximum number of queries it uses as a function of the parameters q, n, and ϵ. When analyzing a q-cover tester, we are particularly interested in its query complexity; namely, we count only the number of distinct queries.

3   An Efficient Tester for Covers and Seeds

Let us fix a string $S \in \Sigma^n$ and a string $C \in \Sigma^q$, for two integers $0 < q < n$. We wish to establish results that help us test whether C is a cover of S.

Observation 1. Let C be a seed of S. Then there is a string $S' = X \cdot S \cdot Y$, with $|X| \in [0, q]$ and $|Y| \in [0, q]$, such that C is a cover of S'.

Observation 2. If C is a cover of S, then C is a seed of any substring of S whose length is at least $|C|$.

Lemma 1. *Let $S_1 = S[i \mathinner{.\,.} j]$ and $S_2 = S[i' \mathinner{.\,.} j']$ be two fragments of S, with $i \leq i'$ and $j \leq j'$, so that they (1) share at least $2q$ positions of S and (2) both have $C \in \Sigma^q$ as a seed. Let $S_3 := S[i \mathinner{.\,.} j']$. Then C is also a seed of S_3.*

Proof. To construct a covering of S_3 with seed C, take such coverings for S_1 and S_2; see Fig. 1 for an illustration. From the covering of S_1, remove the occurrences of C that start after $j - q$. From the covering of S_2, remove the occurrences of C that start before i'. Since $j - i' + 1 \geq 2q$, we have that the union of these coverings now covers all of S_3. Thus, C is a seed of S_3. □

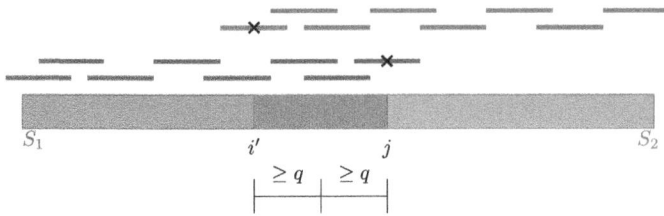

Fig. 1. Combining the seed occurrences of two fragments with a long overlap.

Definition 5 (Period set). *The period set of C, denoted by $\mathrm{PS}(C)$, is the set of periods of C. Thus, $\mathrm{PS}(C) \subseteq [[|C|]]$.*

We are interested in the ways we can combine copies of C to form a longer string. The trivial way is to concatenate C with itself: the length of C is a period of C. The period set tells us which other ways are possible: for every $\lambda \in \mathrm{PS}(C)$, we can construct a string of length $|C| + \lambda$ by overlapping C with itself.

Lemma 2. *Let $\ell \geq 2q^3 + q$ such that $\gcd(\mathrm{PS}(C))|\ell$, with $C \in \Sigma^q$. There is a string of length ℓ for which C is a cover.*

Proof. We construct such a string, denoted by S', by overlapping C with itself. Let $\mathrm{PS}(C) = \{a_1, \ldots, a_k\}$ and let $x_1, \ldots, x_k \in \mathbb{N}$ be such that $\sum_{i=1}^{k} x_i a_i = \ell - q$. These coefficients exist by Corollary 2. Start with $S' := C$. Then for each $i \in [k]$, add a copy of C, overlapping by $(q - a_i)$ positions. This is possible by the definition of $\mathrm{PS}(C)$. This extends the string by a_i letters. Perform this x_i times. At the end of this process, S' is coverable by C and has the correct length. □

The following lemma proves that, when C is a cover of S, the period set $\mathrm{PS}(C)$ determines all the possible positions at which C can occur in S.

Lemma 3. *Let $C \in \Sigma^q$ be a cover of S and $P := \{p_1 < \cdots < p_k\}$ the positions at which C occurs in S. Let $P' := \{p_i - 1 \mid i \in [k]\}$. Then $\gcd(\mathrm{PS}(C))| \gcd(P')$.*

Proof. Let $p_i' \in P'$. We will show that $\gcd(\mathrm{PS}(C))|p_i'$, which implies the claim.

Fix an arbitrary covering of S using C. See Fig. 2 for an example. Let e_i be the ending position of the occurrence of C in this covering that ends first at or after position p_i'. Then C is a cover of the string $S[1 \ldots e_i]$, and therefore, by the definition of $\mathrm{PS}(C)$, there must be $\{x_a \in \mathbb{N}\}_{a \in \mathrm{PS}(C)}$ such that $e_i = \sum_{a \in \mathrm{PS}(C)} x_a a$. Since there is an occurrence of C in S at positions $e_i - q + 1$ and p_i, and $e_i - p_i \leq q$ (by choice of e_i), we have $p_i - e_i + q - 1 \in \mathrm{PS}(C)$ and thus $p_i' - e_i + q \in \mathrm{PS}(C)$. Therefore, p_i' can be written as a conical combination of $\mathrm{PS}(C)$. This implies that $\gcd(\mathrm{PS}(C))|p_i'$. □

Definition 6 (C-Consistent Fragment). *A fragment $S[i \ldots j]$ of S, with $j - i + 1 \geq q$, is C-consistent if and only if (1) C is a seed of $S[i \ldots j]$ and (2) if $p \in [i, j]$ is the position of an occurrence of C in $S[i \ldots j]$, then $\gcd(\mathrm{PS}(C))|p - 1$.*

$$e_i$$
$$p_i' \quad p_i$$

| 1 | 2 | 3 | 4 | 5 | 6 | 7 | 8 | 9 | 10 | 11 | 12 | 13 | 14 |

$S =$ a b b a a b b a b b a b b a

C

$q = |C| = 4$

$\mathrm{PS}(C) = \{3, 4\}$

$p_i - e_i + q - 1 = 3 \in \mathrm{PS}(C)$

$p_i' = 10 = 2 \cdot 3 + 1 \cdot 4$

Fig. 2. $\mathrm{PS}(C)$ determines all the possible positions at which C can occur in S.

Let $\mathcal{S} = \{S_i\}_{i \in [n/(2q^3)]}$ be the set of fragments of S obtained by splitting S into fragments of length $4q^3$ (the last fragment may be shorter than $4q^3$ or empty), each overlapping by $2q^3$ positions. Thus, we have $|\mathcal{S}| = \mathcal{O}(n/q^3)$.

An immediate consequence of Definition 6 is that if a fragment of S is not C-consistent, then C is not a cover of S. Lemma 4 shows that if S is ϵ-far from $\mathrm{QP}_\Sigma(q)$, then this can be detected with high probability by picking a random fragment from set $\mathcal{S}$, which, as we explain next, is what our algorithm does.

Lemma 4. *If $S \in \Sigma^n$ is ϵ-far from $\mathrm{QP}_\Sigma(q)$, for some $q \in \mathbb{N}$, the set $\mathcal{S}' \subseteq \mathcal{S}$ of C-consistent fragments of S, with $C \in \Sigma^q$, has size at most $(1 - \epsilon/2) \cdot n/(2q^3)$.*

Proof. Let $\mathcal{T}$ be the subset of $\mathcal{S}$ that includes all fragments that are not C-consistent. Assume for the sake of contradiction that $|\mathcal{T}| \leq \epsilon n/(4q^3)$. We will show that this implies that S is at Hamming distance less than ϵn from $\mathrm{QP}_\Sigma(q)$, that is, that S is not ϵ-far from $\mathrm{QP}_\Sigma(q)$. In particular, we will show that we can edit S only in fragments that are in the set $\mathcal{T}$ so that C becomes a cover of S. Note that there are at most ϵn positions in $\mathcal{T}$ by the assumption on its size.

Consider any maximal sequence $S_1, \ldots, S_k$ of k consecutive fragments in $\mathcal{T}$. (Note that k could be equal to one.) Let S_{pre} be the fragment in $\mathcal{S}$ before S_1 and S_{post} the one after S_k. Note that we pick S_{pre} and S_{post} so that they have *no overlap* with $S_1, \ldots, S_k$. Since we have assumed that the sequence $S_1, \ldots, S_k$ is maximal, we can assume that $S_{\mathrm{pre}}, S_{\mathrm{post}} \in \mathcal{S}'$, and thus also that C is a seed of S_{pre} and S_{post}. Fix coverings of S_{pre} and S_{post} with seed C; see Fig. 3 for an illustration. For these coverings, let i_{pre} be the first position in S after the covering of S_{pre} by C and i_{post} the last position before the covering of S_{post} by C. By Observation 1, i_{pre} and i_{post} are at most q positions apart from the end and start of S_{pre} and S_{post}, respectively. The fragment $S[i_{\mathrm{pre}} .. i_{\mathrm{post}}]$ can be edited to be coverable by C by replacing it with the string given by Lemma 2. $\square$

S_{pre} $\quad\quad i_{\mathrm{pre}} \quad\quad\quad i_{\mathrm{post}} \quad\quad S_{\mathrm{post}}$

Fig. 3. Editing a string that is not ϵ-far from $\mathrm{QP}_\Sigma(q)$ to be in $\mathrm{QP}_\Sigma(q)$.

Repeating this process, for any maximal sequence of consecutive fragments in $\mathcal{T}$, yields a string that is coverable by C. The coverings of the individual fragments combine to yield a covering of the entire string S by Lemma 1.

Theorem 3. *Algorithm 1 is a tester deciding whether a string $S \in \Sigma^n$ has a cover C of length at most q using $\mathcal{O}(q^3 \epsilon^{-1} \log q)$ queries, where $\epsilon \in \mathbb{R}^+$ is the distance parameter.*

Proof. The query complexity is immediate from the algorithm's definition.

For the correctness, first observe that if S has a cover C of length at most q, then any set of fragments of S is C-consistent (Definition 6). The first condition follows from Observation 2 and the second one from Lemma 3.   □

Algorithm 1: A q-cover tester

Input: $q \in \mathbb{N}, n \in \mathbb{N}, \epsilon \in \mathbb{R}^+$ and oracle access to a string $S \in \Sigma^n$
Output: $b \in \{\text{NO}, \text{YES}\}$
Let $\mathcal{S} := \{S_i\}_{i \in [n/(2q^3)]}$ be the set of fragments of S as defined as above.
Sample $(24 \log q)/\epsilon$ of the length-$(4q^3)$ fragments in $\mathcal{S}$ uniformly at random and also the length-$(4q^3)$ suffix of S. Denote the sampled fragments by set $\mathcal{R}$.
Query the $\mathcal{O}(q^3 \epsilon^{-1} \log q)$ positions of the fragments in $\mathcal{R}$ using the oracle.
Query the $\mathcal{O}(q)$ positions $1, \ldots, q$ and $n - q + 1, \ldots, n$ using the oracle.
For every border C of S, $|C| \leq q$, check whether every $F \in \mathcal{R}$ is C-consistent.
If there is such a border C, output **YES**; otherwise output **NO**.

For the other direction of the correctness, assume that S is ϵ-far from $\text{QP}_\Sigma(q)$. There are q prefixes of S that could be a cover of length at most q. For any such prefix C, the set of fragments in $\mathcal{S}$ that are C-consistent has size at most $(1 - \epsilon/2) \cdot n/(2|C|^3)$ by Lemma 4. Since we sample $(24 \log q)/\epsilon$ of these, we will, with probability at least $3/4$, reject all possible covers. The 24 in this equation comes from the four in our length and a six from a folklore application of Chernoff bounds over the at most q candidate borders; see [16, Theorem 4.4].

Algorithm 1 can be adapted to check if S has a seed by considering the $\mathcal{O}(q^2)$ fragments of $S[1 .. 2q]$ as candidate seeds. Note that, since $2q = \mathcal{O}(q)$, this does not affect the query complexity of the tester. Additionally, the requirement that the first and last fragments in $\mathcal{S}$ must be covered should be slightly relaxed.

Corollary 3. *There is a tester deciding whether a string $S \in \Sigma^n$ has a seed of length at most q using $\mathcal{O}(q^3 \epsilon^{-1} \log q)$ queries, where $\epsilon \in \mathbb{R}^+$ is the distance parameter.*

4   A Simple Streaming Algorithm for Covers via Seeds

Gawrychowski et al. showed a one-pass streaming algorithm for computing the shortest cover of a string of length n that uses $\mathcal{O}(\sqrt{n \log n})$ space and runs in

$\mathcal{O}(n \log^2 n)$ time [8]. One of its routines is a streaming algorithm for computing the length of the shortest cover if it is at most q that uses $\mathcal{O}(q)$ space and runs in $\mathcal{O}(n)$ time. The algorithm underlying Theorem 4 is a fundamentally different and *simple* alternative for the same computation that also uses $\mathcal{O}(q)$ space and runs in $\mathcal{O}(n)$ time w.h.p. It follows from our results in Sect. 3 and can be seen as a different, independently useful consequence of our combinatorial insights.

Theorem 4. *For any string $S \in \Sigma^n$ and any $q \in \mathbb{N}$, there is a one-pass streaming algorithm for computing the shortest cover C of S, if $|C| \leq q$, that uses $\mathcal{O}(q)$ space and runs in $\mathcal{O}(n)$ time w.h.p.*

Proof. We conceptually split S into a set $\mathcal{S}'$ of $\mathcal{O}(n/q)$ fragments of S of length $4q$, each overlapping by $2q$ positions (the last fragment may be shorter). Then, by Lemma 1 and Observation 2, we have that any border C of length at most q of S is a cover of S if and only if C is a seed of all fragments in $\mathcal{S}'$.

   This fact yields a simple streaming algorithm. We start by reading $P := S[1 \,..\, q]$ in memory. The prefixes of P will be our *candidates*. We also insert these q letters in $\mathcal{O}(q)$ total time in a dynamic dictionary $\mathcal{D}$ supporting $\mathcal{O}(1)$-time worst-case look-ups [3]. We then read $S[q+1 \,..\, n]$ from left to right, always storing the last $4q$ letters in memory. By using $\mathcal{D}$, we can assume that $S[q+1 \,..\, n]$ consists only of letters occurring in P (otherwise S cannot be coverable by any prefix of P) and that these letters are mapped onto the range $[q]$. Every time we have a fragment F from $\mathcal{S}'$ in memory, we compute the seeds of F by employing the algorithm of Theorem 1, which takes $\mathcal{O}(q)$ time. In accordance with the $\mathcal{O}(q)$-size representation of the computed seeds [12,19], we can check whether each of the q candidate prefixes is a seed, by first constructing the generalized suffix tree [7] of P and F in $\mathcal{O}(q)$ time (thus finding which prefixes of P occur in F), and then checking whether each of the candidate prefixes $P[1 \,..\, i]$ which occur in F, for $i \in [q]$, is a seed of F in $\mathcal{O}(1)$ time per candidate. In addition to processing all fragments of $\mathcal{S}'$, we also need to check whether any of the remaining prefix candidates is a suffix and thus a border of S. We achieve this simply by computing the borders of string $P\$L$ in $\mathcal{O}(q)$ total time [11], where $\$$ is a letter not in Σ and $L := S[n - q + 1 \,..\, n]$. The total time is thus $\mathcal{O}(q \cdot n/q) = \mathcal{O}(n)$; the algorithm is randomized and works w.h.p. due to the use of dictionary $\mathcal{D}$. $\square$

Acknowledgments. A research visit during which part of the presented ideas were conceived was funded by a Royal Society International Exchanges Award.

References

1. Apostolico, A., Ehrenfeucht, A.: Efficient detection of quasiperiodicities in strings. Theor. Comput. Sci. **119**(2), 247–265 (1993). https://doi.org/10.1016/0304-3975(93)90159-Q
2. Apostolico, A., Farach, M., Iliopoulos, C.S.: Optimal superprimitivity testing for strings. Inf. Process. Lett. **39**(1), 17–20 (1991). https://doi.org/10.1016/0020-0190(91)90056-N

3. Bender, M.A., Conway, A., Farach-Colton, M., Kuszmaul, W., Tagliavini, G.: Iceberg hashing: optimizing many hash-table criteria at once. J. ACM **70**(6), 40:1–40:51 (2023). https://doi.org/10.1145/3625817, https://doi.org/10.1145/3625817
4. Breslauer, D.: An on-line string superprimitivity test. Inf. Process. Lett. **44**(6), 345–347 (1992). https://doi.org/10.1016/0020-0190(92)90111-8
5. Czajka, P., Radoszewski, J.: Experimental evaluation of algorithms for computing quasiperiods. Theor. Comput. Sci. **854**, 17–29 (2021). https://doi.org/10.1016/J.TCS.2020.11.033, https://doi.org/10.1016/j.tcs.2020.11.033
6. Erdős, P., Graham, R.: On a linear Diophantine problem of Frobenius. Acta Arith **21**, 399–408 (1972)
7. Farach, M.: Optimal suffix tree construction with large alphabets. In: 38th Annual Symposium on Foundations of Computer Science, FOCS '97, Miami Beach, Florida, USA, October 19-22, 1997, pp. 137–143. IEEE Computer Society (1997). https://doi.org/10.1109/SFCS.1997.646102, https://doi.org/10.1109/SFCS.1997.646102
8. Gawrychowski, P., Radoszewski, J., Starikovskaya, T.: Quasi-periodicity in streams. In: Pisanti, N., Pissis, S.P. (eds.) 30th Annual Symposium on Combinatorial Pattern Matching, CPM 2019, June 18-20, 2019, Pisa, Italy. LIPIcs, vol. 128, pp. 22:1–22:14. Schloss Dagstuhl - Leibniz-Zentrum für Informatik (2019). https://doi.org/10.4230/LIPICS.CPM.2019.22, https://doi.org/10.4230/LIPIcs.CPM.2019.22
9. Haran, B.: How to order 43 Chicken McNuggets (Frobenius numbers) - Numberphile (2012). https://www.youtube.com/watch?v=vNTSugyS038
10. Iliopoulos, C.S., Moore, D.W.G., Park, K.: Covering a string. Algorithmica **16**(3), 288–297 (1996). https://doi.org/10.1007/BF01955677
11. Knuth, D.E., Jr., J.H.M., Pratt, V.R.: Fast pattern matching in strings. SIAM J. Comput. **6**(2), 323–350 (1977). https://doi.org/10.1137/0206024, https://doi.org/10.1137/0206024
12. Kociumaka, T., Kubica, M., Radoszewski, J., Rytter, W., Walen, T.: A linear time algorithm for seeds computation. In: Rabani, Y. (ed.) Proceedings of the Twenty-Third Annual ACM-SIAM Symposium on Discrete Algorithms, SODA 2012, Kyoto, Japan, January 17-19, 2012, pp. 1095–1112. SIAM (2012). https://doi.org/10.1137/1.9781611973099.86, https://doi.org/10.1137/1.9781611973099.86
13. Kociumaka, T., Kubica, M., Radoszewski, J., Rytter, W., Walen, T.: A linear-time algorithm for seeds computation. ACM Trans. Algorithms **16**(2), 27:1–27:23 (2020). https://doi.org/10.1145/3386369, https://doi.org/10.1145/3386369
14. Lachish, O., Newman, I.: Testing periodicity. Algorithmica **60**(2), 401–420 (2011). https://doi.org/10.1007/S00453-009-9351-Y
15. Mhaskar, N., Smyth, W.F.: Fundam. Info. **190**(1), 17–45 (2022). https://doi.org/10.3233/FI-222164
16. Mitzenmacher, M., Upfal, E.: Probability and computing: randomized algorithms and probabilistic analysis. Cambridge University Press (2005). https://doi.org/10.1017/CBO9780511813603
17. Moore, D.W.G., Smyth, W.F.: An optimal algorithm to compute all the covers of a string. Inf. Process. Lett. **50**(5), 239–246 (1994). https://doi.org/10.1016/0020-0190(94)00045-X
18. Moore, D.W.G., Smyth, W.F.: A correction to "an optimal algorithm to compute all the covers of a string". Inf. Process. Lett. **54**(2), 101–103 (1995). https://doi.org/10.1016/0020-0190(94)00235-Q

19. Radoszewski, J.: Linear time construction of cover suffix tree and applications. In: Gørtz, I.L., Farach-Colton, M., Puglisi, S.J., Herman, G. (eds.) 31st Annual European Symposium on Algorithms, ESA 2023, September 4-6, 2023, Amsterdam, The Netherlands. LIPIcs, vol. 274, pp. 89:1–89:17. Schloss Dagstuhl - Leibniz-Zentrum für Informatik (2023). https://doi.org/10.4230/LIPICS.ESA.2023.89, https://doi.org/10.4230/LIPIcs.ESA.2023.89

KeBaB: *k*-mer Based Breaking for Finding Long MEMs

Nathaniel K. Brown[1], Lore Depuydt[2], Mohsen Zakeri[1], Anas Alhadi[3],
Nour Allam[3], Dove Begleiter[3], Nithin Bharathi Kabilan Karpagavalli[3],
Suchith Sridhar Khajjayam[3], Hamza Wahed[3], Travis Gagie[3]([✉]),
and Ben Langmead[1]

[1] Johns Hopkins University, Baltimore, USA
[2] Ghent University, Ghent, Belgium
[3] Dalhousie University, Halifax, Canada
`travis.gagie@gmail.com`

Abstract. Long maximal exact matches (MEMs) are used in many
genomics applications such as read classification and sequence alignment. Li's ropebwt3 finds long MEMs quickly because it can often ignore
much of its input, skipping matching steps which are redundant to the
final output. In this paper we propose KeBaB, a fast and space efficient *k*-mer filtration step using a Bloom filter. This approach speeds
up MEM-finders such as ropebwt3 even further by letting them ignore
even more, breaking the input into substrings called "pseudo-MEMs"
which are guaranteed to contain all long MEMs. We also show experimentally that KeBaB can accelerate metagenomic classification without
significantly reducing accuracy, either by finding all long MEMs or by
leveraging the filter to find only the long MEMs present in the t longest
pseudo-MEMs.

Keywords: Maximal exact matches · k-mer filtration · Pseudo-MEMs

1 Introduction

A challenge for today's string-matching algorithms is to compute exact matches
with respect to an index over a large, repetitive text. This is a pressing problem in
computational genomics, where databases of reference genomes and pangenomes
are growing very rapidly. One highly practical full-text indexing method for
pangenomes is `ropebwt3` [10], which indexes using a run-length compressed form
of the Burrows-Wheeler Transform of the text. Its strategy for querying the
index involves skipping along the query in the style of Boyer-Moore pattern
matching [3], an idea that was first connected to BWT queries by Gagie [8].

In this paper we propose a fast *k*-mer filtration strategy using a Bloom filter
that allows for more skipping and speeds up MEM-finders such as `ropebwt3` substantially. We call our strategy KeBaB for "*k*-mer based breaking". In Sect. 2 we
briefly review MEM-finding. In Sect. 3, we describe how to break a pattern into

© The Author(s), under exclusive license to Springer Nature Switzerland AG 2026
G. Badkobeh et al. (Eds.): SPIRE 2025, LNCS 16073, pp. 10–17, 2026.
https://doi.org/10.1007/978-3-032-05228-5_2

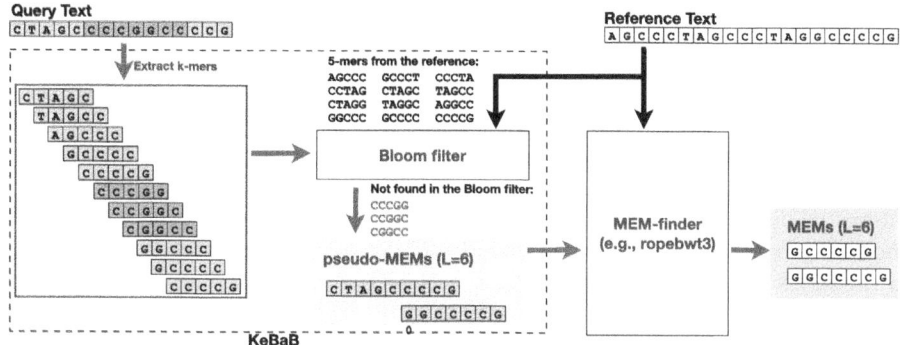

Fig. 1. An example of how to use KeBaB to find pseudo-MEMs.

substrings we call pseudo-MEMs that are guaranteed to contain all sufficiently long MEMs of the pattern with respect to an indexed text. If we are interested only in the t longest MEMs, then we can search in the pseudo-MEMs in non-increasing order by length and stop when we have found t MEMs at least as long as the next pseudo-MEM. This should require modifying existing MEM-finders such as `ropebwt3`, but our experiments in Sect. 4 indicate that simply searching in the t longest pseudo-MEMs and discarding the rest does not significantly affect downstream results in metagenomic classification—even compared to using all the long MEMs. Figure 1 shows an example of how to use KeBaB to find pseudo-MEMs.

2 MEMs, Forward-Backward and BML

A *maximal exact match* (MEM)—also called super-maximal exact match (SMEM)—of a pattern $P[0..m-1]$ with respect to a text $T[0..n-1]$ is a substring $P[i..j]$ such that

- $P[i..j]$ occurs in T,
- $i = 0$ or $P[i-1..j]$ does not occur in T,
- $j = m-1$ or $P[i..j+1]$ does not occur in T.

Finding MEMs is an important step in many bioinformatics pipelines, such as aligning long and error-prone DNA reads to large pangenomic references [18].

For Li's [9] popular forward-backward MEM-finding algorithm, we keep FM-indexes [6] of T and its reverse T^{rev} (see [12,14] for an introduction to FM-indexes). Assuming all the characters in P occur in T, the leftmost MEM starts at $P[0]$. We can therefore find the leftmost MEM $P[0..e_1]$ through forward extension by searching for P^{rev} in the index for T^{rev}. If $e_1 < m-1$ then the second MEM $P[s_2..e_2]$ from the left in P includes $P[e_1+1]$. By definition, no MEM includes $P[s_2-1..e_1+1]$, so we can find s_2 through backward extension by searching for $P[0..e_1+1]$ in the index for T. Conceptually, we can then recurse on $P[s_2..m-1]$

and find e_2 and the remaining MEMs. The number of backward steps this takes in the indexes is proportional to the total length of the MEMs.

For many applications we are interested only in MEMs which are determined to be sufficiently long; these are biologically significant since they are unlikely to be the result of noise. Unfortunately, the total length of the MEMs is often dominated by many short MEMs, which we would like to ignore. Suppose we are interested only in MEMs of length at least L. Gagie [8] recently observed that any such MEM starting in $P[0..L-1]$ includes $P[L-1]$, so if we search for $P[0..L-1]$ in the index for T and find that $P[s..L-1]$ occurs in T but $P[s-1..L-1]$ does not, for some $s > 1$, then we can ignore $P[0..s-1]$ and recurse on $P[s..m-1]$. If we find that all of $P[0..L-1]$ occurs in T then we can still use the first few steps of forward-backward to find the leftmost MEM and the starting position of the second MEM from the left in P, and then recurse. Since this approach is reminiscent of Boyer-Moore pattern matching, we call it *Boyer-Moore-Li* (BML). Li [10] incorporated BML into `ropebwt3` and found it significantly accelerates MEM-finding. Depuydt et al.'s [4] `b-move` also supports accelerated MEM-finding by applying BML to the bidirectional r-index.

3 k-mer Based Breaking Into Pseudo-MEMs

Another technique for speeding up pattern matching is *k-mer filtration*. In contrast to BML, this requires scanning the whole input and deciding which parts can be ignored because they cannot contain significant-length matches. If the alphabet's size is polylogarithmic in n and BML uses a sublinear number of backward steps, then in the word-RAM model filtration is asymptotically slower; however, the filtration scan is sequential, incurring few cache misses and allowing it to be fast in practice compared to FM-index queries, which tend to incur many cache misses.

Suppose we are given k when we index T and we build a Bloom filter [2] for the distinct k-mers in T. Bloom filters can give false positive results but not false negative ones, so if the filter answers "no" for a k-mer $P[i..i+k-1]$ then no MEM of length at least k includes that k-mer. It follows that when we are given P and $L > k$, we can break P up into maximal substrings—which can overlap by $k-2$ characters but cannot nest—containing only k-mers for which the filter answers "yes", that contain all the MEMs of length at least L. We call these substrings *pseudo-MEMs* because they are our best guesses at the MEMs of length at least L based on the information we can glean from the filter. Further, any filter which cannot give false negatives can be used.

Definition 1. *Let a filter be a function $f : \Sigma^k \to \{0,1\}$ with respect to a given k, alphabet Σ, and text T, such that if k-mer x occurs in T, then $f(x) = 1$. Thus, a k-mer x appears in f if and only if $f(x) = 1$ (the filter answers "yes").*

Definition 2. *A pseudo-MEM of a pattern $P[0..m-1]$ with respect to a text $T[0..n-1]$, an integer $k \geq 1$, a filter f for the distinct k-mers in T, and an integer $L > k$, is any maximal non-empty substring $P[i..j]$ of P of length at least L such that all the k-mers in $P[i..j]$ appear in the filter f.*

Proposition 1. *All the MEMs of P with respect to T of length at least $L > k$ are contained in the pseudo-MEMs of P with respect to T and any filter for the distinct k-mers in T.*

Our experiments in Sect. 4 show that computing the pseudo-MEMs and searching in them is in practice already faster than searching in all of P. Further, they show that if T is highly repetitive then the Bloom filter tends to be smaller than the respective indexes for the MEM-finders we tested.

If we seek only the top-t longest MEMs of length at least L, however, then we can leverage the filter step to accelerate the MEM-finding step even further through early stopping.

Proposition 2. *If we are finding the top-t longest MEMs of length at least L and we are searching the pseudo-MEMs in non-increasing order by length, we can stop when we have already found t MEMs longer than the next pseudo-MEM.*

We can compute and sort the pseudo-MEMs independently of the MEM-finding algorithm in use. While modifying a MEM-finder to track the t longest MEMs found so far and stop when the next pseudo-MEM becomes shorter would enable early stopping optimizations, KeBaB fits seamlessly into any MEM-finding pipeline without requiring such changes and preserves exact MEM output. As a result, we have not yet modified any MEM-finder, even though doing so could provide further speedups.

Without modifying a MEM-finder, we can estimate how long it would take to find the top-t long MEMs by finding them ourselves ahead of time and giving it only the pseudo-MEMs it would search in before stopping. Our experiments in Sect. 4 show that for reasonable values of t, this should be much faster than running `ropebwt3` on all the pseudo-MEMs; moreover, at least for the metagenomic classifier using `b-move` we tested, it does not significantly hurt the accuracy. In fact, we found that using the long MEMs we found in only the top-t pseudo-MEMs—which are not guaranteed to be the top-t long MEMs but which we can find without modifying existing MEM-finders—is even faster and still results in nearly identical classification accuracy.

4   Experiments

Our `kebab` implementation in `C++` is available at github.com/drnate-brown/kebab. It streams over k-mers using a rolling nucleotide hash defined by ntHash supporting both the forward and reverse complement DNA strands [13]. We use HyperLogLog [7] to estimate the cardinality of a text collection to initialize the Bloom filter size, which is optimized with respect to the number of filter hashes used. We then add canonical k-mers (the smaller of each k-mer and its reverse complement by hash value) to the filter. Given a pattern, we query its canonical k-mers and extract the pseudo-MEMs. Further optimizations such as Fibonacci hashing, parallelization, and latency hiding are described in the full paper. For all experiments, the Bloom filter uses a desired false positive rate of $\epsilon = 1/10$.

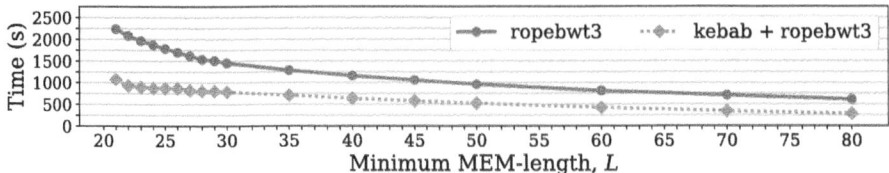

Fig. 2. Total runtime in seconds for MEM-finding methods, searching in a microbial pangenome with different minimum MEM-length values L.

4.1   MEM-Finding

We tested the speed of MEM-finding on a mock community dataset of 7 microbial species (5867 genomes, $\sim$27 GB) used in Ahmed et al.'s [1] SPUMONI 2 study. Patterns consist of long ONT *null reads* (10245 yeast reads with average length 19693) and *positive reads* (581802 microbial reads with average length 25378). Experiment runtimes were measured using GNU time on a server with an Intel(R) Xeon(R) Gold 6248R CPU running at 3.00 GHz with 48 cores and 1.5TB DDR4 memory, averaged over 10 runs using 16 threads. Constructing ropebwt3 took 162.88 minutes with an 0.7988 GB index. Building kebab with $k = 20$ and one hash function took 4.02 minutes with an 0.2684 GB filter (about a third of the size of ropebwt3's index).

We compared the time to find MEMs with ropebwt3 alone with default settings, to the time to first generate pseudo-MEMs with kebab and then search them with ropebwt3. We also simulated early stopping to find the 10 longest MEMs as explained in Sect. 3. Figure 2 shows the total times for different choices of L, and Fig. 3 shows times for null and positive reads with $L = 40$. For $L \geq 30$, the running-time of only the kebab step on the reads was at most about 3 times more than the time to copy them to another file, which is a rough lower bound on file I/O for a filtering step. For $L = 40$, the pseudo-MEMs for positive reads had an average length of 70.97, while the long MEMs themselves had an average length of 70.46.

4.2   Metagenomic Classification

To see how using a few long MEMs affects downstream applications, we replicated the metagenomic classification experiment in Depuydt et al.'s [5] tagger study, using the same dataset of 8 microbial species (8165 genomes, $\sim$37GB) and 50000 simulated long ONT reads with average length 5236. By default, tagger uses b-move with BML to find long MEMs together with sample species containing occurrences of them, then classifies the reads based on the sample species containing each read's long MEMs. We note that b-move is usually larger but faster than ropebwt3, so speedups with kebab are not as dramatic. Experiment runtimes were measured using GNU time on a server with an Intel(R) Xeon(R) E5-2698 v3 CPU running at 2.30 GHz with 32 cores (two threads per core) and 270 GB memory, using a single thread.

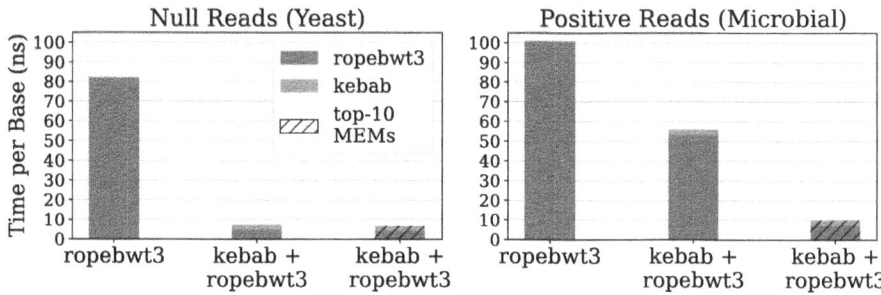

Fig. 3. For $L = 40$, time per base in the original input to find all long MEMs or only the 10 longest MEMs.

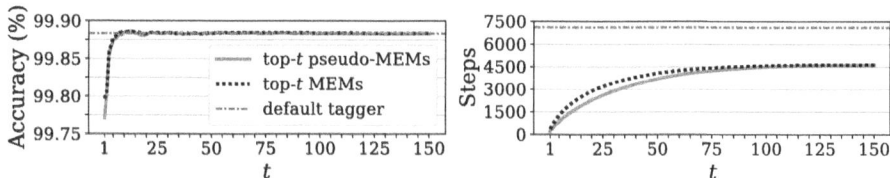

Fig. 4. tagger's accuracy (**left**) and the average number of steps b-move takes (**right**) for MEM-finding.

We computed tagger's accuracies (Fig. 4, on the left)—that is, its percentages of true-positive classifications—and the average number of steps b-move takes (Fig. 4, on the right) when finding and classifying based on

- all the reads' MEMs of length at least L ("default"),
- only the longest t MEMs from each read ("top-t MEMs"),
- only the MEMs of length at least L in the longest t pseudo-MEMs from each read ("top-t pseudo-MEMs"),

for $L = 25$ and various values of t. We ran tagger with default settings and $L = 25$ because Depuydt et al. found it gave good results. Clearly, for t greater than about 10, using only the t longest MEMs in each read or the MEMs of length at least $L = 25$ in the t longest pseudo-MEMs, does not noticeably hurt tagger's accuracy but significantly reduces the number of steps b-move takes for MEM-finding.

We also computed the total times, shown in Fig. 5, to classify the reads with tagger after first

- finding all the MEMs of length at least L with b-move ("tagger"),
- finding all the pseudo-MEMs with kebab and then finding all the MEMs of length least L in them with b-move ("kebab + tagger"),
- finding all the pseudo-MEMs with kebab and then finding the MEMs of length at least L in the t longest pseudo-MEMs from each read with b-move ("kebab + tagger, $t = 30, 20, 10$").

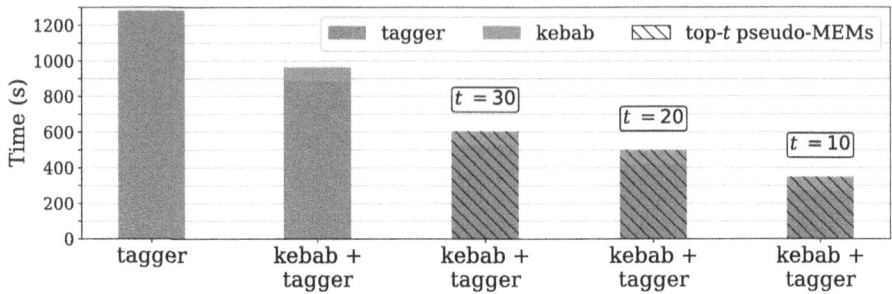

Fig. 5. The time to classify using the MEMs of length at least $L = 25$ using only `tagger`, or `kebab` followed by `tagger`, or all the MEMs of length at least $L = 25$ in the t longest pseudo-MEMs in each read using `kebab` followed by `tagger`.

We ran `tagger` with default settings and $L = 25$, and `kebab` with $k = 20$ and one hash function. The index for `b-move` took 7.869 GB and the filter for `kebab` took an additional 0.2684 GB. Clearly, `kebab` can also speed up `tagger`'s pipeline.

5   Conclusion

KeBaB substantially speeds up MEM-finding, and can further speed up analyses like metagenomic classification by considering only the longest MEMs or long MEMs present in the longest pseudo-MEMs. Parameter tuning and further optimizations (blocked Bloom filters [16], SIMD acceleration [11], and sub-k-mer queries [15,17]) could make `kebab` even more efficient. In fact, adopting the approach of sub-k-mer methods such as Robidou and Peterlongo's `findere` may reduce Bloom filter queries by skipping certain k-mers. Future work includes implementing early stopping in `ropebwt3`, exploring other early-stopping approaches, and applying `kebab` to scenarios that can particularly benefit from filtration such as sequences with hypervariable regions.

Acknowledgments. Many thanks to Finlay Maguire for pointing out the similarities between BML and Boyer-Moore, and to the reviewers for insightful comments. NKB, MZ, and BL were funded by NIH grants R01HG011392 and R56HG013865 to BL. NKB was also funded by a JHU CS PhD Fellowship and an NSERC PGS-D. LD was funded by a PhD Fellowship FR (1117322N), Research Foundation Flanders (FWO). TG was funded by NSERC grant RGPIN-07185-2020.

Disclosure of Interests. The authors declare no competing interests.

References

1. Ahmed, O.Y., Rossi, M., Gagie, T., Boucher, C., Langmead, B.: SPUMONI 2: improved classification using a pangenome index of minimizer digests. Genome Biol. **24**(1), 122 (2023)

2. Bloom, B.H.: Space/time trade-offs in hash coding with allowable errors. Commun. ACM **13**(7), 422–426 (1970)
3. Boyer, R.S., Moore, J.S.: A fast string searching algorithm. Commun. ACM **20**(10), 762–772 (1977)
4. Depuydt, L., Renders, L., Van de Vyver, S., Veys, L., Gagie, T., Fostier, J.: b-move: Faster lossless approximate pattern matching in a run-length compressed index. Algorithms for Molecular Biology (accepted)
5. Depuydt, L., Ahmed, O.Y., Fostier, J., Langmead, B., Gagie, T.: Run-length compressed metagenomic read classification with SMEM-finding and tagging. bioRxiv, pp. 2025–02 (2025)
6. Ferragina, P., Manzini, G.: Indexing compressed text. J. ACM **52**(4), 552–581 (2005)
7. Flajolet, P., Fusy, É., Gandouet, O., Meunier, F.: HyperLogLog: the analysis of a near-optimal cardinality estimation algorithm. Discrete Mathematics & Theoretical Computer Science (2007)
8. Gagie, T.: How to find long maximal exact matches and ignore short ones. In: Proc. 28th Conference on Developments in Language Theory (DLT), pp. 131–140 (2024)
9. Li, H.: Exploring single-sample SNP and INDEL calling with whole-genome de novo assembly. Bioinformatics **28**(14), 1838–1844 (2012)
10. Li, H.: BWT construction and search at the terabase scale. Bioinformatics **40**(12), btae717 (2024)
11. Lu, J., et al.: Ultra-fast bloom filters using SIMD techniques. IEEE Trans. Parallel Distrib. Syst. **30**(4), 953–964 (2018)
12. Mäkinen, V., Belazzougui, D., Cunial, F., Tomescu, A.I.: Genome-scale algorithm design: bioinformatics in the era of high-throughput sequencing. Cambridge University Press (2023)
13. Mohamadi, H., Chu, J., Vandervalk, B.P., Birol, I.: ntHash: recursive nucleotide hashing. Bioinformatics **32**(22), 3492–3494 (2016)
14. Navarro, G.: Compact data structures: A practical approach. Cambridge University Press (2016)
15. Pellow, D., Filippova, D., Kingsford, C.: Improving bloom filter performance on sequence data using k-mer bloom filters. J. Comput. Biol. **24**(6), 547–557 (2017)
16. Putze, F., Sanders, P., Singler, J.: Cache-, hash-, and space-efficient bloom filters. J. Exp. Alg. (JEA) **14**, 4–4 (2010)
17. Robidou, L., Peterlongo, P.: findere: fast and precise approximate membership query. In: International Symposium on String Processing and Information Retrieval, pp. 151–163. Springer (2021)
18. Varki, R., et al.: Accurate short-read alignment through r-index-based pangenome indexing. Genome Res. **35**(7), 1609–1620 (2025)

Analysing New Entropy Measures for Tries

Lorenzo Carfagna[1]([⊠])(iD) and Carlo Tosoni[2]([⊠])(iD)

[1] Department of Computer Science, University of Pisa, Pisa, Italy
lorenzo.carfagna@phd.unipi.it
[2] DAIS, Ca' Foscari University of Venice, Venice, Italy
carlo.tosoni@unive.it

Abstract. We introduce new entropy measures for tries taking into account the distribution of the edge labels. To do that, we study the combinatorial problem of counting the number of tries with a given symbol distribution. We provide an alternative proof for the closed formula counting the tries belonging to such a class. This formula allows us to directly define the worst-case entropy for the aforementioned set of tries. Moreover, we propose a new notion of k-th order empirical entropy for tries and we show that the relationships between these two entropy measures are similar to those between the corresponding well-known measures for strings. Contrary to the label entropy [FOCS '05], which was designed to compress the labels of a node-labeled ordered tree, our empirical entropy considers not only the labels, but the entire structure of the trie. Finally, we relate our empirical entropy to the repetitiveness measure r proposed by Prezza [SODA '21], which counts the number of runs in the XBWT of the trie. We show that these measures exhibit relations analogous to those of their string counterparts.

Keywords: Worst-case entropy · Empirical entropy · Cardinal trees · Tries · Digital trees · XBWT

1 Introduction

A *cardinal tree* is a particular type of ordered tree where every node is represented by an array of σ positions, and each position in that array is either null or stores a pointer to a child node. Cardinal trees are equivalent to tries, that is, edge-labeled ordered trees over an alphabet of size σ, in which (i) the labels of the edges outgoing from the same node are distinct, and (ii) sibling nodes are ordered by their incoming label. In the literature, it is known that the number of tries with n nodes over an alphabet Σ of size σ is equal to $\mathcal{S}(n, \sigma) = \frac{1}{n}\binom{\sigma n}{n-1}$ [7,13]. This formula directly leads to the *worst-case entropy* $\mathcal{C}(n, \sigma)$ for tries [2,19] defined as $\mathcal{C}(n, \sigma) = \log_2 \mathcal{S}(n, \sigma)$. This worst-case entropy is particularly famous in the scientific community and the space complexity of many trie representations is expressed in terms of $\mathcal{C}(n, \sigma)$. For instance, Benoit

© The Author(s), under exclusive license to Springer Nature Switzerland AG 2026
G. Badkobeh et al. (Eds.): SPIRE 2025, LNCS 16073, pp. 18–27, 2026.
https://doi.org/10.1007/978-3-032-05228-5_3

et al. proposed a trie representation taking $\mathcal{C}(n,\sigma) + \Omega(n)$ bits [2]. This space occupation was later improved to $\mathcal{C}(n,\sigma) + o(n) + O(\log\log\sigma)$ bits by Raman et al. [19] and Farzan et al. [9]. All these data structures were devised to support queries on the trie topology plus the cardinal query of retrieving the child labeled with the i-th character of the alphabet in constant time. While the previous data structures are static, we also mention solutions proposing dynamic representations of tries with space complexity related to $\mathcal{C}(n,\sigma)$ [1,5,6]. In particular, these data structures take $\mathcal{C}(n,\sigma) + \Omega(n)$ bits of space. However, to the best of our knowledge, contrary to what happens for strings, there exist no notions of empirical entropy for tries that take into account the distribution of the edge labels and the context of the nodes.

Our Contribution. Motivated by the above reasons, in this paper we aim to refine the formula $\mathcal{C}(n,\sigma)$ to consider also how many times a character of the alphabet occurs as a label in the trie. To devise this formula, we consider the combinatorial problem of finding the number $|\mathcal{U}|$ of tries having a given *symbol distribution* $\{n_c \mid c \in \Sigma\}$, that is, tries having n_c edges labeled with symbol c. During a bibliographic search, we found a recent technical report [18] showing $|\mathcal{U}| = \frac{1}{n}\prod_{c\in\Sigma}\binom{n}{n_c}$, with n being the number of nodes, by sketching a proof based on generating functions [12]. However, in this paper we prove this closed formula by showing a simple bijection between these tries and a particular class of binary matrices. As an immediate consequence, we obtain the aforementioned worst-case entropy $\mathcal{H}^{wc}$ for this refined class of tries. In addition to that, we define a new notion of k-th order empirical entropy $\mathcal{H}_k$ for tries. Similarly to the label entropy of Ferragina et al. [10,11], defined for arbitrary ordered labeled trees, our $\mathcal{H}_k$ divides the nodes based on their k-length incoming string. However, while the label entropy can encode only the labels of the tree, our empirical entropy can encode both the labels and the topology of the trie *simultaneously*. Moreover, we show that the formulas $\mathcal{H}^{wc}$ and $\mathcal{H}_k$ for tries share similar properties of their corresponding string counterparts. In particular, it is $n\mathcal{H}_0 = \mathcal{H}^{wc} + O(\sigma\log n)$ and for every $k \geq 0$ it holds that $\mathcal{H}_{k+1} \leq \mathcal{H}_k$. Finally, we show that, due to [17, Theorem 3.1], for every trie and integer $k \geq 0$ it holds that $r \leq n\mathcal{H}_k + \sigma^{k+1}$ with r being the number of runs in the XBWT of a trie [17]. See the technical report [3] for an extended version of the paper.

2 Notation

In the following we refer with Σ a finite alphabet of size σ, totally ordered according to a given relation $\preceq$. For every integer $k \geq 0$ we define Σ^k as the set of all length-k strings with symbols in Σ where $\Sigma^0 = \{\epsilon\}$, i.e., the singleton set containing the empty string ϵ. With $[n]$ we denote the set $\{1,..,n\} \subseteq \mathbb{N}$ and given a finite set X, $|X|$ is the cardinality of X. Unless otherwise specified, all logarithms are base 2 and denoted by $\log x$, furthermore we assume that $0\log(x/0) = 0$ for every $x \geq 0$. Given a matrix M we denote the i-th row/column of M as $M[i][-]$ and $M[-][i]$, respectively. In the following we denote with $\mathrm{ones}(M')$ the number of entries equal to 1 in a submatrix M' of M. In this paper, we work with *cardinal trees*, also termed tries, i.e., edge-labeled ordered trees in which sibling nodes have different incoming labels and are ordered according to that label. We denote by $\mathcal{T} = (V, E)$ a trie over an alphabet Σ, where V is set of nodes with $|V| = n$ and E is the set of edges with labels drawn from Σ. We denote with n_i the number of edges labeled by the i-th character $c_i \in \Sigma$ according to $\preceq$. Note that since $\mathcal{T}$ is a tree we have that $\sum_{i=1}^{\sigma} n_i = |E| = n-1$. We denote by $\lambda(u)$ the label of the incoming edge of a node $u \in V$. If u is the root, we define $\lambda(u) = \#$, where $\#$ is a special character of Σ not labeling any edge. Given a node $u \in V$, the function $\pi(u)$ returns the parent node of u in $\mathcal{T}$, where $\pi(u) = u$ if u is the root. Furthermore, the function $out(u)$ returns the set of labels of the edges outgoing from node u, i.e., $out(u) = \{c \in \Sigma : (u, v, c) \in E \text{ for some } v \in V\}$.

3   On the Number of Tries with a Given Symbol Distribution

This section is dedicated to prove the following result.

Theorem 1. *Let $\mathcal{U}$ be the set of tries with n nodes and labels drawn from an alphabet $\Sigma = \{c_1, \ldots, c_\sigma\}$, where each symbol c_i labels n_i edges, then $|\mathcal{U}| = \frac{1}{n} \prod_{i=1}^{\sigma} \binom{n}{n_i}$*

The same problem has already been addressed by Prezza [17] and Prodinger [18]. Prezza claimed that $|\mathcal{U}|$ is equal to $\prod_{i=1}^{\sigma} \binom{n}{n_i}$, however, we point out that this formula overestimates the correct number of tries by a multiplicative factor n. On the other hand, Prodinger deduced the correct number in a technical report using the Lagrange Inversion Theorem [12]. In this paper, we give an alternative proof using a simple bijection between these tries and a class of binary matrices. For the more general problem of counting tries with n nodes and alphabet of size σ, the corresponding formula $\mathcal{S}(n, \sigma)$ is well-known and in particular it is $\mathcal{S}(n, \sigma) = \frac{1}{n} \binom{n\sigma}{n-1}$ [7,13]. We now introduce the definition of set $\mathcal{M}$, which depends on $\mathcal{U}$.

Definition 1 (Set $\mathcal{M}$). *We define $\mathcal{M}$ as the set of all $\sigma \times n$ binary matrices M, such that $\mathrm{ones}(M[i][-]) = n_i$ for every $i \in [\sigma]$.*

Note that $\mathrm{ones}(M) = n - 1$, trivially follows from $\sum_{i=1}^{\sigma} n_i = n - 1$. In order to count the elements in the set $\mathcal{U}$, we now define a function f mapping each trie $\mathcal{T}$ into an element of $M \in \mathcal{M}$. According to our function f, the number of ones in the columns of M encodes the topology of the trie, while the fact that M is binary encodes the standard trie labeling constraint.

Definition 2 (Function f) . *We define the function $f : \mathcal{U} \to \mathcal{M}$ as follows: given a trie $\mathcal{T} \in \mathcal{U}$ consider its nodes $u_1, \ldots, u_n$ in pre-order visit. Then $M = f(\mathcal{T})$ is the $\sigma \times n$ binary matrix such that $M[i][j] = 1$ if and only if $c_i \in out(u_j)$.*

Note that, given a trie $\mathcal{T}$, the i-th column $M[-][i]$ of the corresponding matrix $M = f(\mathcal{T})$ is the characteristic bitvector of the outgoing labels of node u_i (under the specific ordering on Σ) and in particular it holds that $\mathrm{ones}(M[-][i])$ is the out-degree of u_i. We observe that the function f is injective. To see that, consider two distinct tries $\mathcal{T}$ and $\mathcal{T}'$ and their matrices $M = f(\mathcal{T})$ and $M' = f(\mathcal{T}')$. If the trie topologies differ, there exists i such that $M[-][i] \neq M'[-][i]$ since $\mathrm{ones}(M[-][i]) \neq \mathrm{ones}(M'[-][i])$. Otherwise, $\mathcal{T}$ and $\mathcal{T}'$ differ by an outgoing label, and consequently again $M[-][i] \neq M'[-][i]$ necessarily holds. However, f is not surjective in general: some matrices $M \in \mathcal{M}$ may not belong to the image of f because during the inversion, connectivity constraints could be violated (see Fig. 1). To characterize the image of f, we define the following two sequences.

Definition 3 (Sequences D *and* L). *For every $M \in \mathcal{M}$ we define the integer sequences D and L of length n, such that $\mathrm{D}[i] = \mathrm{ones}(M[-][i]) - 1$ and $\mathrm{L}[i] = \sum_{j=1}^{i} \mathrm{D}[j]$ for every $i \in [n]$.*

We can observe that if $M \in f(\mathcal{U})$, then if we consider the unique trie $\mathcal{T} = f^{-1}(M)$, D is the out-degree of the i-th node in pre-order visit minus one, namely $\mathrm{D}[i] = |out(u_i)| - 1$. Consequently, we have that $\mathrm{L}[i] = \sum_{j=1}^{i} |out(u_j)| - i$, and therefore $\mathrm{L}[i] + 1$ denotes the total number of "pending" edges immediately after the pre-order visit of the node u_i, where the $+1$ term stems from the fact that the root has no incoming edges. In other words, $\mathrm{L}[i] + 1$ denotes the number of edges whose source has already been visited, while their destination has not. We note that for every $M \in \mathcal{M}$, it holds that $\mathrm{L}[n] = -1$, since $\mathrm{ones}(M) = n - 1$. In the following, we show that the sequence L can be used to determine if $M \in f(\mathcal{U})$ holds for a given matrix $M \in \mathcal{M}$. In fact, we will see that if $M \in f(\mathcal{U})$, the array L is a so-called *Lukasiewicz path* as defined in the book of Flajolet and Sedgewick [12] (see Section I. 5. "Tree Structures"). Specifically, a Lukasiewicz path $\mathcal{L}$ is a sequence of n integers satisfying the following conditions. (i) $\mathcal{L}[n] = -1$, and for every $i \in [n-1]$, it holds that (ii) $\mathcal{L}[i] \geq 0$, and (iii) $\mathcal{L}[i+1] - \mathcal{L}[i] \geq -1$. Lukasiewicz paths can be used to encode the topology of a trie since it is known that their are in bijection with unlabeled ordered trees of n nodes [12]. Specifically, given an ordered unlabeled tree $\mathcal{T}'$ the i-th point in its corresponding Lukasiewicz path $\mathcal{L}$ is $\mathcal{L}[i] = \sum_{j=1}^{i}(|out(u_j)| - 1)$, i.e., $\mathcal{L}$ is obtained by prefix-summing the sequence formed by the nodes out-degree minus 1 in pre-order. The reverse process consists in scanning $\mathcal{L}$ left-to-right and appending the current node u_i of out-degree $\mathcal{L}[i] - \mathcal{L}[i-1] + 1$ to the deepest pending edge on the left; obviously u_1 is the root of $\mathcal{T}'$ where we assume $\mathcal{L}[0] = 0$. Since f is injective, the next result proves that the class of tries $\mathcal{U}$ is in bijection with the matrices of $\mathcal{M}$ whose corresponding array L is a *Lukasiewicz path*. Similar results have been proved in other articles [7, 8, 20].

Lemma 1. *A binary matrix $M \in \mathcal{M}$ belongs to $f(\mathcal{U})$ if and only if, the corresponding array L is a Lukasiewicz path.*

Proof. $(\Rightarrow)$: Consider the trie $\mathcal{T} = (V, E) = f^{-1}(M)$ and its underlying unlabeled ordered tree $\mathcal{T}'$, i.e., $\mathcal{T}'$ is the ordered tree obtained by deleting the labels from $\mathcal{T}$. Obviously, the corresponding nodes of $\mathcal{T}$ and $\mathcal{T}'$ in pre-order have the same out-degree. Therefore, by Definition 3, the array L of M is equal to the Łukasiewicz path of $\mathcal{T}'$. ($\Leftarrow$) : Given $M \in \mathcal{M}$ and its corresponding array L, we know by hypothesis that L is a Łukasiewicz path. Therefore, we consider the unlabeled ordered tree $\mathcal{T}'$ obtained by inverting L. Let u_i be the i-th node of $\mathcal{T}'$ in pre-order, we label the j-th left-to-right edge outgoing from u_i with the symbol corresponding to the j-th top-down 1 in $M[-][i]$. The resulting ordered labeled tree is a trie $\mathcal{T}''$ and by construction it follows that $f(\mathcal{T}'') = M$.   □

To count the number of tries in $\mathcal{U}$, we now introduce the concept of *rotations*.

Definition 4 (Rotations) . *For every matrix $M \in \mathcal{M}$ and integer $r \geq 0$, we define the r-th rotation of M, denoted by M^r, as $M^r[-][j] = M[-][(j + r - 1) \bmod n + 1]$ for every $j \in [n]$.*

Informally, a rotation M^r of a matrix $M \in \mathcal{M}$ is obtained by moving the last r columns of M at its beginning, and therefore $M^0 = M$. Moreover, since $M^r = M^{r \bmod n}$, in the following we consider only rotations M^r with $r \in [0, n-1]$. Note also that, for every $r \geq 0$, we know that $M^r \in \mathcal{M}$ as rotations do not change the number of entries equal to 1 in a row. Moreover, we can observe that the array D of a rotation M^r corresponds to a cyclic permutation of the array D' of M. Next we show that all n rotations of a matrix $M \in \mathcal{M}$ are distinct and exactly one rotation has an array L which is Łukasiewicz. To do this, we recall a known result from [20] and [12, Section I. 5. "Tree Structures", Note I.47].

Lemma 2 *[20, Lemma 2]. Consider a sequence of integers $A = a_1, \ldots, a_n$ with $\sum_{i=1}^{n} a_i = -1$. Then (i) all n cyclic permutations of A are distinct, and (ii) there exists a unique cyclic permutation $A' = a'_1, \ldots, a'_n$ of A such that $\sum_{i=1}^{j} a'_i \geq 0$, for every $j \in [n - 1]$*

The following corollary is a consequence of the above lemma and an example is shown in Fig. 1.

Corollary 1. *For every matrix $M \in \mathcal{M}$ all its rotations are distinct and only one of them belongs to $f(\mathcal{U})$.*

Proof. Consider the sequence D of matrix M. We know that the sequence of integers $D[1], \ldots, D[n]$ sums to -1. Consequently, by Lemma 2, it follows that all the cyclic permutations of D are distinct and thus all the corresponding rotations of M are distinct too. By Lemma 1, to conclude the proof it remains to be shown that there exists a unique rotation M^r whose array L is a Łukasiewicz path. By Lemma 2, we know that there exists a unique rotation M^r such that its array L is nonnegative excluding the last position, which is a necessary condition for L to be a Łukasiewicz path. Therefore M^r is the only candidate that may belong to $f(\mathcal{U})$. Moreover, we know that $L[n] = -1$ and by Definition 3 it holds that $L[i + 1] - L[i] = \mathbf{ones}(M^r[-][i]) - 1 \geq -1$ for every $i \in [n - 1]$. Since these observations prove that L is Łukasiewicz, it follows that $M^r \in f(\mathcal{U})$.   □

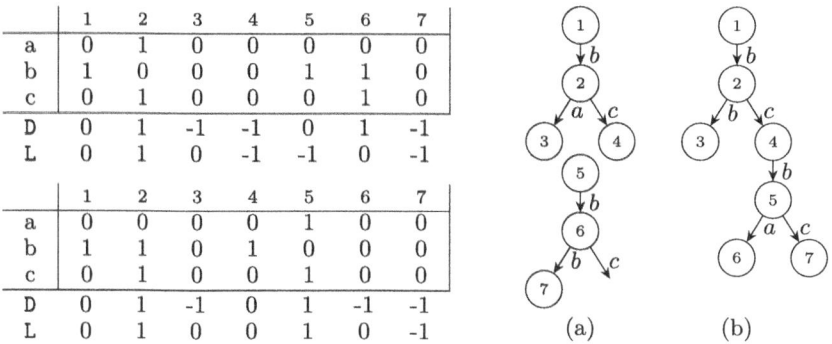

	1	2	3	4	5	6	7
a	0	1	0	0	0	0	0
b	1	0	0	0	1	1	0
c	0	1	0	0	0	1	0
D	0	1	-1	-1	0	1	-1
L	0	1	0	-1	-1	0	-1

	1	2	3	4	5	6	7
a	0	0	0	0	1	0	0
b	1	1	0	1	0	0	0
c	0	1	0	0	1	0	0
D	0	1	-1	0	1	-1	-1
L	0	1	0	0	1	0	-1

Fig. 1. In the top-left corner, the figure shows a 3×7 binary matrix M with $n_1 = 1$, $n_2 = 3$, and $n_3 = 2$, and its sequences D and L. Since L is not Łukasiewicz, by inverting M, we obtain the object (a) which is not a trie. This happens because there are no pending edges to which the pre-order node 5 can be attached since $L[4] = -1$. However, the rotated matrix M^3 showed in the bottom-left corner produce the valid trie shown in (b). In this case L is Łukasiewicz.

In the following, we prove the main result of this section.

Proof (Theorem 1). Let $\sim$ be the relation over $\mathcal{M}$ defined as follows. For every $M, \bar{M} \in \mathcal{M}$ we have that $M \sim \bar{M}$ holds if and only if, there exists an integer i such that $M^i = \bar{M}$. It is easy to see that $\sim$ is an equivalence relation over $\mathcal{M}$, indeed, $\sim$ satisfies reflexivity, symmetry, and transitivity. By Corollary 1, we know that for every $M \in \mathcal{M}$ the equivalence class $[M]_\sim$ contains n elements with a unique matrix $\bar{M} \in [M]_\sim$ such that $\bar{M} \in f(\mathcal{U})$. Therefore, since f is injective, we deduce that $|\mathcal{U}| = |f(\mathcal{U})| = |\mathcal{M}|/n$. Finally, since it is easy to observe that $|\mathcal{M}| = \prod_{i=1}^{\sigma} \binom{n}{n_i}$ it follows that $|\mathcal{U}| = \frac{1}{n} \prod_{i=1}^{\sigma} \binom{n}{n_i}$. □

4   Entropy of a Trie

We now use Theorem 1 from the previous section to introduce the *worst-case entropy* of a trie with given symbol frequencies $n_1, \ldots, n_\sigma$.

Definition 5 (Worst-case entropy). *The worst-case entropy $\mathcal{H}^{wc}$ of a trie $\mathcal{T}$ with symbol distribution $n_1, \ldots, n_\sigma$ is $\mathcal{H}^{wc}(\mathcal{T}) = \log|\mathcal{U}| = \sum_{i=1}^{\sigma} \log \binom{n}{n_i} - \log n$*

Clearly the above worst-case entropy is a lower-bound for the minimum number of bits (in the worst-case) required to encode a trie with a given number of symbol occurrences. Note that $\mathcal{H}^{wc}(\mathcal{T})$ is always smaller than or equal to the worst-case entropy $\mathcal{C}(n, \sigma) = \log \frac{1}{n} \binom{n\sigma}{n-1}$ [19] of tries having n nodes over an alphabet of size σ and it is much smaller when the symbol distribution is skewed. Next we introduce the notion of 0-*th order empirical entropy* of a trie $\mathcal{T}$.

Definition 6 (0-th order empirical entropy). *The 0-th order empirical entropy $\mathcal{H}_0(\mathcal{T})$ of a trie $\mathcal{T}$ with symbol distribution $n_1, \ldots, n_\sigma$ is defined as $\mathcal{H}_0(\mathcal{T}) = \frac{1}{n} \sum_{i=1}^{\sigma} n_i \log(n/n_i) + (n - n_i) \log(n/(n - n_i))$*

Given a bitvector $B_{n,m}$ of size n with m ones, the worst-case entropy is defined as $\mathcal{H}^{wc}(B_{n,m}) = \log\binom{n}{m}$ while the corresponding 0-th order empirical entropy is $\mathcal{H}_0(B_{n,m}) = \frac{m}{n}\log(\frac{n}{m}) + \frac{n-m}{n}\log(\frac{n}{n-m})$ [4,16]. Furthermore, the following known inequalities relate the above entropies of a binary sequence: it holds that $n\mathcal{H}_0(B_{n,m}) - \log(n+1) \le \mathcal{H}^{wc}(B_{n,m}) \le n\mathcal{H}_0(B_{n,m})$ [4, Equation 11.40]. The next result is a direct consequence of these inequalities.

Lemma 3. *For every trie $\mathcal{T}$ with n nodes, the following formula holds $n\mathcal{H}_0(\mathcal{T}) - \sigma\log(n+1) - \log n \le \mathcal{H}^{wc}(\mathcal{T}) \le n\mathcal{H}_0(\mathcal{T}) - \log n$*

The following corollary directly follows from the above lemma.

Corollary 2. *For every trie $\mathcal{T}$ it holds that $n\mathcal{H}_0(\mathcal{T}) = \mathcal{H}^{wc}(\mathcal{T}) + O(\sigma\log n)$*

Note that the results of Lemma 3 and Corollary 2 are valid also if we consider the effective alphabet formed by all the symbols labeling at least one edge in $\mathcal{T}$, as the other characters do not affect $\mathcal{H}^{wc}$ or $n\mathcal{H}_0$. The result of Corollary 2 is consistent with known results for strings: given a string S of length n it is known that $n\mathcal{H}_0(S) = \mathcal{H}^{wc}(S) + O(\sigma\log n)$ (see [16, Section 2.3.2]) where $\mathcal{H}^{wc}(S) = \log\binom{n}{n_1,\dots,n_\sigma}$ is the worst-case entropy for the set of strings with character occurrences n_i and $H_0(S)$ is the 0-th order empirical entropy of S as defined in [16]. We now aim to extend the 0-th order trie entropy to any arbitrary integer $k > 0$. To this purpose, similarly as in [10,11,17], for every integer $k \ge 0$, we define the *k-th order context* of a node u, denoted by $\lambda_k(u)$, as the string formed by the last k symbols of the path from the root to u. Formally, we have $\lambda_0(u) = \epsilon$ and $\lambda_k(u) = \lambda_{k-1}(\pi(u)) \cdot \lambda(u)$. By definition of π (see Sect. 2), if a node u has a depth d with $d < k$, then such string is left-padded by the string $\#^{k-d}$. We now define the integers n_w and $n_{w,c}$, for every $w \in \Sigma^k$ and $c \in \Sigma$ as follows: n_w is the number of nodes having w as their k-th order context and $n_{w,c}$ is the number of nodes having w as their k-th order context and an outgoing edge labeled by c. Formally, $n_w = |\{u \in V : w = \lambda_k(u)\}|$ and $n_{w,c} = |\{u \in V : w = \lambda_k(u) \text{ and } c \in out(u)\}|$. With the next definition we generalize our notion of empirical entropy of a trie to higher orders in a similar way of what has been done in [10,11] for ordered labeled trees.

Definition 7 (k-th order empirical entropy). *For every integer $k \ge 0$, the k-th order empirical entropy $\mathcal{H}_k(\mathcal{T})$ of a trie $\mathcal{T}$ is defined as follows;*

$$\mathcal{H}_k(\mathcal{T}) = \sum_{w \in \Sigma^k} \sum_{c \in \Sigma} \frac{n_{w,c}}{n} \log\left(\frac{n_w}{n_{w,c}}\right) + \frac{n_w - n_{w,c}}{n} \log\left(\frac{n_w}{n_w - n_{w,c}}\right)$$

Note that for $k = 0$ this formula coincides with that of Definition 6 since in this case $\Sigma^0 = \{\epsilon\}$ and $n_\epsilon = n$ and for every character $c_i \in \Sigma$, we have $n_{\epsilon,c_i} = n_i$. Moreover, analogously to strings, by the log-sum inequality (see [4, Eq. (2.99)]), for every trie $\mathcal{T}$ and integer $k \ge 0$ it holds that $H_{k+1}(\mathcal{T}) \le H_k(\mathcal{T})$. In the extended version of the paper [3], we show that known data structures [14,19] can represent and index a trie within $n\mathcal{H}_k(\mathcal{T}) + o(n)$ bits of space, for every

$k = o(\log_\sigma n)$ simultaneously, assuming $\sigma \leq n^\varepsilon$ for some positive constant $\varepsilon < 1$. Specifically, the above index is based on the XBWT [10,11,17] of a trie and can efficiently count the number of nodes reached by a given string pattern.

A different notion of entropy for ordered node-labeled trees is the *label entropy* introduced by Ferragina et al. [10]. This entropy is a measure of the repetitiveness of the labels in the tree knowing the context of the nodes. In the following we report a trivial adaptation of their formula to ordered edge-labeled trees. Let $cover(w)$ be the string obtained by concatenating (in any order) the labels outgoing from the nodes having the same length-k context $w \in \Sigma^k$. The (unnormalized) k-th order label entropy $\mathcal{H}_k^{label}$ for a tree $\mathcal{T}$ is defined as $\mathcal{H}_k^{label}(\mathcal{T}) = \sum_{w \in \Sigma^k} |cover(w)| \mathcal{H}_0(cover(w))$ [10, Section 6] where $\mathcal{H}_0(cover(w))$ is the 0-th order empirical entropy of the string $cover(w)$. We observe the following main differences between $\mathcal{H}_k$ and $\mathcal{H}_k^{label}$. While $\mathcal{H}_k^{label}$ is defined for labeled ordered trees and considers only the labels appearing in the tree, our entropy $\mathcal{H}_k$ of Definition 7 is specifically designed for the sub-class of tries and differently from $\mathcal{H}_k^{label}$ it encodes also the topology of the trie. Now we relate our trie entropy $\mathcal{H}_k$ to another repetitiveness measure for tries, namely the number r of XBWT runs of Prezza [17]. Consider the string $s(u)$ reaching the node u of a trie from the root, i.e., $s(u) = \lambda_d(u)$, where d is the depth of u, with the root having depth 0. We now recall the definition of the XBWT$(\mathcal{T})$ of a trie $\mathcal{T}$ [17, Definition 3.1]. Consider the nodes $u_1, \ldots, u_n$ of the trie sorted according to co-lexicographic ordering $\preceq$ of the strings $s(u_1) \preceq \ldots \preceq s(u_n)$ reaching them, then XBWT$(\mathcal{T})$ is defined as the sequence of n sets, where the i-th set contains the labels outgoing from u_i. Note that u_1 is the root of the trie since $s(u_1) = \epsilon$. We say that an integer $i \in [n]$ is a c-run break if $c \in out(u_i)$ and either $i = n$ or $c \notin out(u_{i+1})$ holds. Then the number r of *XBWT runs*, is $r = \sum_{c \in \Sigma} r_c$, where r_c is the total number of c-run breaks in XBWT$(\mathcal{T})$ [17]. In the same paper, Prezza proposed an index for tries, termed r-index, whose space usage is $O(r \log n) + o(n)$ bits[1]. In [17, Theorem 3.1] Prezza proved that for every integer $k \geq 0$, the following inequality holds $r \leq \sum_{w \in \Sigma^k} \sum_{c \in \Sigma} \log \binom{n_w}{n_{w,c}} + \sigma^{k+1}$ [17, Theorem 3.1], which by Equation 11.40 of [4] in turn implies that $r \leq n\mathcal{H}_k(\mathcal{T}) + \sigma^{k+1}$ for every integer k. We observe that a similar relation is known also for strings [15, Theorem 1]. In [3, Proposition 5.5] we also prove that for a family of tries over a binary alphabet it holds that $r = \Theta(n\mathcal{H}_0(\mathcal{T})) = \Theta(n)$.

Acknowledgments. We would like to thank Sung-Hwan Kim for remarkable suggestions concerning the combinatorial problem addressed in the article.

Lorenzo Carfagna: partially funded by the INdAM-GNCS Project CUP E53C24001950001, and by the PNRR ECS00000017 Tuscany Health Ecosystem, Spoke 6, CUP I53C22000780001, funded by the NextGeneration EU programme.

Carlo Tosoni: funded by the European Union (ERC, REGINDEX, 101039208). Views and opinions expressed are however those of the authors only and do not necessarily reflect those of the European Union or the European Research Council Executive

---

[1] This is the space usage to represent the trie and support count queries on it.

Agency. Neither the European Union nor the granting authority can be held responsible for them.

References

1. Arroyuelo, D., Davoodi, P., Satti, S.R.: Succinct dynamic cardinal trees. Algorithmica **74**(2), 742–777 (2015). https://doi.org/10.1007/s00453-015-9969-x
2. Benoit, D., Demaine, E.D., Munro, J.I., Raman, R., Raman, V., Rao, S.S.: Representing trees of higher degree. Algorithmica **43**(4), 275–292 (2005). https://doi.org/10.1007/S00453-004-1146-6
3. Carfagna, L., Tosoni, C.: Indexing tries within entropy-bounded space (2025). https://arxiv.org/abs/2507.02728
4. Cover, T.M., Thomas, J.A.: Elements of Information Theory. Wiley-Interscience, USA (2006)
5. Darragh, J.J., Cleary, J.G., Witten, I.H.: Bonsai: a compact representation of trees. Softw. Pract. Exper. **23**(3), 277–291 (1993). https://doi.org/10.1002/spe.4380230305
6. Davoodi, P., Rao, S.S.: Succinct dynamic cardinal trees with constant time operations for small alphabet. In: Ogihara, M., Tarui, J. (eds.) Theory and Applications of Models of Computation, pp. 195–205. Springer Berlin Heidelberg, Berlin, Heidelberg (2011). https://doi.org/10.1007/978-3-642-20877-5_21
7. Dershowitz, N., Zaks, S.: The cycle lemma and some applications. Eur. J. Comb. **11**(1), 35–40 (1990). https://doi.org/10.1016/S0195-6698(13)80053-4
8. Dvoretzky, A., Motzkin, T.: A problem of arrangements. Duke Math. J. **14**(2), 305–313 (1947). https://doi.org/10.1215/S0012-7094-47-01423-3
9. Farzan, A., Raman, R., Rao, S.S.: Universal succinct representations of trees? In: Albers, S., Marchetti-Spaccamela, A., Matias, Y., Nikoletseas, S., Thomas, W. (eds.) ICALP 2009. LNCS, vol. 5555, pp. 451–462. Springer, Heidelberg (2009). https://doi.org/10.1007/978-3-642-02927-1_38
10. Ferragina, P., Luccio, F., Manzini, G., Muthukrishnan, S.: Structuring labeled trees for optimal succinctness, and beyond. In: 46th Annual IEEE Symposium on Foundations of Computer Science (FOCS 2005), 23-25 October 2005, Pittsburgh, PA, USA, Proceedings, pp. 184–196. IEEE Computer Society (2005). https://doi.org/10.1109/SFCS.2005.69
11. Ferragina, P., Luccio, F., Manzini, G., Muthukrishnan, S.: Compressing and indexing labeled trees, with applications. J. ACM **57**(1), 4:1–4:33 (2009). https://doi.org/10.1145/1613676.1613680
12. Flajolet, P., Sedgewick, R.: Analytic Combinatorics. Cambridge University Press (2009)
13. Graham, R.L., Knuth, D.E., Patashnik, O.: Concrete Mathematics: A Foundation for Computer Science, 2nd Ed. Addison-Wesley (1994). https://www-cs-faculty.stanford.edu/%7Eknuth/gkp.html
14. Kosolobov, D., Sivukhin, N.: Compressed multiple pattern matching. In: Pisanti, N., Pissis, S.P. (eds.) 30th Annual Symposium on Combinatorial Pattern Matching, CPM 2019, June 18-20, 2019, Pisa, Italy. LIPIcs, vol. 128, pp. 13:1–13:14. Schloss Dagstuhl - Leibniz-Zentrum für Informatik (2019). https://doi.org/10.4230/LIPICS.CPM.2019.13

15. Mäkinen, V., Navarro, G.: Succinct suffix arrays based on run-length encoding. In: Apostolico, A., Crochemore, M., Park, K. (eds.) CPM 2005. LNCS, vol. 3537, pp. 45–56. Springer, Heidelberg (2005). https://doi.org/10.1007/11496656_5
16. Navarro, G.: Compact Data Structures - A Practical Approach. Cambridge University Press (2016)
17. Prezza, N.: On locating paths in compressed tries. In: Marx, D. (ed.) Proceedings of the 2021 ACM-SIAM Symposium on Discrete Algorithms, SODA 2021, Virtual Conference, January 10 - 13, 2021, pp. 744–760. SIAM (2021). https://doi.org/10.1137/1.9781611976465.47
18. Prodinger, H.: Counting edges according to edge-type in t-ary trees (2022). https://doi.org/10.48550/arXiv.2205.13374
19. Raman, R., Raman, V., Satti, S.R.: Succinct indexable dictionaries with applications to encoding k-ary trees, prefix sums and multisets. ACM Trans. Algorithms **3**(4), 43–es (2007). https://doi.org/10.1145/1290672.1290680
20. Rote, G.: Binary trees having a given number of nodes with 0, 1, and 2 children. Séminaire Lotharingien de Combinatoire, B38b (1997). https://www.emis.de/journals/SLC/wpapers/s38proding.html

Depth First Representations of k^2-trees

Gabriel Carmona$^{(\boxtimes)}$ and Giovanni Manzini

Department of Computer Science, University of Pisa, Pisa, Italy
gabriel.carmona@phd.unipi.it, giovanni.manzini@unipi.it

Abstract. The k^2-tree is a compact data structure designed to efficiently store sparse binary matrices leveraging both sparsity and clustering of nonzero elements. This representation efficiently supports naviga-
tional operations and complex binary operations, such as matrix-matrix
multiplication, while maintaining space efficiency. The standard k^2-tree
follows a level-by-level representation, which, while effective, prevents
further compression of identical subtrees and it is not cache friendly
when accessing individual subtrees. In this work, we introduce some novel
depth-first representations of the k^2-tree and propose an efficient linear-
time algorithm to identify and compress identical subtrees within these
structures. Our experimental results show that the use of a depth-first
representation is a strategy worth pursuing: for the adjacency matrix of
web graphs exploiting the presence of identical subtrees does improve
both compression ratio and peak memory usage, and for some matrices, depth-first representations turn out to be faster than the standard
k^2-tree in computing the matrix-matrix multiplication.

Keywords: Web graphs · Sparse binary matrices · Succinct tree
representations · Compact data structure

1 Introduction

The growing need to manage massive datasets has driven the development of
compressed data structures [20], which enable operations on data without full
decompression. The k^2-*tree* [4] is a compact structure for representing sparse
binary matrices that has proven effective in various domains, including Web
Graphs [5,6], Graph Databases [1,9], Geographic Information Systems [7,16],
and Social Networks [7].

This paper explores further compressing the k^2-tree by identifying and
reusing identical subtrees. While challenging in a level-by-level representation,
a depth-first representation appears to be more suitable for this purpose. We
introduce several new depth-first encodings of the k^2-tree and present a linear-
time algorithm to identify identical subtrees. Our evaluation on Web Graph,
Graph Database and random matrices shows that this technique improves both
compression and peak memory usage, and in some cases, also speeds up matrix-
matrix multiplication due to better cache locality. In general, our results show
that the use of depth-first representations is a strategy worth pursuing, and in
Sect. 7 we describe possible lines of further investigation.

© The Author(s), under exclusive license to Springer Nature Switzerland AG 2026
G. Badkobeh et al. (Eds.): SPIRE 2025, LNCS 16073, pp. 28–44, 2026.
https://doi.org/10.1007/978-3-032-05228-5_4

2   Notation

Let Σ be a finite ordered alphabet of constant size σ. A *string*, (or sequence or array), of length n over alphabet Σ is denoted with $S[1,n] \in \Sigma^n$. We write $S[i..j]$ to denote the substring $S[i]S[i+1]\cdots S[j]$ if $1 \le i \le j \le n$ or the empty string otherwise. The Suffix Array [17] of S is a permutation of the integers $\{1,\ldots,n\}$ such that for $i = 2,\ldots,n$, $S[SA[i-1],n] \prec S[SA[i],n]$, where $\prec$ denotes the lexicographic ordering. The LCP Array [17] $LCP[2,n]$ is an array of integers such that $i = 2,\ldots,n$, $LCP[i]$ is the length of the longest common prefix between $S[SA[i-1],n]$ and $S[SA[i],n]$. Both SA and the LCP array can be computed in $O(n)$ time [12–15].

Given a sequence S and a symbol $c \in \Sigma$, $\mathtt{rank}_c(S,i)$ returns the number of occurrences of c in $S[1,i]$, and $\mathtt{select}_c(S,i)$ returns the position of the i-th occurrence of c in S or -1 if such occurrence does not exist. Using additional $o(n)$ bits of space we can preprocess S so that both $\mathtt{rank}$ and $\mathtt{select}$ operations can be computed in $O(1)$ time [18]. If $\Sigma = \{(,)\}$, the sequence S contains the same number of (and) symbols and no prefix $S[1,i]$ contains more) than (then S is called a *balanced parenthesis* sequence. Given a balanced parenthesis sequence S, $\mathtt{find\_close}(S,i)$ (resp. $\mathtt{find\_open}(S,i)$) returns the index of the closing (opening) parenthesis that matches a given opening (closing) parenthesis $S[i]$. Such operations can be supported in $O(1)$ time [19].

3   The Canonical Representation of k^2-trees

Given a square binary matrix M, the corresponding k^2-tree [4] is built by recursively dividing M using an *MX-Quadtree* strategy [21], splitting it into k^2 equal-sized submatrices, each containing n^2/k^2 cells. Each submatrix corresponds to a child of the root node, with its value set to 1 if at least one cell in the submatrix is 1; otherwise, its value is 0. The process is recursively applied to all submatrices with at least one 1 until all cells have been processed, see Fig. 1. This approach works seamlessly when n is a power of k. In cases where n is not a power of k, the matrix can be padded with additional rows and columns of zeros until its size becomes a power of k. In the following we will assume k is a constant, so the size of the padded matrix is still $O(n)$.

By construction, the k^2-tree is structured as a k^2-ary tree of height $h = \lceil \log_k n \rceil$, where each node, except the root, stores a single bit of information. For all levels $\ell < h$ internal nodes store a 1 while leaf nodes store a 0 meaning that the corresponding $k^{h-\ell} \times k^{h-\ell}$ submatrix only contains zero elements. At level h all nodes are leaves, each one storing an actual bit value of the input matrix. In [4] the authors propose a representation of the above tree, and therefore of the associated matrix, consisting of two bit arrays: T (tree) stores all the bits that are not in the last level of the tree. The bits are stored following a top-down level-wise traversal; L (leaves) stores all the bits of the last level of the tree.

In the following we call the above the *canonical representation*. For example, for the k^2-tree in Fig. 1, the corresponding T and L bit arrays are as follows (colors are used to represent tree levels following Fig. 1):

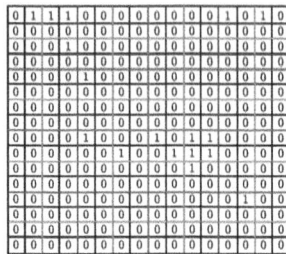

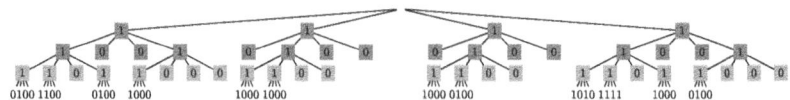

Fig. 1. A 16×16 matrix split recursively into 2×2 submatrices (top) and the corresponding k^2-tree with $k = 2$ (bottom). The color of each the internal nodes is related to the size of the submatrix it represents.

T: 1111 1001 0100 0100 1001 1101 1000 1100 1100 1101 1000
L: 0100 1100 0100 1000 1000 1000 1000 0100 1010 1111 1000 0100

The reason for which the two bit arrays T and L are considered separately is technical. To access specific elements of the input matrix it is necessary to efficiently navigate the k^2-tree. The authors observed that we can compute the i-th child of a node at position x using the formula $\mathtt{child}(x, i) = \mathtt{rank}_1(x) \cdot k^2 + i$ [4]. Hence, adding an $o(|T|)$ bits data structure supporting the $\mathtt{rank}_1$ operation over T in constant time, it is possible to efficiently navigate the tree. Later, Brisaboa et al. [5] extended the functionalities of the k^2-tree with operations still relying only the $\mathtt{rank}$ operation on T. There is no need to support the $\mathtt{rank}$ operation over L, so it can be stored using a different (simpler) representation.

For an $n \times n$ matrix with m nonzeros, [4] shows that the number of bits in this representation is bounded by $k^2 m \left(\log_{k^2}(n^2/m) + O(1) \right)$. However, such value is obtained for pathological inputs and the behavior in practice is much better. Experiments with web graphs [5] show that the actual space usage is significantly lower and competitive with other graph compression schemes.

4   Depth-First Traversal Representations

The canonical representation is extremely space efficient since it essentially uses a single bit for each node. Considering that it supports tree navigation in constant time it seems hard to improve. However, such representation appears ill suited to exploit the possible presence of identical submatrices in the input matrix. Such submatrices corresponds to identical subtrees but with the canonical representation the information on every subtree is partitioned into the different tree levels. It is therefore non trivial to detect equal subtrees within the canonical representation. Another issue of the canonical representation is that nodes which are

$P = 1111$
1001 1101 **0100** **1100** **0100** 1000 **1000**
0100 1100 **1000** **1000**
0100 1100 **1000** **0100**
1001 1101 **1010** **1111** **1000** 1000 **0100**

Fig. 2. The Plain Depth-First representation of the same tree as Fig. 1. The representation consists of the bit array P containing the concatenation of the 4-tuples of bits shown above: the first row represents the children of the root node. Each of the subsequent four rows corresponds to the subtrees rooted at each child of the root, listed in depth-first traversal order; different colors represent different tree levels.

close in the tree can be stored far apart: for some computation this layout can generate a large number cache misses. As a possible solution to the above shortcomings, in this section we describe alternative representations based on a depth first traversal of the k^2-tree: such approach stores together the information of a given subtree thus improving the locality of references and simplifying the task of detecting identical subtrees.

4.1   Plain Depth-First Representation

The simplest Depth-First representation consists in traversing the tree in depth-first order; when the visit reaches a non-leaf node u, we write the k^2 bits associated to u's children. The resulting bit array P has length mk^2, where m is the number of internal nodes of the k^2-tree. See Fig. 2 for an example. It is easy to see that P is a permutation of the bits in $T \cup L$ of the canonical representation; more precisely both P and $T \cup L$ can be partitioned into m blocks of k^2 bits, and the blocks of P are a permutation of the blocks of $T \cup L$.

We call this representation the Plain Depth First (PDF) representation since we do not add any additional information to speed up navigation. Note that given a pair of indices i, j we are still able to determine the entry $M[i][j]$ of the matrix represented by the k^2-tree, but this requires a depth first traversal of the tree up to the node representing the largest non-empty submatrix containing position i, j. Although the access to a single entry is inefficient, some operations involving the whole matrix (such as computing the vector product $y = Mx$) do require visiting the whole k^2-tree and this can be done with a simple left-to-right scan of the bit array P. Such visit is much more cache friendly than a visit of the canonical representation of the same tree

4.2   Enriched Depth-First Representation

The main drawback of the PDF representation is that in order to find the starting position in P of the subtree corresponding to the, say, third child of the root note, we must execute a visit of the first two subtrees. The cost of the visit is proportional to the number of nodes in such subtrees so this is a major problem when such subtrees are large: since the visit consists of a linear scan of a subarray of P, small subtrees are less of a problem.

The above observation suggests to "enrich" to plain depth-first representation, with additional information that allow us to skip large subtrees without visiting them. The simplest approach is to associate to the block of k^2 bits representing an internal node, the size of its subtrees with the exception of the last one. In the example of Fig. 2, with the block 1111 we would store the size of the first three subtrees, e.g. 7 (first subtree), 4 (second subtree), and 4 (third subtree). If we need to access the third subtree we skip $7 + 4 = 11$ blocks, while to access the fourth subtree we skip $7 + 4 + 4 = 15$ blocks. We do not store the size of the last subtree (7 in our example), since it is not used for skipping any subtree.

Storing this information for all nodes would be too expensive, so we choose a threshold τ and store the "skip" values only for subtrees of size larger than τ. For example if we set $\tau = 6$ we use "skip" values only for the root and for the first and last children of the root, whose subtrees have size 7. Since those children have two subtrees, they only need to store a single skip value (4 in our example). If N is the total number of nodes and we choose $\tau = f(N)$ we can estimate the overhead of storing the skip values as follows. Consider the subtrees which have more than $f(N)$ nodes and do not contain any subtree with more than $f(N)$ nodes. Clearly there are at most $N/f(N)$ such subtrees, and each such subtree will have at most $\log_k n$ ancestors. Not all ancestors are distinct but we can state that the number of node containing skip values is at most $O(N \log n / f(N))$. Assuming we use $O(\log N)$ bits for each skip value, the total overhead for skip values is $O(Nk^2 \log N \log n / f(N))$ bits. In our experiments we set $\tau = \sqrt{N}$ and $k = 4$ so the overhead is $O(\sqrt{N} \log N \log n)$ bits. Since $N = |P| + 1$, this is $o(|P|)$ when $N = \Omega(\log^{2+\epsilon} n)$, that is when the matrix is not pathologically sparse.

We call the above representation Enriched Depth First (EDF). There are a few details of the representation that has to be decided during the implementation. The first one is how to encode the skip values: to save space one could use a variable length code (eg Elias, Rice, Variable-Byte, etc. see for example [8, Chapter 11]) depending on the desired time-space tradeoff. In our experiments we found that setting $\tau = O(\sqrt{N})$ the number of skip values is relatively small so we do not compress them to favor speed.

The second implementation choice is where to store the skip values. A first alternative is to interleave them in the bit array P. This approach is cache friendly since the whole representation still consists of a single bit array which is mostly accessed in sequential scans. However, the size of the subtrees, hence the skip values, now depend also on the skip values stored at the lower levels: this makes the construction of the representation more complex since now it has to be done, bottom up and left to right. We therefore decided to store the skip values in a separate array S. This is also non trivial since, in order to skip a subtree T_i we need to skip the portion of the P array containing the encoding of T_i nodes *and* the portion of the S array containing the skip values for T_i. Hence the size of such portion of S has to be saved as well: this can be done in the array S itself. For the matrix-matrix multiplication algorithm (Sect. 6), we combined the above *static* set of skip values with a *dynamic* set: when the recursive multiplication algorithm reaches a submatrix whose corresponding subtree has no skip values such values are computed on the fly for all nodes in the subtree and stored in a temporary array S'. Since by construction the subtrees with no skip values have less than τ nodes computing S' takes $O(\tau)$ time and space.

4.3   Balanced Parenthesis Representation

In this section we introduce another depth-first based representation of k^2-trees based on Balanced Parenthesis (BP) encoding, following the ideas introduced by Munro and Raman [19]. Our representation follows the classical approach used for general trees; however we exploit the special structure of the k^2-tree introducing an optimization for the last-level nodes. Instead of representing these nodes using parenthesis, we store their actual values in a separate bit array L'.

Let $h = \lceil \log_k n \rceil$ denote the tree height. The construction of our BP representation is done as follows. We visit the k^2 tree in depth first order and we write to a vector B a (every time we start the visit of a node and we write a) every time the visit of a node is complete and we go back to the parent node. However, when we reach a level $h-1$ node which is not a leaf, instead of visiting its children we write a pair () and we write the k^2 bits associated to its children to the bit array L'. See Fig. 3 for an example.

Note that the subtrees rooted at a level $h-1$ node are represented by either (), if the have no children, or by (()) if the have children. Since in a k^2 tree every node has either k^2 children or none, the sequence (()) cannot represent a subtree rooted at a level $< h-1$. Hence there is a one-to-one correspondence between the occurrences of the pattern (()) in B, the subtrees with children at level $h-1$ and the blocks of k^2 bits in the L' array. By construction, the correspondence is order preserving in the sense that the i-th occurrence of (()) corresponds to the i-th block in L'. This means that to access the bit values stored in a level-h leaf, we need to locate the position p of its parent, and count the number x of occurrences of the pattern (()) in B up to position p; the desired bit values are stored starting from position xk^2 in L'.

Let t denote the number of nodes in levels $0, \ldots, h-1$ and ℓ denotes the number of level-h leaves. It is immediate to see that the parenthesis vector B and the bit vector L' have length

$$|B| = 2t + \frac{2\ell}{k^2}, \qquad |L'| = \ell. \tag{1}$$

$$B = ($$
$$(((())(())()(()))()()(((())()()()))$$
$$(()(((())(()))()()()()()))$$
$$(()(((())(()))()()()()))$$
$$(((())(())()(()))()()(((())()()()))$$
$$)$$

$L' = 0100\ 1100\ 0100\ 1000\ 1000\ 1000\ 1000\ 0100\ 1010\ 1111\ 1000\ 0100$

Fig. 3. The balanced parenthesis representation of the k^2-tree of Fig. 1. The resulting parenthesis vector B is the concatenation of six rows of parentheses: the first and last row are the open and close parenthesis for the root; and the other four rows represent the subtrees rooted at the four children of the root. Also shown is the bit vector L' containing the values in the bottom level stored according to the depth-first order visit.

The BP representation therefore takes roughly twice the space used by the canonical and (Enriched) Depth-First representations that take $t + \ell$ bits plus possibly lower order terms to support fast navigation. The advantages of the BP representation are the locality of reference and the possibility, explored in Sect. 5 to compress identical subtrees. To ensure constant time navigation with the BP representation we need to support the `find_close` operation in constant time that, as recalled in Sect. 2, can be supported using $o(|B|)$ bits of auxiliary space. In addition, when we reach an internal node at level $h - 1$, to read the corresponding bits in L' we need to support the constant time rank operation for the pattern (()) on the array B; this can be achieved by a straightforward modification of the rank data structures for the `rank` data structure still using $o(|B|)$ bits of auxiliary space.

5   Subtree Compression

Identical submatrices in the original input matrix M can lead to identical subtrees in the k^2-tree representing M. We now show how to detect identical subtrees and replace them with a "pointer" to a previous occurrence, and how to perform navigation operations on this compressed representation. The algorithm for detecting identical subtrees are based on the Suffix Array and LCP array and takes linear time, i.e. time proportional to the size of k^2-tree representation.

5.1   Compressed Balanced Parenthesis

For the BP representation, instead of considering identical subtrees in the traditional sense, we will consider the wider concept of interchangeable subtrees according to the following definition, that, for our convenience, excludes the pathological case of subtrees consisting of a single node.

Definition 1. *Let T_1 and T_2 be subtrees of height at least 1 of the k^2-tree T. We say that T_1 and T_2 are interchangeable if their respective balanced parenthesis sequences generated during the visit described in Sect. 4.3 are identical.*

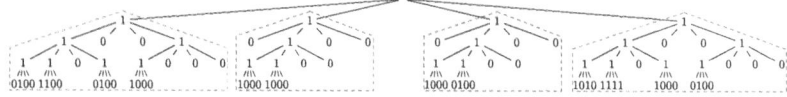

Fig. 4. The subtrees surrounded by dashed lines of the same color are interchangeable according to Definition 1. Note that the subtrees surrounded in red are interchangeable even if the values stored in some of the leaves at the last level are different.

We point out that the balanced parenthesis sequences generated during the visit in Sect. 4.3 does not include any information regarding the last level leaves since they are stored in binary form in the vector L'. Hence, some subtrees are interchangeable even if some of the values stored in the last level differ, see for example the subtrees surrounded in red in Fig. 4.

Lemma 1. *The subtree starting at position i in B is interchangeable with the subtree starting in position j, if and only if, setting $c = $ find_close(i), the substring $B[i..c]$ is equal to $B[j..j + c - i]$.*

Because of the above lemma, we can detect all interchangeable subtrees in linear time as follows. First we compute the suffix array SA and LCP array of the parenthesis sequence B. For $i = 1, \ldots |B| - 1$, if $B[SA[i]] = ($, we set $c = $ find_close$(SA[i])$. Then, if $c - SA[i] \leq LCP[i]$ by the above lemma the subtree starting at position $SA[i]$ is interchangeable with the subtree starting at $SA[i - 1]$. During this procedure we ignore positions i when $LCP[i] \leq 32$ (or some other larger threshold) since in that case an interchangeable subtree would be so small that replacing it with a pointer would offer no benefit.

The above algorithm will find *all* interchangeable subtrees but for our purposes this is too much. The problem is that subtrees of interchangeable subtrees are still interchangeable but we are only interested in finding *maximal* interchangeable subtrees, i.e. subtrees which are not contained in larger interchangeable subtrees. Therefore, instead of scanning the SA array we scan the parenthesis sequence B: for $j = 1, \ldots, |B|$, if $B[j] = ($ we set $i = SA^{-1}[j]$ and we check whether $LCP[i] > $ find_close$(j) - j$. If the subtree starting at j is interchangeable with a previous subtree, we restart the scanning of B at position find_close$(j) + 1$ therefore skipping all non-maximal interchangeable subtrees contained in the subtree starting at j. Note that if we have k consecutive suffixes $SA[i], \ldots, SA[i + k - 1]$ each representing interchangeable subtrees, then they are all interchangeable among themselves. In this case we select the subtree with the leftmost starting position as the reference subtree and the other subtrees will point to this reference subtree.

Having identified which (maximal) subtrees should be replaced by a pointer, we call them *pruned* subtrees, the construction of the Compressed Balanced Parenthesis (CBP) representation is done as follows. Let h denote the number of tree levels, ℓ denote the number of level h leaves (whose value are stored in L'), and p denote the number of pruned subtrees. Given the uncompressed sequence of balanced parenthesis B we build the compressed sequence B_c by replacing the

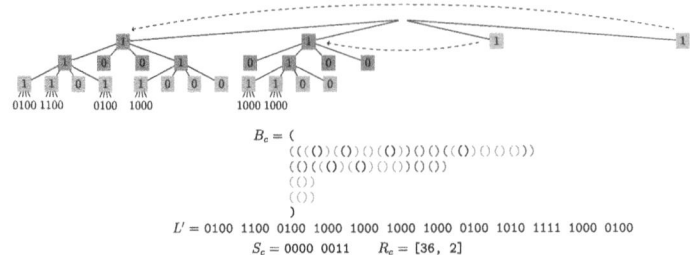

$$B_c = ($$
$$((()))(())()(())()()((())()()))$$
$$(())(())(())()()()()()()()$$
$$(())$$
$$(())$$
$$)$$

$L' = 0100\ 1100\ 0100\ 1000\ 1000\ 1000\ 1000\ 0100\ 1010\ 1111\ 1000\ 0100$

$S_c = 0000\ 0011 \quad R_c = [36, 2]$

Fig. 5. The Compressed Balanced Parenthesis representation for the k^2-tree shown in Fig. 3 with two subtrees (in orange) pruned and replaced with pointers to interchangeable subtrees. The two pointers are explicitly stored in the array R_c

subsequence representing each pruned subtree with the pattern (()) (note B_c is still balanced). Since this is the same pattern used to represent internal nodes at level $h - 1$, we also introduce a binary vector $S_c[1, \ell + p]$ such that for each occurrence of the pattern (()) in B_c we store a 0 in S_c if it represent an internal node level $h-1$, and we store a 1 if it represents a pruned subtree. We also use an integer array $R_c[1, p]$ such that $R_c[v]$ is the starting position in B_c of the reference subtree for the v-th pruned subtree. When the navigation reaches at position i of B_c a (()) pattern which is not at level $h - 1$, we compute $r = \mathtt{rank}_{(())}(B_c, i)$, $v = \mathtt{rank}_1(S_c, r)$. By construction, the pruned subtree starting at position i is the v-th pruned subtree, hence $j = R_c[v]$ is the starting position in B_c of the reference subtree for the one starting at i. Finally, since interchangeable subtrees can have different values stored at level h (i.e. in the bitarray L'), when jumping from position i to position j we need to compute the length of the portion of L' corresponding to the tree nodes between position j and position i. Such lengths can be stored in another integer array $L_c[1, p]$ or computed on the fly using an auxiliary binary array containing the unary encoding of the number of level-h for each pruned subtree (details will be given in the full paper). Figure 4 shows the k^2-tree of Fig. 3 with two subtrees pruned and the resulting B_c sequence (Fig. 5).

5.2   Compressed Plain/Enriched Depth-First Representations

Finding repeated subtrees in the Plain Depth-First (PDF) representation is conceptually similar to the BP case. The main difference is that in PDF, leaf values are mixed with internal nodes; hence when compressing such representations we are interested in finding *maximal identical subtrees* i.e. subtrees that match completely from root to leaves.

Let P denote the binary sequence representing the PDF encoding of the input k^2-tree, and assume $k = 2$. To find identical subtrees we still use the SA and LCP arrays. However, a substring $P[a, b]$ encoding a subtree t now can be identical to a substring $P[c, d]$ not representing a subtree; in this case the subtree t cannot be compressed (replaced by a pointer). To avoid this spurious matches and detect identical subtrees, we build an auxiliary sequence P', such that $P'[i]$ is a pair containing the depth of node i and the 4-bit block associated to node i. For example, for the tree encoding in Fig. 2 the sequence P' starts with:

(1 1111) (2 1001) (3 1101) (4 0100) (4 1100) (4 0100) (3 1000) ...

Using the SA and LCP array of P' we can compute the maximal identical subtrees as in Sect. 5.1, finding equal subsequences $P'[i, i+k]$ and $P'[j, j+k]$ of length at least equal to the number of nodes in the subtree rooted at $P'[i]$. Each subsequence representing a pruned subtree is replaced by the 4-bit block 0000 which is a block that cannot appear in the original sequence P. As for the BP case, we maintain an integer array R_c storing pointers to the subtree identical to the pruned one. When we reach a node associated to a 0000 block, to find the corresponding subtree we only need to count (using a rank data structure) the number of 0000 blocks up to the current one and read the corresponding pointer from R_c (there is no need of the bitarray S_c since 0000 can only denote a pruned subtree).

We call the resulting representation CPDF (Compressed PDF). As in Sect. 4.2, we can enrich it by computing the skip values on the pruned tree, treating the roots of pruned subtrees as leaves. This data is stored separately to speed up navigation in the upper tree levels. The resulting structure is called CEDF (Compressed EDF). Currently, CEDF does not support dynamic computation of skip values, but we plan to add this soon.

6 Experimental Results

In this section, we present preliminary results on our depth-first representations of k^2-trees. We evaluate not only compression performance but also the ability to support matrix operations efficiently. Specifically, we benchmark matrix-matrix multiplication, for which an efficient recursive implementation exists for canonical k^2-trees, as described by Arroyuelo et al. [1]. Such implementation is based on the classical divide-and-conquer recursive procedure: Given two matrices A and B, if we partition them into four submatrices of equal size $A = \begin{pmatrix} A_0 & A_1 \\ A_2 & A_3 \end{pmatrix}$, $B = \begin{pmatrix} B_0 & B_1 \\ B_2 & B_3 \end{pmatrix}$, then the product can be obtained by the recursive formula $A \times B = \begin{pmatrix} A_0 \times B_0 + A_1 \times B_2 & A_0 \times B_1 + A_1 \times B_3 \\ A_2 \times B_0 + A_3 \times B_2 & A_2 \times B_1 + A_3 \times B_3 \end{pmatrix}$. This approach aligns well with k^2-trees for $k = 2$, where each submatrix maps to a root's child. Since the resulting matrix is compressed, the algorithm also reflects construction efficiency. Its recursive nature and arithmetic demands make it ideal for comparing k^2-tree variants. Experiments were run on an Ubuntu machine with 394 GiB RAM and an Intel Xeon Gold 6132 @ 2.60 GHz. We measured running time and peak memory with /usr/bin/time across five implementations: (1) K2-TREE[1] the canonical level-by-level k^2 implementation by Arroyuelo et al. [1]; (2) K2-EDF, K2-CPDF and K2-CEDF[2] implementations based on the depth-first representations

---

[1] https://github.com/adriangbrandon/rpq-matrix/tree/main.:   https://github.com/adriangbrandon/rpq-matrix

[2] https://github.com/acubeLab/k2tree/tree/compressionk2tree. K2-CPDF & K2-CEDF: https://github.com/acubeLab/k2tree/

Table 1. Size, number of nonzero, number of nodes (canonical representation) per nonzero, and density of nonzero elements for the matrices in the Web Graph dataset.

File	Matrix size	# Nonzero	Nodes/Nonz	Density
CN-2000	325,557	3,216,152	0.87	$3 \cdot 10^{-5}$
AM-2008	735,323	5,158,388	2.32	$9 \cdot 10^{-6}$
EU-2005	862,664	19,235,140	1.02	$2 \cdot 10^{-5}$
IN-2004	1,382,908	16,917,053	0.73	$8 \cdot 10^{-6}$
ID-2004	7,414,866	194,109,311	0.60	$3 \cdot 10^{-6}$
AR-2005	22,744,080	639,999,458	0.68	$1 \cdot 10^{-6}$
UK-2005	39,459,925	936,364,282	0.68	$6 \cdot 10^{-7}$

described in Sects. 4.2 and 5.2 respectively. For the enriched representations we used as threshold the square root of the number of nodes; (3) K2-BP and K2-CBP[3] implementations based on the BP representations described in Sects. 4.3 and 5 and implemented using the sdsl-lite library [10]. We tested them using three different datasets: Web Graph [2,3], Graph Database [1] and Random Matrices [1]. For the matrices of Web Graph, we tested seven web graphs reported in Table 6. Since matrices have different sizes, in this case matrix multiplication was performed by squaring each matrix. For the Graph Database and Random Matrices, a description and results of both datasets can be found in the appendix (Table 1).

We first present the space usage results of each variant. As shown in Table 2, K2-BP uses the most space in bits per nonzero, as expected since BP representation uses 2 bits per node. K2-CBP reduces this by roughly a factor of three by compressing the BP representation. Interestingly, Table 7 in the appendix shows that on random matrices K2-CBP consistently requires fewer bits than K2-BP across all densities except for matrices with a density of 10^{-4}. At such a low density, the number of nonzero elements is minimal, making it unlikely to find identical subtrees within a matrix of that size (10^3). K2-CBP outperforms K2-TREE only in ID-2004, AR-2005, and UK-2005; in the remaining web graphs, the matrices are too small for compression to offset the BP overhead. K2-CPDF consistently achieves the best compression across all Web Graph datasets. In nearly all cases, it improves upon K2-TREE by at least one bit per nonzero, with AM-2008 being the only exception. Comparing K2-CPDF and K2-CEDF, we find that the space used by skip values is relatively modest. K2-CEPF also uses less space than K2-TREE, except in AM-2008. The comparison between K2-TREE and K2-EDF shows that eliminating support for the rank operation and replacing it with skip values values results, for the largest matrices, in lower space usage. For sparser matrices, skip values can use more space, as shown in Table 5 for the Graph Database dataset. Here, K2-EDF uses more space than K2-TREE; however

---

[3] https://github.com/Yhatoh/k2tree/tree/experimental. & K2-CBP: https://github.com/Yhatoh/k2tree

Table 2. Disc space usage of Web Graph; table reports the disc space usage in bits per nonzero for each data structure

File	K2-TREE	K2-EDF	K2-CPDF	K2-CEDF	K2-BP	K2-CBP
CN-2000	3.50	3.92	**2.68**	3.07	9.61	3.92
AM-2008	9.28	9.88	**9.26**	9.85	26.25	12.59
EU-2005	4.07	4.27	**3.18**	3.36	11.34	4.47
IN-2004	3.12	3.11	**2.00**	2.16	7.80	2.90
ID-2004	2.56	2.45	**1.31**	1.34	6.20	1.87
AR-2005	2.92	2.75	**1.69**	1.69	7.24	2.37
UK-2005	2.89	2.75	**1.66**	1.68	7.16	2.35

Table 3. Results for matrix multiplication for web graph matrices; table reports the peak memory usage in bits per nonzero and running time in minutes.

File	K2-TREE		K2-EDF		K2-CPDF		K2-CEDF		K2-BP		K2-CBP	
	bits	min	bits	min	bits	min	bits	min	bits	min	bits	min
CN-2000	49.26	**0.1**	32.92	0.2	**32.39**	0.3	32.68	0.3	108.82	0.6	102.59	1.7
AM-2008	61.76	**1.3**	**40.95**	4.5	42.30	9.95	42.83	8.2	156.40	9.5	141.68	38.8
EU-2005	58.46	**2.5**	38.57	4.8	**37.79**	8.9	37.96	7.2	177.31	13.6	169.64	36.0
IN-2004	40.67	1.1	28.17	**0.9**	**27.40**	2.0	27.55	1.1	104.19	3.4	98.83	4.4
ID-2004	28.01	125	20.44	**26.4**	**19.41**	47.6	19.44	33.1	91.61	146	86.65	242
AR-2005	37.54	84.6	25.16	**76.4**	**24.26**	230	24.28	117	125.98	246	120.73	652
UK-2005	25.72	**89.4**	16.70	215	**15.76**	657	15.77	300	85.70	499	80.31	2090

K2-CPDF and K2-CEDF still achieve significant subtree compression and use less space. For this dataset, we excluded K2-BP and K2-CBP due to the matrices being larger but significantly sparser, rendering those variants non-competitive.

For matrix multiplication we measured both running time and peak memory usage. Peak memory usage is reported in terms of bits per total number of nonzero elements in the input matrices; note however that all algorithms compute the product in-memory so the space usage also depends on the structure of the product matrix. In terms of peak memory usage, Table 3 shows that K2-BP and K2-CBP are the ones using more memory. This is expected since both algorithms produce the output in the space inefficient K2-BP format. Table 3 also shows that K2-EDF, K2-CPDF, and K2-CEDF use much less memory than K2-TREE; we believe the reason is that K2-TREE uses a queue to traverse both input trees at the moment of merging two trees, while the depth-first variants process inputs sequentially and avoid this. Table 6 in the appendix confirms this phenomenon, though the memory gap is smaller. This is likely because in Web Graph dataset, the output matrices are sometimes much denser than inputs, leading to larger trees and higher memory use. Graph Database matrices are sparser, producing smaller trees and reducing queue overhead in K2-TREE. Table 3 also reports running times for matrix multiplication for the Web Graph dataset. Using skip values, K2-CEDF and K2-EDF outperform K2-CPDF in all cases except CNR-2000, where the matrix is smaller and skip values are less effective. K2-EDF is faster than K2-TREE on IN-2004, ID-2008, and AR-2005, but slower on CNR-2000, AM-2008, EU-2005, and UK-2005. A possible explanation is that CNR-2000, AM-2008, and EU-2005 have the most nodes per nonzero, increasing K2-EDF's cost, especially when logically splitting subtrees to implement the splitting of submatrices. For UK-2005, K2-TREE is significantly faster than the other algorithms; we conjecture this is because of the lower density of the matrix. Indeed, Table 6 in the appendix confirms that in terms of speed K2-TREE outperforms K2-EDF at smaller densities. For the random matrices, Table 8 in appendix shows that at higher densities, K2-EDF outperforms K2-TREE, and K2-CPDF is faster than K2-EDF due to pruning small subtrees, which increases disk usage (see Table 7 in appendix) but lowers traversal overhead. At lower densities, K2-TREE remains faster. Although more experiments are clearly needed for a complete understanding, the above results show that it is indeed possible to significantly save space exploiting repeated subtrees and that in some setting this can lead even to a reduction of the running time for an important operation such as matrix multiplication.

7   Conclusions and Further Works

In this paper we introduced new k^2-tree representations using a depth-first layout, which is more cache friendly and enables subtree compression. Experiments on web graph adjacency matrices confirm the presence of repeated subtrees. Our Compressed Plain Depth-First format achieves the best compression in disk space and peak memory, though with slower matrix-matrix multiplication. When the input matrix is not too sparse, our Enriched Depth-First format is both smaller and faster than the standard k^2-tree. Although still preliminary, our results show that depth first layouts can bring advantages for some classes of matrices and operations, and we plan to pursue this idea along the following lines: (1) we plan to improve the performance of our balanced parenthesis representations by implementing our own version of the find_close operation and re-implementing the current one; (2) we plan to experiment with the DFUDS representation with possibly some of the improvements from [11]; (3) we plan to extend our comparison to a multithread setting; (4) we plan to investigate whether subtree compression can lead to a speed-up for others matrix operations.

Acknowledgments. The authors whish to thank Paolo Ferragina and Francesco Tosoni for the many fruitful discussions on k^2-tree representations. This work has been partially funded by the INdAM-GNCS Project CUP E53C24001950001, and by the PNRR ECS00000017 Tuscany Health Ecosystem, Spoke 6, CUP I53C22000780001, funded by the NextGeneration EU programme.

Disclosure of Interests. The authors declare no competing interests.

Appendix: Experiments on additional datasets

In this appendix, we report experiments on two additional datasets. (1) Graph Database: We use the dataset from [1], which consists of squared matrices of size $348{,}945{,}080^2$. Our experiments focus on matrices with at least 10^6 ones (density $\geq 10^{-11}$). Since all matrices share the same dimensions, we pair them and perform matrix multiplication. We tested 18 matrices, as detailed in Table 4. (2) Random Matrices: Following Arroyuelo et al. [1], we generated $10^3 \times 10^3$ matrices with densities 2×10^{-1}, 10^{-1}, 10^{-2}, 10^{-3}, and 10^{-4}. For each density, 10 matrices with uniformly distributed values were created. We then conducted matrix-matrix multiplication on all pairs.

Table 4. Number of nonzero, nodes (canonical representation) per nonzero, and density of nonzero elements for the matrices used in Graph Database dataset.

Matrix ID	# Nonzero	Nodes/Nonz	Density	Matrix ID	# Nonzero	Nodes/Nonz	Density
0001	105,901,917	2.80	$9 \cdot 10^{-10}$	2831	78,222,146	10.25	$6 \cdot 10^{-10}$
0004	53,883,532	8.79	$4 \cdot 10^{-10}$	3744	18,499,841	12.52	$2 \cdot 10^{-10}$
0006	53,883,271	6.48	$4 \cdot 10^{-10}$	0889	19,341,438	11.52	$2 \cdot 10^{-10}$
0007	53,883,271	3.71	$4 \cdot 10^{-10}$	3868	14,895,138	10.58	$1 \cdot 10^{-10}$
0008	39,609,445	10.29	$3 \cdot 10^{-10}$	2670	166,682,725	8.93	$1 \cdot 10^{-9}$
0012	49,516,226	4.46	$4 \cdot 10^{-10}$	3936	16,611,881	6.71	$1 \cdot 10^{-10}$
0159	10,089,284	4.54	$8 \cdot 10^{-11}$	4660	17,457,608	11.21	$1 \cdot 10^{-10}$
0606	19,961,090	10.56	$2 \cdot 10^{-10}$	3867	14,895,138	10.54	$1 \cdot 10^{-10}$
1619	19,430,075	9.99	$2 \cdot 10^{-10}$	3935	18,421,701	8.58	$2 \cdot 10^{-10}$

Table 5. Disc space usage of Graph Database matrices; table reports the disc space usage in bits per nonzero by the data structure.

Matrix ID	K2-TREE	K2-EDF	K2-CPDF	K2-CEDF	Matrix ID	K2-TREE	K2-EDF	K2-CPDF	K2-CEDF
0001	11.21	11.30	**5.56**	5.59	2831	35.70	35.88	**31.43**	31.53
0004	35.16	35.46	**29.56**	29.75	3744	50.07	50.71	**43.93**	44.27
0006	25.92	26.16	**19.68**	19.81	0889	46.09	46.68	**40.75**	41.10
0007	14.83	14.99	**9.57**	9.65	3868	42.31	42.93	**37.36**	37.73
0008	41.18	41.57	**36.15**	36.39	2670	40.99	41.26	**36.15**	36.31
0012	17.86	18.08	**15.02**	15.17	3936	26.84	27.31	**37.26**	37.60
0159	18.15	18.63	**16.29**	16.67	4660	44.84	45.44	**39.47**	39.83
0606	42.24	42.77	**38.30**	38.63	3867	42.17	42.78	**37.23**	37.60
1619	39.96	40.53	**36.37**	36.72	3935	34.32	34.81	**32.23**	32.59

Table 6. Results for matrix multiplication for graph database matrices; table reports the peak memory usage in bits per nonzero and running time in minutes.

Matrix ID	K2-TREE		K2-EDF		K2-CPDF		K2-CEDF	
	bits	min	bits	min	bits	min	bits	min
0001×2831	64.16	**0.03**	60.64	0.65	**51.51**	15.0	60.64	0.40
3744×0004	40.83	**13.0**	38.80	97.9	**34.41**	127	34.61	88.6
0006×0889	64.66	**58.1**	54.42	352	**49.40**	380	49.59	306
3868×0007	21.97	**0.3**	20.94	3.9	**16.35**	18.4	16.49	3.2
0008×2670	42.87	**84.3**	40.62	582	**37.23**	917	37.41	596
3936×0012	21.31	**2.5**	20.31	17.6	**18.33**	64.8	18.53	23.3
0159×4660	5.59	**0.01**	4.93	0.03	**5.33**	1.52	4.97	0.04
0606×3867	44.64	**0.2**	42.51	1.4	**39.54**	7.5	39.91	1.3
1619×3935	39.80	**9.6**	37.88	61.0	**36.58**	180	36.93	86.0

Table 7. Disc space usage for random matrices; table reports average total disc space usage by the data structure in bits per nonzero.

Dens.	K2-TREE	K2-EDF	K2-CPDF	K2-CEDF	K2-BP	K2-CBP
0.2	**4.86**	4.91	10.37	10.69	13.63	8.46
0.1	**6.73**	6.95	18.50	19.13	19.10	12.55
10^{-2}	**13.58**	14.25	33.90	35.54	42.53	19.17
10^{-3}	**22.14**	30.80	40.72	46.28	55.41	36.14
10^{-4}	**44.03**	58.17	48.77	80.51	113.74	125.18

Table 8. Results for matrix multiplication for random matrices; table reports average peak memory usage by the multiplication in bits per nonzero and average running time in seconds.

Dens.	K2-TREE		K2-EDF		K2-CPDF		K2-CEDF		K2-BP		K2-CBP	
	bits	sec	bits	sec	bits	sec	bits	sec	bits	sec	bits	sec
0.2	88	9.457	82.25	3.568	**81.63**	2.526	82.39	**2.487**	123	18.387	121	27.80
0.1	180	4.373	163	2.631	**161**	2.009	163	**1.992**	237	12.382	235	18.59
10^{-2}	1631	**0.190**	1547	0.288	1563	0.372	1561	0.373	1974	0.890	1967	2.74
10^{-3}	15005	**0.022**	14706	0.032	14778	0.034	14743	0.034	17751	0.055	17764	0.12
10^{-4}	149707	**0.006**	144942	0.008	146101	0.008	145330	0.008	175762	0.011	175989	0.014

References

1. Arroyuelo, D., Gómez-Brandón, A., Navarro, G.: Evaluating regular path queries on compressed adjacency matrices. VLDB J. **34**(1), 2 (2025)

2. Boldi, P., Rosa, M., Santini, M., Vigna, S.: Layered label propagation: a multires-olution coordinate-free ordering for compressing social networks. In: Proceedings of the 20th International Conference on World Wide Web. ACM Press (2011)
3. Boldi, P., Vigna, S.: The WebGraph framework I: Compression techniques. In: Proc. of the Thirteenth International World Wide Web Conference (WWW 2004), pp. 595–601. ACM Press, Manhattan, USA (2004)
4. Brisaboa, N.R., Ladra, S., Navarro, G.: k2-trees for compact web graph representa-tion. In: International Symposium on String Processing and Information Retrieval, pp. 18–30. Springer (2009)
5. Brisaboa, N.R., Ladra, S., Navarro, G.: Compact representation of web graphs with extended functionality. Inf. Syst. **39**, 152–174 (2014). https://doi.org/10.1016/j.is.2013.08.003,https://www.sciencedirect.com/science/article/pii/S0306437913001051
6. Claude, F., Navarro, G.: Fast and compact web graph representations. ACM Trans. Web (TWEB) **4**(4), 1–31 (2010)
7. de Bernardo, G., Gagie, T., Ladra, S., Navarro, G., Seco, D.: Faster com-pressed quadtrees. J. Comput. Syst. Sci. **131**, 86–104 (2023). https://doi.org/10.1016/j.jcss.2022.09.001, https://www.sciencedirect.com/science/article/pii/S0022000022000629
8. Ferragina, P.: Pearls of algorithm engineering. Cambridge University Press (2023). https://doi.org/10.1017/9781009128933
9. García, S.A., Brisaboa, N.R., Bernardo, G. D., Navarro, G.: Interleaved k2-tree: indexing and navigating ternary relations. In: 2014 Data Compression Conference, pp. 342–351 (2014). https://doi.org/10.1109/DCC.2014.56
10. Gog, S., Beller, T., Moffat, A., Petri, M.: From theory to practice: plug and play with succinct data structures. In: Gudmundsson, J., Katajainen, J. (eds.) SEA 2014. LNCS, vol. 8504, pp. 326–337. Springer, Cham (2014). https://doi.org/10.1007/978-3-319-07959-2_28
11. Jansson, J., Sadakane, K., Sung, W.K.: Ultra-succinct representation of ordered trees with applications. J. Comput. Syst. Sci. **78**(2), 619–631 (2012). https://doi.org/10.1016/j.jcss.2011.09.002
12. Kärkkäinen, J., Sanders, P., Burkhardt, S.: Linear work suffix array construction. J. ACM **53**(6), 918–936 (2006)
13. Kärkkäinen, J., Manzini, G., Puglisi, S.J.: Permuted longest-common-prefix array. In: Kucherov, G., Ukkonen, E. (eds.) CPM 2009. LNCS, vol. 5577, pp. 181–192. Springer, Heidelberg (2009). https://doi.org/10.1007/978-3-642-02441-2_17
14. Kim, D.K., Sim, J.S., Park, H., Park, K.: Linear-time construction of suffix arrays. In: Proc. 14th Symposium on Combinatorial Pattern Matching (CPM '03), pp. 186–199. Springer-Verlag LNCS n. 2676 (2003)
15. Ko, P., Aluru, S.: Space efficient linear time construction of suffix arrays. In: Proc. 14th Symposium on Combinatorial Pattern Matching (CPM '03), pp. 200–210. Springer-Verlag LNCS n. 2676 (2003)
16. Ladra, S., Paramá, J.R., Silva-Coira, F.: Scalable and queryable compressed storage structure for raster data. Inf. Syst. **72**, 179–204 (2017)
17. Manber, U., Myers, G.: Suffix arrays: a new method for on-line string searches. SIAM J. Comput. **22**(5), 935–948 (1993)
18. Munro, J.I.: Tables. In: Proceeding of the 16th Conference on Foundations of Soft-ware Technology and Theoretical Computer Science, pp. 37–42. Lecture Notes in Computer Science, vol. 1180, Springer (1996)

19. Munro, J.I., Raman, V.: Succinct representation of balanced parentheses and static trees. SIAM J. Comput. **31**(3), 762–776 (2002). https://doi.org/10.1137/S0097539799364092
20. Navarro, G.: Compact Data Strucures. Cambridge University Press (2016)
21. Samet, H.: Foundations of multidimensional and metric data structures. Morgan Kaufmann (2006)

Dorst–Smeulders Coding for Arbitrary Binary Words

Alessandro De Luca[1] and Gabriele Fici[2]([✉])

[1] DIETI, Università di Napoli Federico II, Naples, Italy
alessandro.deluca@unina.it
[2] Dipartimento di Matematica e Informatica, Università di Palermo, Palermo, Italy
gabriele.fici@unipa.it

Abstract. A binary word is Sturmian if the occurrences of each letter are balanced, in the sense that in any two factors of the same length, the difference between the number of occurrences of the same letter is at most 1. In digital geometry, Sturmian words correspond to discrete approximations of straight line segments in the Euclidean plane. The Dorst–Smeulders coding, introduced in 1984, is a 4-tuple of integers that uniquely represents a Sturmian word w, enabling its reconstruction using $|w|$ modular operations, making it highly efficient in practice. In this paper, we present a linear-time algorithm that, given a binary input word w, computes the Dorst–Smeulders coding of its longest Sturmian prefix. This forms the basis for computing the Dorst–Smeulders coding of an arbitrary binary word w, which is a minimal decomposition (in terms of the number of factors) of w into Sturmian words, each represented by its Dorst–Smeulders coding. This coding could be leveraged in compression schemes where the input is transformed into a binary word composed of long Sturmian segments. Although the algorithm is conceptually simple and can be implemented in just a few lines of code, it is grounded in a deep analysis of the structural properties of Sturmian words.

Keywords: Sturmian word · Factorization · Dorst–Smeulders coding

1 Introduction

Sturmian (finite) words are binary balanced words. They can be used to describe straight line segments in the discrete plane. In 1984, Dorst and Smeulders introduced a coding for uniquely representing a Sturmian word [9]. Although other codings were proposed in the literature (see, e.g., [13]) the coding of Dorst and Smeulders has a clear interpretation in terms of combinatorics on words. In fact, Sturmian words are factors of (lower primitive) Christoffel words, and Christoffel

Gabriele Fici is partly supported by MUR project PRIN 2022 APML – 20229BCXNW, funded by the European Union – Mission 4 "Education and Research" C2 - Investment 1.1.

G. Badkobeh et al. (Eds.): SPIRE 2025, LNCS 16073, pp. 45–53, 2026.
https://doi.org/10.1007/978-3-032-05228-5_5

words have lots of interesting combinatorial properties. For every length $\ell > 0$ and every height (number of 1's) h coprime with ℓ, there is exactly one Christoffel word $u_{h,\ell}$. Moreover, Christoffel words are precisely the Lyndon Sturmian words. Let w be a Sturmian word of minimum period p. The Lyndon conjugate of the prefix of length p of w is therefore a Christoffel word $u_{h,p}$, and $w_1 \cdots w_p = \sigma^s(u_{h,p})$ for some $s \geq 0$, where σ is the usual (right) shift operator. The word w is then uniquely determined by its length n, its period p, the height h, and the shift s. The 4-tuple (n, p, h, s) is the *Dorst–Smeulders coding* of w. In this paper, we extend the Dorst–Smeulders coding to an arbitrary binary word w by considering a particular factorization of w in Sturmian words and encoding each term with its Dorst–Smeulders coding.

Let Σ be an alphabet, and let $L \subseteq \Sigma^*$ be a language such that $\Sigma \subseteq L$. Then every word w in Σ^* can be factored in words of L, i.e., there exist $x_1, \ldots, x_k \in L$ such that $w = x_1 \cdots x_k$. We call k the *length* of the factorization. We call a factorization $w = x_1 \cdots x_k$ *minimal* if any other factorization in words of L has length at least k. For example, if $L = Pal$ is the language of palindromes, the length of a minimal factorization of w is also called the *palindromic length* of w [11].

For some languages, a minimal factorization can be obtained by the *greedy algorithm*, which consists in taking the longest prefix of w that belongs to L and recursing on what remains after removing this prefix from w. This is the case, for example, for the language Lyn of Lyndon words, where the minimal factorization is also called the *Chen–Fox–Lyndon factorization* [4], but it is not the case for the language of palindromes. For example, the greedy factorization of $w = 0010$ is $w = 00 \cdot 1 \cdot 0$ and it is not minimal, since $w = 0 \cdot 010$.

In particular, if the language L is *factorial* (i.e., it is closed under taking factors), then the greedy algorithm always produces a minimal factorization— this can be easily proved by contradiction. Actually, it is sufficient that the language is closed under taking suffixes [5]. By symmetry, for languages closed under taking prefixes, the right-to-left greedy factorization produces a minimal factorization.

The complexity of the problem of determining a minimal factorization depends on the language L. For example, it is known that this complexity is linear in the length of the input word for Pal [12] and Lyn [10].

For those languages such that the greedy algorithm produces a minimal factorization, this complexity reduces to the complexity of computing the longest prefix of w that belongs to L.

In this paper, we are particularly interested in the language $Sturm$ of Sturmian words. A binary word is Sturmian if the occurrences of each letter are balanced, in the sense that in any two factors of the same length, the difference between the number of occurrences of the same letter is at most 1. The language $Sturm$ is a factorial language; hence, a minimal Sturmian factorization can be obtained by the greedy algorithm. We then call *Dorst–Smeulders coding* of an arbitrary binary word w the list of Dorst–Smeulders codings of the terms in the minimal Sturmian factorization of w obtained from the greedy algorithm.

We present a linear-time algorithm to compute the Dorst–Smeulders coding of an arbitrary binary word. The core of the algorithm is a procedure that, given an arbitrary binary input word w, computes the longest prefix of w that is Sturmian (a different linear time algorithm for this, based on geometric considerations, is described in [2]). Our procedure scans w from left to right, letter by letter, checking for an arithmetic property that ensures balance, and updates the 4 parameters of the Dorst–Smeulders coding, thus yielding as output the Dorst–Smeulders coding of the longest Sturmian prefix. Our algorithm is straightforward to implement since it consists of only elementary arithmetic operations.

2   Preliminaries

Let Σ be a finite alphabet. A *language* L over Σ is a set of words over Σ, i.e., a subset of Σ^*, the free monoid generated by Σ. A language L is *factorial* (or factor-closed) if it contains all the factors of its words.

Let $w = w_1 w_2 \cdots w_n$, $w_i \in \Sigma$, be a word of length $n = |w|$. An integer $p > 0$ is a *period* of w if $w_i = w_j$ whenever $i = j \mod p$. The prefix $\rho(w)$ of w whose length is the minimum period of w is called the *fractional root* of w. A *border* of w is a factor that occurs as a prefix and as a suffix in w. The word w has a border of length b if and only if $|w| - b$ is a period of w. An *unbordered word* is a word that coincides with its fractional root, i.e., such that $|w|$ is its minimum period.

The *reversal* $\widetilde{w}$ of w is the word $\widetilde{w} = w_n w_{n-1} \cdots w_1$. A word is a *palindrome* if it coincides with its reversal.

Let $w = w_1 \cdots w_{n-1} w_n$ be a word of length $n > 0$. The *shift* of w is the word $\sigma(w) = w_n w_1 \cdots w_{n-1}$. Two words w and w' are *conjugates* if $w = uv$ and $w' = vu$ for some words u and v. The conjugacy class of a word w can be obtained by repeatedly applying the shift operator, and contains $|w|$ distinct elements if and only if w is *primitive*, i.e., $w \neq v^k$ for any nonempty word v and $k > 1$.

A nonempty word is *Lyndon* if it is lexicographically smaller than all its nonempty suffixes. Lyndon words are unbordered. Every primitive word w has a (unique) Lyndon conjugate, i.e., $w = \sigma^s(w')$ for a Lyndon word w' and an integer $s \geq 0$.

From now on, we suppose $\Sigma = \{0,1\}$, and we use the lexicographic order on Σ^* induced by $0 < 1$. Occasionally, we will also consider the letters of Σ as numbers, i.e., make use of their arithmetic value.

A word $w \in \Sigma^*$ is Sturmian (or 1-balanced) if for every two factors u and v of w of the same length, one has $||u|_0 - |v|_0| \leq 1$ (or, equivalently, $||u|_1 - |v|_1| \leq 1$), where $|w|_x$ denotes the number of occurrences of the letter x in the word w. For example, 01001, 010101 and 110101 are Sturmian words, whereas 0011 is not. Famous examples of Sturmian words are the *Fibonacci words*, defined recursively by $f_1 = 1$, $f_2 = 0$ and $f_n = f_{n-1} f_{n-2}$ for each $n > 2$.

For any Sturmian word w, at least one between $0w$ and $1w$ (resp. between $w0$ and $w1$) is Sturmian. A Sturmian word w is *left (resp. right) special* if $0w$ and $1w$ (resp. $w0$ and $w1$) are both Sturmian, and it is *bispecial* if it is left and right special. Of course, a word is left (resp. right) special if and only if it is a prefix (resp. a suffix) of a bispecial word. A bispecial word is *strictly bispecial* if $0w0$, $1w0$, $0w1$ and $1w1$ are in *Sturm*. For example, 00 is strictly bispecial; 10, instead, is bispecial but not strictly bispecial, since $1 \cdot 10 \cdot 0$ is not Sturmian.

A *central word* is a word having two coprime periods, p and q, and length $p + q - 2$. It is well-known that central words are Sturmian and palindromes.

Proposition 1 ([8]). *Let w be a binary word. The following are equivalent:*

1. *w is a central word;*
2. *w is a strictly bispecial Sturmian word;*
3. *w is a power of a single letter or there exist palindromes (actually, central words) P, Q such that $w = P01Q = Q10P$.*

A (lower) *Christoffel word* is either a single letter or a word of the form $0c1$, where c is a central word. Christoffel words are precisely the Lyndon Sturmian words.

Every Christoffel word $u = 0c1$ has a unique palindromic factorization $u = \alpha\beta$. Moreover, the lengths of α and β are the two coprime periods of c, as well as the multiplicative inverses of $|u|_0$ and $|u|_1$ modulo $|u|$, respectively. This factorization is straightforward if c is a power of a single letter; otherwise, by Proposition 1, one has $\alpha = 0P0$ and $\beta = 1Q1$.

For example, 010010010 is a central word, with coprime periods $p = 3$ and $q = 8$, and $00100100 \cdot 101$ is the palindromic factorization of the corresponding Christoffel word.

For more details on Christoffel and Sturmian words, the reader is pointed to [14, Chap. 2] and [1].

3   The Coding of Dorst and Smeulders

We now describe how to code any Sturmian word using 4 integers of size bounded by the length of the word. This coding is due to Dorst and Smeulders [9].

Every Sturmian word w occurs as a factor of length $n = |w|$ starting at some position s' in the infinite periodic Christoffel word $u_{h,p}^\omega = u_{h,p}u_{h,p} \cdots$ of slope h/p, where $p = |\rho(w)|$ is the minimum period of w and $h = |\rho(w)|_1$ is the height of the fractional root of w. The number $s = (1 - s') \bmod p$, called shift, is in fact the distance of the root of w from its Christoffel conjugate. In other words, the root $w_1 w_2 \cdots w_p$ of w has Christoffel conjugate $w_{s+1} w_{s+2} \cdots w_s$. In particular, $s = 0$ if and only if the root of w is a Christoffel word.

For example, $w = 101001$ occurs in the infinite periodic Christoffel word $(00101)^\omega$ of slope $2/5$, starting at position 3. Notice that w has the same minimum period $p = 5$ of 00101, since 00101 is a conjugate of the fractional root 10100 of w, and the height $h = 2$ of 00101 is the height of the fractional root of w.

Therefore, every Sturmian word w is completely determined by its length n, its period p, the height h of its fractional root, and the shift s. The 4-tuple (n, p, h, s) is called the *Dorst–Smeulders coding* of w.

Since for every $i \geq 1$, the ith letter of the infinite periodic Christoffel word $u_{h,p}^{\omega}$ is

$$\left\lfloor i\frac{h}{p} \right\rfloor - \left\lfloor (i-1)\frac{h}{p} \right\rfloor$$

(or equivalently, the prefix of $u_{h,p}^{\omega}$ of length i has height $\lfloor ih/p \rfloor$), we have that the ith letter of w, for every $1 \leq i \leq n$, is

$$w_i = \left\lfloor (i-s)\frac{h}{p} \right\rfloor - \left\lfloor (i-s-1)\frac{h}{p} \right\rfloor.$$

This leads to a reconstruction of w from its Dorst–Smeulders coding that is very fast in practice, since it performs only arithmetic operations.

4   An Optimal Online Algorithm

We now give a linear-time algorithm that computes the longest Sturmian prefix w of an input word and returns the Dorst–Smeulders coding of w.

It scans the input word from left to right, character by character, and maintains four integer variables, which represent the Dorst–Smeulders coding of the current balanced prefix. In order to check whether the next prefix is balanced, we make use of the following result.

Lemma 1. *Let $w \neq \varepsilon$ be a Sturmian word with Dorst–Smeulders coding (n, p, h, s), and $a \in \{0, 1\}$. If wa has period p, then it is Sturmian. Otherwise, wa is Sturmian (and w is right special) if and only if*

$$(-h(n+1-s)) \bmod p \in \{0, 1\}. \tag{1}$$

Proof. Since w is Sturmian if and only if its fractional root $\rho(w)$ is a conjugate of a Christoffel word (cf. [7]), clearly wa is balanced if $|\rho(wa)| = p$, that is, if $\rho(wa) = \rho(w)$.

On the other hand, if wa does not have period p, then necessarily wb does, where $\{a, b\} = \{0, 1\}$. Thus, in this case wa is balanced if and only if w is right special, which in turn is equivalent to $\widetilde{w}$ being *left* special. Now, a Sturmian word is left special if and only if its fractional root is either a single letter or cxy for a central word c and $\{x, y\} = \{0, 1\}$ [6]. The single letter case means $p = 1$, trivially satisfying (1). Otherwise, either $10c$ or $01c$ is a suffix of w.

The first option holds if and only if $w = \lambda u^k 0c$, where $k \geq 0$ and λ is the suffix of $u := u_{h,p} = 0c1$ of length s (recall that the first occurrence of the lower Christoffel word u in $\rho(w)^{\omega}$ begins at position $s+1$). Equivalently, considering the lengths of the words involved, we have $n + 1 - s \equiv 0 \pmod{p}$; as $p > 1$ and $\gcd(h, p) = 1$, this is equivalent to $-h(n+1-s) \equiv 0 \pmod{p}$.

Finally, let us examine the second possibility, i.e., w ending in $01c$. Recall that $u = 0c1$, being a primitive Christoffel word of length $p > 1$, has a unique factorization $u = \alpha\beta$ in two palindromes α, β, whose lengths verify (see [3])

$$|\alpha| \cdot |u|_0 \equiv |\beta| \cdot |u|_1 \equiv 1 \pmod{p}. \tag{2}$$

Now, w ends in $01c$ if and only if $w0 = \lambda u^m \alpha$ for some $m \geq 0$. Since we have $|u|_1 = |\rho(w)|_1 = h$ and $|\beta| \equiv -|\alpha| \pmod{p}$, the above equation is equivalent to $-h|\alpha| \equiv 1 \pmod{p}$, and hence to $-h(n + 1 - s) \equiv 1 \pmod{p}$, as $n + 1 - s = |u^m \alpha| = mp + |\alpha| \equiv |\alpha| \pmod{p}$.   □

Remark 1. The same ideas from the previous lemma lead to a characterization of Sturmian words in terms of their *period array* (i.e., the array whose i-th entry is the minimum period of the prefix of length i, for $1 \leq i \leq n$; note the connection with the *border* array). Namely, a binary word w with period array

$$(\underbrace{p_1, \ldots, p_1}_{k_1}, \underbrace{p_2, \ldots, p_2}_{k_2}, \ldots, \underbrace{p_m, \ldots, p_m}_{k_m})$$

(with $1 = p_1 < p_2 < \cdots < p_m = |\rho(w)|$) is balanced if and only if p_j divides either k_j or $k_j + p_{j-1}$, whenever $1 < j < m$.

Lemma 2. *Let $w \in \{0, 1\}^*$ be a right special Sturmian word and (n, p, h, s) be its Dorst–Smeulders coding, with $p > 1$. If $a \in \{0, 1\}$ is such that $|\rho(wa)| > p$, then the coding $(n + 1, p', h', s')$ of wa satisfies the following:*

- $p' = n + 1 - ((n + 1 - (-1)^a \bar{h}) \bmod p)$,
- $h' = \left\lfloor \frac{n+1-(-1)^a \bar{h}}{p} \right\rfloor h + (-1)^a \left\lfloor \frac{h\bar{h}}{p} \right\rfloor$,
- $s' = as + (1 - a)(n + 1 - p)$,

where $\bar{h}$ is the multiplicative inverse of h modulo p.

Proof. By Lemma 1, the number $-h(n + 1 - s)$ is, modulo p, either 0 or 1.

1. As seen in the proof of Lemma 1, the first case occurs when $w = \lambda(0c1)^k 0c$ for some lower Christoffel word $u = 0c1$ of length p, and λ its proper suffix of length s. As $|\rho(wa)| \neq p$, we have $a = 0$ in this case.
   Now, $0c0$ occurs only as a suffix in $wa = w0$, and all other factors of length p are conjugates of the Lyndon word $0c1$, thus lexicographically greater. Therefore, $0c0$ is the lexicographically least factor of length p, so that the Lyndon conjugate of the root $\rho(wa)$ (whose length is larger than p, by hypothesis) must start with $0c0$. In other words, we have

$$s' = |\lambda(0c1)^k| = |w0| - |0c0| = n + 1 - p,$$

   the desired value for $a = 0$. Now let $0c1 = \alpha\beta$ be the palindromic factorization. The unique occurrence of $0c0$ also implies that the longest border of wa is the longest proper suffix of $0c0$ (or equivalently, of $1c0 = \beta\alpha$) that is also a prefix of $\lambda 0c$.
   If $s < |\beta|$, then the word λ, being a suffix of $0c1 = \alpha\beta$ in general, must be a suffix of β. Hence, by the above observation, the longest border of $wa = \lambda(\alpha\beta)^k 0c0$ is also the longest border of $\lambda\alpha\beta\alpha$, which is $\lambda\alpha$ (as $\beta\alpha$ is unbordered). Thus, $p' = |wa| - |\lambda\alpha| = n + 1 - (s + |\alpha|)$; this is the desired

value, since in this case we have $a = 0$, $n+1 \equiv s$ (mod p), and $\bar{h} = |\beta| \equiv -|\alpha|$ (mod p) by Eq. (2). We also have

$$h' = |\beta(\alpha\beta)^{k-1}0c0|_1 = |\beta(0c1)^k|_1 - 1 = |\beta|_1 - 1 + kh$$
$$= \left\lfloor \frac{h\bar{h}}{p} \right\rfloor + \left(\frac{n+1-s}{p} - 1 \right) h,$$

where for the last equality we used the fact that β is the suffix of length $\bar{h}$ of the Christoffel word $0c1$, and so its height minus 1 equals the height of the *prefix* of $0c1$ of the same length (recall that c is a palindrome). The value of h' again satisfies our thesis since $s < \bar{h} < p$.

If $s \geq |\beta| = \bar{h}$, instead, let $\lambda = \gamma\beta$, so that γ is a proper suffix of α. The longest border of wa then coincides with the longest border of $\lambda\alpha = \gamma\beta\alpha$, which is γ. Thus, we again obtain $p' = n+1 - ((n+1-\bar{h}) \bmod p)$. Moreover, we have

$$h' = |\beta(0c1)^{k+1}|_1 - 1 = |\beta|_1 - 1 + (k+1)h = \left\lfloor \frac{h\bar{h}}{p} \right\rfloor + \frac{n+1-s}{p}h,$$

again as desired, since $0 < \bar{h} < s < p$ here.

2. By Lemma 1, the second case $-h(n+1-s) \equiv 1$ (mod p) occurs when $w0 = \lambda(\alpha\beta)^k\alpha$ but $a = 1$, so that $1c1$ occurs only as a suffix in $wa = w1$. Therefore, the longest border of wa is the longest proper suffix of $1c1$ (and hence also of $0c1 = \alpha\beta$) that is also a prefix of $\lambda\alpha\beta$; in other words, it coincides with the longest border of $\lambda\alpha\beta$, which is λ. Thus, $p' = n+1-s$ in this case; as $n+1-s+\bar{h} = |(\alpha\beta)^{k+1}| \equiv 0$ (mod p), the formula in the statement is again verified. Now, the lower Christoffel conjugate of $\rho(wa)$ must end with the factor of length p that is lexicographically greatest, that is, $1c1$. This implies $s' = s$.

Finally, we have

$$h' = |(\alpha\beta)^k\alpha|_1 + 1 = |(\alpha\beta)^{k+1}|_1 - (|\beta|_1 - 1) = \left\lfloor \frac{n+1+\bar{h}}{p} \right\rfloor h - \left\lfloor \frac{h\bar{h}}{p} \right\rfloor.$$

$\square$

The total running time is clearly linear, and the working space is constant, assuming the Word-RAM model. The pseudocode is shown in Algorithm 1.

Example 1. Let $w = 0101001101010000010010010101001001000101$, of length 40. The Dorst–Smeulders coding of w is

$$(7,5,2,4), (7,7,3,5), (11,10,3,0), (11,11,4,3), (4,2,1,0)$$

corresponding to the factorization

$$0101001 \cdot 1010100 \cdot 00010010010 \cdot 10100100100 \cdot 0101.$$

Algorithm 1. Dorst–Smeulders coding of the longest Sturmian prefix
___

Input: Binary word $w = w_0 w_1 \ldots w_{N-1}$, where $w_i \in \{0, 1\}$
Output: Tuple (n, p, h, s) that is the Dorst–Smeulders coding of the longest Sturmian
    prefix of w
1: **if** all letters in w are equal **then**
2:     **return** $(N, 1, w_0, 0)$
3: $p \leftarrow$ index of first occurrence of $1 - w_0$, plus 1
4: $h \leftarrow (-1)^{w_0} \bmod p$
5: $s \leftarrow (-w_0) \bmod p$
6: **for** $n = p$ to $N - 1$ **do**
7:     **if** $w_n \neq w_{n-p}$ **then**
8:         $h^{-1} \leftarrow$ modular inverse of h modulo p
9:         **if** $(n - s + 1) \bmod p = 0$ **then**
10:             $h \leftarrow \left\lfloor \frac{n+1-h^{-1}}{p} \right\rfloor \cdot h + \left\lfloor \frac{h \cdot h^{-1}}{p} \right\rfloor$
11:             $s \leftarrow n + 1 - p$
12:             $p \leftarrow n + 1 - ((n + 1 - h^{-1}) \bmod p)$
13:         **else if** $(n - s + 1 + h^{-1}) \bmod p = 0$ **then**
14:             $h \leftarrow \left\lfloor \frac{n+1+h^{-1}}{p} \right\rfloor \cdot h - \left\lfloor \frac{h \cdot h^{-1}}{p} \right\rfloor$
15:             $p \leftarrow n + 1 - ((n + 1 + h^{-1}) \bmod p)$
16:         **else**
17:             **return** (n, p, h, s)
18: **return** (N, p, h, s)
___

5   Conclusions and Future Work

We described a simple algorithm that, on a given binary input string, returns the
Dorst–Smeulders coding of the Sturmian factors in the greedy decomposition.
Our algorithm is conceptually simple and needs only constant working space
in the Word-RAM model, since it only needs to maintain a constant number
of integers whose size is bounded by n. Our algorithm could be used as part
of a compression scheme after applying some preprocessing that maps repetitive
strings to strings with long Sturmian factors.

References

1. Berstel, J., Lauve, A., Reutenauer, C., Saliola, F.: *Combinatorics on Words: Christoffel Words and Repetition in Words*, vol. 27 of *CRM monograph series*. American Mathematical Society (2008)
2. Berstel, J., Pocchiola, M.: Random generation of finite sturmian words. Discret. Math. **153**(1–3), 29–39 (1996)
3. Berthé, V., De Luca, A., Reutenauer, C.: On an involution of Christoffel words and Sturmian morphisms. Eur. J. Comb. **29**(2), 535–553 (2008)
4. Chen, K. T., Fox, R. H., Lyndon, R. C.: Free differential calculus, IV. the quotient groups of the lower central series. *Ann. Math.***68**(1):81–95 (1958)

5. Cohn, M., Khazan, R. Parsing with prefix and suffix dictionaries. In: Storer, J. A., Cohn, M., (eds.) *Proceedings of the 6th Data Compression Conference (DCC '96), Snowbird, Utah, USA, March 31 - April 3, 1996*, pp. 180–189. IEEE Computer Society (1996)
6. De Luca, A., De Luca, A.: Pseudopalindrome closure operators in free monoids. Theoret. Comput. Sci. **362**(1–3), 282–300 (2006)
7. De Luca, A., De Luca, A.: Some characterizations of finite Sturmian words. Theor. Comput. Sci. **356**(1–2), 118–125 (2006)
8. De Luca, A., Mignosi, F.: Some combinatorial properties of Sturmian words. Theoret. Comput. Sci. **136**(2), 361–385 (1994)
9. Dorst, L., Smeulders, A.W.M.: Discrete representation of straight lines. IEEE Trans. Pattern Anal. Mach. Intell. **6**(4), 450–463 (1984)
10. Duval, J.: Factorizing words over an ordered alphabet. J. Algorithms **4**(4), 363–381 (1983)
11. Frid, A.E., Puzynina, S., Zamboni, L.Q.: On palindromic factorization of words. Adv. Appl. Math. **50**(5), 737–748 (2013)
12. Kosolobov, D., Shur, M.A., Rubinchik, M., Borozdin, K.: Palindromic length in linear time. In: *Annual Symposium on Combinatorial Pattern Matching: 28th Annual Symposium, CPM 2017*, pp. 1–23. Leibniz-Zentrum für Informatik (2017)
13. Lindenbaum, M., Koplowitz, J.: A new parameterization of digital straight lines. IEEE Trans. Pattern Anal. Mach. Intell. **13**(8), 847–852 (1991)
14. Lothaire, M.: Algebraic Combinatorics on Words. Encyclopedia of Mathematics and its Applications. Cambridge Univ, Press (2002)

Prefix-Free Parsing for Merging Big BWTs

Diego Díaz-Domínguez[1]📷, Travis Gagie[2(✉)]📷, Veronica Guerrini[3]📷,
Ben Langmead[4]📷, Zsuzsanna Lipták[5]📷, Giovanni Manzini[3]📷,
Francesco Masillo[6]📷, and Vikram Shivakumar[4]📷

[1] University of Helsinki, Helsinki, Finland
[2] Dalhousie University, Halifax, Canada
travis.gagie@gmail.com
[3] University of Pisa, Pisa, Italy
[4] Johns Hopkins University, Baltimore, USA
[5] University of Verona, Verona, Italy
[6] Dortmund Technical University, Dortmund, Germany

Abstract. When building Burrows-Wheeler Transforms (BWTs) of truly huge datasets, prefix-free parsing (PFP) can use an unreasonable amount of memory. In this paper we show how if a dataset can be broken down into small datasets that are not very similar to each other—such as collections of many copies of genomes of each of several species, or collections of many copies of each of the human chromosomes—then we can drastically reduce PFP's memory footprint by building the BWTs of the small datasets and then merging them into the BWT of the whole dataset.

Keywords: Burrows-Wheeler Transform · Prefix-free parsing · Low-memory algorithms · Pangenomics

1 Introduction

The discovery of the r-index [12,13] was an exciting time in bioinformatics but, as Paolo Ferragina [8] likes to say, to use an index one must first *build* it! There were at that point no algorithms for building Burrows-Wheeler Transforms (BWTs) that could handle dozens of human genomes in reasonable time and memory but, inspired by their desire to use the r-index, Boucher et al. [5,6] quickly proposed prefix-free parsing (PFP). Although lacking good worst-case guarantees, PFP is fast and fairly small in practice, and easy to implement: on Feb. 21st, 2018 Gagie sent his final design [11] to Manzini, who had a working implementation a week later. These features have made PFP popular, with several groups (for example, [1–4,9,10,14,15,17,19,24,25]) using and extending it, mostly without Manzini's help. It is even covered in the new edition of Mäkinen et al.'s [20] textbook (which also covers r-indexes, BWTs, etc.)!

© The Author(s), under exclusive license to Springer Nature Switzerland AG 2026
G. Badkobeh et al. (Eds.): SPIRE 2025, LNCS 16073, pp. 54–63, 2026.
https://doi.org/10.1007/978-3-032-05228-5_6

Sadly, PFP seems to have reached the limit of its usefulness now that we have truly huge pangenomic datasets to index: it ran out of memory when Li [18] tried to build the BWT of AllTheBacteria [16], and when the fourth and eighth authors of this paper tried to build the BWT of the latest HPRC reference of 231 human genomes ("Assemblies Release 2" at https://humanpangenome.org/data). These datasets are not repetitive enough for recursive PFP [9] to work well, so it seems we must turn to newer and more sophisticated algorithms, such as Díaz-Domínguez and Navarro's [7] grlBWT, Li's [18] ropebwt3, Masillo's [22] CMS-BWT or Olbrich's [23] algorithm. In this paper, however, we show there is life in PFP yet!

Specifically, we show if a dataset can be broken down into small datasets that are not very similar to each other—such as collections of many copies of genomes of each of several species, or collections of many copies of each of the human chromosomes—then we can drastically reduce PFP's memory footprint by building the BWTs of the small datasets and then merging them into the BWT of the whole dataset. In Sect. 2 we briefly review how PFP works, in Sect. 3 we describe how it can be used for merging BWTs, and in Sect. 4 we demonstrate experimentally that our idea is practical.

2 PFP

PFP is based on a modification of rsync [27]: we run a sliding window of a given length w over the dataset and insert a phrase break whenever the Karp-Rabin hash of the contents of the window are 0 modulo a given value p (which need not be prime). Unlike rsync, however, we include the contents of the window that triggered the break—called a *trigger string*—in both the preceding and next phrases. If we treat the string as circular then

- consecutive phrases overlap by exactly w characters;
- every character in the dataset is contained in exactly one phrase in which it is not among the last w characters;
- every phrase starts with a trigger string, ends with a trigger string and contains no other trigger string;
- no proper phrase suffix of length at least w—called a *valid* phrase suffix—is a proper prefix of any other proper phrase suffix (otherwise there would be a trigger string in the middle of a phrase).

It follows that if we know the lexicographic order of the dataset's suffixes starting at phrase boundaries—and we can find that by building the BWT of the parse—then we can easily determine the order in the BWT of any two characters in the dataset. To do this, we consider the unique phrases in which those characters appear but not among the last w characters, and compare the valid phrase suffixes immediately following the characters. If those phrase suffixes differ, then the characters' order in the BWT is the same as the lexicographic order of the phrase suffixes; otherwise, it is the same as the lexicographic order of the dataset's suffixes starting at the ends of the phrases.

```
int pos = 0;
for (int i = 0; i < dSize; i++) {
    if (valid(SAD[i])) {
        if (unique(SAD[i])) {
            memset(&BWT[pos], dict[SAD[i]] - 1], occ(SAD[i]));
        }
        pos += occ(SAD[i]);
    } }
```

Fig. 1. A code fragment for filling in the characters in the BWT that precede in the dataset valid phrase suffixes always preceded by the same distinct character. By itself, this usually fills in most of the BWT.

To compare valid phrase suffixes quickly, we can build the suffix array SAD of the concatenation dict of the phrases in the dictionary. In fact, if

- dSize is the number of characters in dict,
- valid(SAD[i]) indicates whether the phrase suffix starting at dict[SAD[i]] is valid,
- unique(SAD[i]) indicates whether all occurrences in dict of the phrase suffix starting at dict[SAD[i]] are preceded by copies of the same character,
- occ(SAD[i]) is the frequency in the parse of the phrase containing dict[SAD[i]],
- BWT is a string with the same length as the dataset,

then the code fragment shown in Fig. 1 fills in the characters in the BWT that precede in the dataset valid phrase suffixes always preceded by the same distinct character. It considers only the dictionary and not the parse, usually fills in most of the BWT by itself—and it uses some ideas that will be useful to us in this paper.

To see why the code fragment is correct, consider that all the characters that precede occurrences of a valid phrase suffix α in the dataset are consecutive in the BWT. If α is always preceded by the same distinct character then the code fragment just needs to fill in the right number of copies of that character; otherwise, it should leave enough blank cells to hold them.

The code fragment scans through the valid phrase suffixes in lexicographic order, by scanning through the suffixes of the dictionary and checking if each starts with a valid phrase suffix. For each phrase β in dict ending with α,

- if α is always preceded by the same distinct character, then the code fragment fills into the BWT all the characters that precede α in occurrences of β in the dataset;
- otherwise, it leaves enough blank cells to hold all the characters that precede α in occurrences of β in the dataset.

Since the number of characters that precede occurrences of α in the dataset is the sum over the phrases β ending with α of the number of characters that precede α in occurrences of β in the dataset, the code fragment is correct.

3   Merging

Suppose we have a dataset that can be broken down into small datasets that are not very similar to each other, in the sense that most w-mers that appear in the whole dataset appear in only one of the small datasets. (In our experiments, we found setting $w = 20$ is usually enough.) We find the set of trigger strings that appear in more than one small dataset, then perform a modified parsing of the small datasets treating as trigger strings only the ones appearing in exactly one small dataset.[1] Once we have the small datasets' BWTs we will be interested only in their dictionaries, so we can discard the parses.

Each valid phrase suffix now appears in the dictionary of exactly one of the small datasets. The characters in the whole dataset's BWT appear in the lexicographic order of the valid phrases suffixes they immediately precede in the whole dataset. If two characters precede the same valid phrase suffix then they must be in the same small dataset and—assuming the BWT of the whole dataset is an extended BWT (eBWT) [21] with the small datasets treated as separate strings or collections of strings—their order in the whole dataset's BWT is the same as in that small dataset's BWT. This is illustrated in Fig. 2.

This means that if SAD is now the suffix array of the concatenation dict of the small dataset's dictionaries, dSize, valid(SAD[i]) and occ(SAD[i]) are defined as before but BWT is now a file, and

- oldBWT[j] is a file containing the BWT of the jth small dataset,
- DS(SAD[i]) is the number of the small dataset containing the phrase suffix starting at dict[SAD[i]],
- buffer is an empty string at least as long as the frequency of the most common valid phrase suffix,

then the code fragment in Fig. 3 merges the small datasets' BWTs into the BWT of the whole dataset. Once we have concatenation of the small dictionaries, we can build SAD with a standard algorithm for suffix-array construction.

To see why the code fragment is correct, consider that for each valid phrase suffix α and phrase β ending with α, it copies as many characters from the BWT of the small dataset containing copies of α as there are occurrences of β in that dataset. The number of occurrences of α in the whole dataset is the total number of occurrences of phrases ending with α, so the code fragment is correct.

The code fragment needs enough memory to store SAD and buffer and data structures to support valid(SAD[i]), occ(SAD[i]) and DS(SAD[i]). The latter are easy to make small: we store

- a table saying which phrase in dict contains every, say, 50th character of dict;

---

[1] We note that if someone has mistakenly labelled a Salmonella genome as an E. coli genome, for example, then the E. coli dataset will probably include almost all the trigger strings in the Salmonella dataset, so the Salmonella genomes will be parsed into only a few very long phrases and our method will work poorly overall. On the other hand, it could be useful to know about such unexpected similarities.

$T_1 = \texttt{GATTACAT; GATACAT; GATTAGATA\$}$ | $T_2 = \texttt{AGCCGTGC; AGCGTGC; AGCCGAGCG\$}$

phrase	occurrences		phrase	occurrences
AT;GAT	2		GC;AGC	2
ATA\$GAT	1		GCCGAGC	1
ATACAT	1		GCCGTGC	1
ATTACAT	1		GCG\$AGC	1
ATTAGAT	1		GCGTGC	1

BWT$_1$	phrase suffix	BWT$_1$	phrase suffix		BWT$_2$	phrase suffix	BWT$_2$	phrase suffix
A	\$GAT	A;\$;	GAT		G	\$AGC	C	CGAGC
TT	;GAT	AA	T;GAT		CC	;AGC	GC	CGTGC
T	A\$GAT	A	TA\$GAT		;\$G;	AGC	C	G\$AGC
TT	ACAT	TA	TACAT		GG	C;AGC	C	GAGC
T	AGAT	T	TAGAT		G	CCGAGC	TTAAAA	GC
CCGGGG	AT	A	TTACAT		G	CCGTGC	CC	GTGC
AA	CAT	A	TTAGAT		G	CG\$AGC	GG	TGC

phrase suffix	freq.	dataset	BWT		phrase suffix	freq.	dataset	BWT
\$AGC	1	2	G		CGAGC	1	2	C
\$GAT	1	1	A		CGTGC	2	2	GC
;AGC	2	2	CC		G\$AGC	1	2	C
;GAT	2	1	TT		GAGC	1	2	C
A\$GAT	1	1	T		GAT	4	1	A;\$;
ACAT	2	1	TT		GC	6	2	TTAAAA
AGAT	1	1	T		GTGC	2	2	CC
AGC	4	2	;\$G;		T;GAT	2	1	AA
AT	6	1	CCGGGG		TA\$GAT	1	1	A
C;AGC	2	2	GG		TACAT	2	1	TA
CAT	2	1	AA		TAGAT	1	1	T
CCGAGC	1	2	G		TGC	2	2	GG
CCGTGC	1	2	G		TTACAT	1	1	A
CG\$AGC	1	2	G		TTAGAT	1	1	A

Fig. 2. Suppose we have computed the BWTs of T_1 and T_2 and we build their PFP dictionaries with trigger strings AT and GC, each of which occurs in only one dataset. Then the combined BWT consists of interleaved blocks from the two BWTs, with each block corresponding to a valid phrase suffix α: the length of the block is the total frequency of phrases ending with α and the source of the block is the BWT for the dataset containing those phrases. For example, for $\alpha = \texttt{AT}$ the block is CCGGGG from the BWT of T_1.

```
for (int i = 0; i < dSize; i++) {
    if (valid(SAD[i])) {
        fread(&buffer, 1, occ(SAD[i]), oldBWT[DS(SAD[i])]);
        fwrite(&buffer, 1, occ(SAD[i]), BWT);
    } }
```

Fig. 3. A code fragment that merges the BWTs of the small datasets into the BWT of the whole dataset.

– a table saying where each phrase ends in `dict`;
– a table saying how many times each phrase occurs in the whole dataset;
– a table saying which dataset contains each phrase in `dict`.

The BWT of the whole dataset is written sequentially to disk as it is generated, so we do not need to store it in memory.

Given $SAD[i]$, we use the first table to get a close lower bound on which phrase contains $dict[SAD[i]]$, then start at the corresponding number in the second table and scan until we find $SAD[i]$'s successor to determine the exact phrase that contains $dict[SAD[i]]$. We check that $dict[SAD[i]]$ is not the first character in the phrase (if so the phrase suffix starting at $dict[SAD[i]]$ is not proper) nor among the last $w - 1$ characters, to determine $valid[SAD[i]]$. We look up $occ[SAD[i]]$ in the third table and $DS[SAD[i]]$ in the fourth table.

In practice, storing SAD takes less memory than storing the parse—so our merging approach uses less memory than building the BWT for the whole dataset directly with PFP—but it still takes a lot. We can simulate scanning SAD by building the suffix arrays of the small dataset's dictionaries and streaming them from disk, merging them to obtain SAD (and discarding each entry $SAD[i]$ when we have finished the corresponding pass through the code fragment's loop). To merge the suffix arrays, we can store the small dataset's dictionaries and perform `strcmp()` queries to decide which suffix array points to the lexicographically next phrase suffix. Storing the dictionaries instead of SAD reduces the space by a factor of about 4 or 8, depending on how large the suffix-array entries are.

If the BWT of the whole dataset will be used to support searches for queries over only $\{A, C, G, T\}$ then we can do somewhat better. We replace all the non-$\{A, C, G, T\}$ characters in the small datasets by copies of X, which does not change the multiset of substrings over $\{A, C, G, T\}$ in the whole dataset. PFP uses 0x0 as an end-of-file symbol, 0x1 to mark the ends of phrases in its dictionaries and 0x2 to separate strings, but we can still store the small datasets' dictionaries in 3 bits per character. Because phrase suffixes' encodings may start offset by different amounts into bytes, we can no longer use `strcmp()`; with careful casting and bit-shifting we can compare 20-character blocks with one operation, however, so comparing phrase suffixes is still fairly fast.

Currently we keep a list of the small datasets' IDs, sorted into the lexicographic order of the phrase suffixes pointed to by those datasets' next suffix-array entries, and after we process a suffix-array entry we bubble its dataset's ID down the list to its new position. When dealing with many small datasets, however, a min-heap should be more efficient.

We may be able to parallelize our merging algorithm efficiently, although we have not tried this yet. We can choose a position partway through one of the small dataset's suffix array, say, then use binary search to find positions in the other small datasets' suffix arrays such that at some time in its execution, our algorithm be at all those positions in the suffix arrays simultaneously. We can sum the previous $occ(SAD[i])$ values for each suffix array to find where our algorithm would be in the small datasets' BWTs at that time. Therefore, we can start our algorithm from that time in its execution, and it follows we can

Table 1. The time and peak memory used by each program on 30 bacterial pangenomes totalling 9.39 GB.

	time (mm:ss)	memory (GB)
`pfp-merge`	23:50	0.91
`bigbwt`	16:26	13.84
`glrbwt`	9:01	1.71
`ropebwt3`	27:10	5.12
`lg`	7:41	7.00

parallelize our merging algorithm. Notice the dictionaries and tables are static and can be shared between threads, so parallelization should not significantly increase our memory footprint.

4   Experiments

We downloaded genomes of 30 species of bacteria from the AllTheBacteria project[2], with sizes ranging from 309.93 MB to 314.53 MB and totalling 9.39 GB. We built their BWTs or eBWTs with our `pfp-merge`[3], Manzini's `bigbwt`[4] (based on plain PFP), Díaz's `grlbwt`[5], Li's `ropebwt3`[6] and Olbrich's `lg`[7] (for "Lyndon grammar"). We omitted Masillo's CMT-BWT because it requires a single reference. We used default parameter settings for all the programs (except we used `ropebwt3`'s -R flag to prevent it indexing also the datasets' reverse complements) and, due to time constraints, 16 threads for each program on similar nodes of a cluster at the University of Helsinki. For `pfp-merge`, `bigbwt`, and `lg`, we removed newlines and converted non-$\{A, C, G, T\}$ characters to Xs. We ran `ropebwt3` on the original FASTA files and `grlbwt` on the concatenation of the datasets (one string per line). Table 1 shows the wall-clock time and peak memory used by each program, measured with `usr/bin/time -v`. Although `pfp-merge` was almost the slowest—which may improve with additional parallelization—it used the least memory by a noticeable margin. The actual merging used only 0.77 GB of memory, suggesting we may perform relatively even better on more datasets.

5   Future Work

For the full version of this paper, we plan further experiments to explore when PFP-merge is practical. For the case when we have a dataset of many sim-

---

[2] https://allthebacteria.org/.
[3] https://gitlab.com/manzai/pfp-merge.
[4] https://gitlab.com/manzai/Big-BWT.
[5] https://github.com/ddiazdom/grlBWT.
[6] https://github.com/lh3/ropebwt3.
[7] https://gitlab.com/qwerzuiop/lyndongrammar.

ilar strings that are not internally repetitive, we are also working on using Mumemto [26] to "shred" the strings at unique landmark substrings that match across all the strings ("multi-MUMs"; see [26] for details). This yields many smaller repetitive datasets that are not very similar to each other.

For example, suppose we are given a single dataset of 1000 copies of the human chromosome 1 s, each of which is about 249 million characters. We may shred it into 25 datasets, each consisting of 1000 copies of a distinct region of chromosome 1 of length about 10 million. Then we can build the BWTs of the small datasets and—with some new ideas to reverse the shredding—merge them into the BWT of the original dataset.

Finally, we need the small datasets' dictionaries in memory only to be able to determine the lexicographically next valid phrase suffix occurring in each dictionary. We noticed after submitting this paper that if we encode fairly compactly the valid phrase suffixes in each dictionary such that we can decode them sequentially, then we can store those phrase suffixes on disk and avoid storing even the dictionaries in memory. For the first dictionary in the toy example in Fig. 2, we want to store compactly $GAT, ;GAT, A$GAT, ACAT, AGAT, etc., such that we can decode them sequentially. For each consecutive pair of valid phrase suffixes in a dictionary, we can store the lengths of a common prefix and a common suffix and the substring between them in the second phrase suffix in the pair. The encoding for the list above is then $GAT, (0, 3, ";"), (0, 3, "A$"), (1, 2, "C"), (1, 2, "G"), etc., which is not smaller than the list itself but suggests this technique may be useful for larger datasets.

Acknowledgments. Many thanks to the reviewers for their suggestions, to Lavinia Egidi, Felipe Louza and Giovanna Rosone for helpful discussions, and (belatedly) to Paweł Gawrychowski, Dmitry Kosolobov and Tatiana Starikovskaya for suggestions that sped up PFP in the past. TG funded by NSERC grant RGPIN-07185-2020. BL and VS funded by NIH grant R01HG011392 to BL. ZsL partially funded by MUR PRIN project no. 2022YRB97K and INdAM-GNCS project no. E53C24001950001. VG and GM funded by the Next Generation EU PNRR MUR M4 C2 Inv 1.5 project ECS00000017 Tuscany Health Ecosystem Spoke 6 CUP I53C22000780001. DD funded by EU Horizon Europe grant No 101060011.

Disclosure of Interests. The authors declare no competing interests.

References

1. Ahmed, O., et al.: Pan-genomic matching statistics for targeted nanopore sequencing. iScience **24**(6) (2021)
2. Ahmed, O.Y., Rossi, M., Gagie, T., Boucher, C., Langmead, B.: SPUMONI 2: improved classification using a pangenome index of minimizer digests. Genome Biol. **24**(1), 122 (2023)

3. Boucher, C., Cenzato, D., Lipták, Z., Rossi, M., Sciortino, M.: Computing the original eBWT faster, simpler, and with less memory. In: Proceedings 28th Symposium on String Processing and Information Retrieval (SPIRE), pp. 129–142 (2021)
4. Boucher, C., et al.: PFP compressed suffix trees. In: Proceedings 23rd Workshop on Algorithm Engineering and Experiments (ALENEX), pp. 60–72 (2021)
5. Boucher, C., Gagie, T., Kuhnle, A., Langmead, B., Manzini, G., Mun, T.: Prefix-free parsing for building big BWTs. Algorithms Mol. Biol. **14**, 1–15 (2019)
6. Boucher, C., Gagie, T., Kuhnle, A., Manzini, G.: Prefix-free parsing for building big BWTs. In: Proceedings 18th Workshop on Algorithms in Bioinformatics (WABI), pp. 2:1–2:16 (2018)
7. Díaz-Domínguez, D., Navarro, G.: Efficient construction of the BWT for repetitive text using string compression. Inf. Comput. **294**, 105088 (2023)
8. Ferragina, P., Gagie, T., Manzini, G.: Lightweight data indexing and compression in external memory. Algorithmica **63**(3), 707–730 (2012)
9. Ferro, E., Oliva, M., Gagie, T., Boucher, C.: Building a pangenome alignment index via recursive prefix-free parsing. iScience **27**(10) (2024)
10. Gagie, T.I.T., Manzini, G., Navarro, G., Sakamoto, H., Takabatake, Y.: Rpair: rescaling RePair with Rsync. In: Brisaboa, N.R., Puglisi, S.J. (eds.) SPIRE 2019. LNCS, vol. 11811, pp. 35–44. Springer, Cham (2019). https://doi.org/10.1007/978-3-030-32686-9_3
11. Gagie, T., Manzini, G.: Prefix-free parsing for building big BWTs. arxiv (2018)
12. Gagie, T., Navarro, G., Prezza, N.: Optimal-time text indexing in BWT-runs bounded space. In: Proceedings 29th Symposium on Discrete Algorithms (SODA), pp. 1459–1477 (2018)
13. Gagie, T., Navarro, G., Prezza, N.: Fully functional suffix trees and optimal text searching in BWT-runs bounded space. J. ACM **67**(1), 1–54 (2020)
14. Goga, A., Baláž, A.: Prefix-free parsing for building large tunnelled Wheeler graphs. In: Proceedings 22nd Workshop on Algorithms in Bioinformatics (WABI 2022), pp. 18:1–18:12 (2022)
15. Hong, A., Rossi, M., Boucher, C.: LZ77 via prefix-free parsing. In: Proceedings 25th Symposium on Algorithm Engineering and Experiments (ALENEX), pp. 123–134 (2023)
16. Hunt, M., et al.: AllTheBacteria - all bacterial genomes assembled, available and searchable. bioRxiv (2024)
17. Kim, J., Varki, R., Oliva, M., Boucher, C.: Re^{2}pair: increasing the scalability of RePair by decreasing memory usage. In: Proceedings 32nd European Symposium on Algorithms (ESA), pp. 78–1 (2024)
18. Li, H.: BWT construction and search at the terabase scale. Bioinformatics **40**(12), btae717 (2024)
19. Lipták, Z., Lucà, S., Masillo, F.: Measuring genomic data with PFP. bioRxiv (2025)
20. Mäkinen, V., Belazzougui, D., Cunial, F., Tomescu, A.I.: Genome-Scale Algorithm Design: Bioinformatics in the Era of High-Throughput Sequencing, 2nd edn. Cambridge University Press (2023)
21. Mantaci, S., Restivo, A., Rosone, G., Sciortino, M.: An extension of the Burrows-Wheeler transform. Theoret. Comput. Sci. **387**(3), 298–312 (2007)
22. Masillo, F.: Matching statistics speed up BWT construction. In: Proceedings 31st European Symposium on Algorithms (ESA), pp. 83:1–83:15 (2023)
23. Olbrich, J.: Fast and memory-efficient BWT construction of repetitive texts using Lyndon grammars. arXiv (2025)
24. Olbrich, J., Büchler, T., Ohlebusch, E.: Generating multiple alignments on a pangenomic scale. Bioinformatics **41**(3), btaf104 (2025)

25. Rossi, M., Oliva, M., Langmead, B., Gagie, T., Boucher, C.: MONI: a pangenomic index for finding maximal exact matches. J. Comput. Biol. **29**(2), 169–187 (2022)
26. Shivakumar, V.S., Langmead, B.: Mumemto: efficient maximal matching across pangenomes. Genome Biol. **26**(1), 169 (2025)
27. Tridgell, A., Mackerras, P.: The Rsync algorithm. Technical report, TR-CS-96-05, The Australian National University (1996)

RLZ-r and LZ-End-r: Enhancing Move-r

Patrick Dinklage$^{(\boxtimes)}$, Johannes Fischer, Lukas Nalbach, and Jan Zumbrink

TU Dortmund University, Dortmund, Germany
{patrick.dinklage,johannes.fischer,lukas.nalbach,
jan.zumbrink}@cs.tu-dortmund.de

Abstract. In pattern matching on strings, a locate query asks for an enumeration of all the occurrences of a given pattern in a given text. The r-index [Gagie et al., 2018] is a recently presented compressed self index that stores the text and auxiliary information in compressed space. With some modifications, locate queries can be answered in optimal time [Nishimoto & Tabei, 2021], which has recently been proven relevant in practice in the form of Move-r [Bertram et al., 2024]. However, there remains the practical bottleneck of evaluating function Φ for every occurrence to report. This motivates enhancing the index by a compressed representation of the suffix array featuring efficient random access, trading off space for faster answering of locate queries [Puglisi & Zhukova, 2021].

In this work, we build upon this idea considering two suitable compression schemes: Relative Lempel-Ziv [Kuruppu et al., 2010], improving the work by Puglisi and Zhukova, and LZ-End [Kreft & Navarro, 2010], introducing a different trade-off where compression is better than for Relative Lempel-Ziv at the cost of slower access times. We enhance both the r-index and Move-r by the compressed suffix arrays and evaluate locate query performance in an experiment.

We show that locate queries can be sped up considerably in both the r-index and Move-r, especially if the queried pattern has many occurrences. The choice between two different compression schemes offers new trade-offs regarding index size versus query performance.

Keywords: compression · compressed text index · algorithm engineering

1 Introduction

Searching for occurrences of a pattern in a text or a collection of texts is a ubiquitous problem. A common use case is to query different patterns against the same text, e.g., picture different users searching for different terms on (a snapshot of) the internet or DNA reads being matched in a genomic database. This scenario is typically tackled by building an *index* on the text, a data structure that allows for efficient pattern matching queries. Since the text can be prohibitively large to be indexed plainly, we are very much interested in compressed indexes. Arguably an important milestone in this area was the invention of the

G. Badkobeh et al. (Eds.): SPIRE 2025, LNCS 16073, pp. 64–78, 2026.
https://doi.org/10.1007/978-3-032-05228-5_7

r-index by Gagie et al. [6] that can be stored in space $\mathcal{O}(r)$, where r is the number of runs in the text's Burrows-Wheeler transform – a well-accepted measure of compressibility. Augmented by the move data structure by Nishimoto et al. [16], pattern matching queries can be answered in optimal time. Bertram et al. [1] recently implemented this and presented Move-r, achieving a very good practical time/space trade-off.

We differentiate between two types of queries: while *count* queries tell us how often a pattern occurs in the text, *locate* queries ask for an enumeration of all positions at which the pattern occurs. For this, in the r-index, we need to evaluate a function Φ for every occurrence, which turns out to be the main performance bottleneck in practice. In independent work (Move-r was not around yet), Puglisi and Zhukova [18,19] considered storing a compressed representation of the suffix array alongside the r-index that features efficient random access. For locate queries, we can now directly decode the relevant portion of the suffix array instead of evaluating Φ for every step. This resulted in a new trade-off where locate queries could be answered much faster, at the cost of having to store a compressed representation of the suffix array alongside the index.

Our Contributions. We transfer the idea of [19] and explore enhancing Move-r by a compressed representation of the suffix array with efficient random access, expecting this to be a practical trade-off for speeding up locate queries. For this, we consider two different compression schemes. First, like [19], we consider Relative Lempel-Ziv, where we greatly improved the reference construction as well as the parsing procedure. While the source code of [19] remains closed, we publish our reimplementation under an open source license. Second, we consider LZ-End [13], a different Lempel-Ziv compression scheme that allows for efficient random access. Here, we give a competitive generalized and simplified algorithm to compute the LZ-End parsing of a string over an integer alphabet based on the (suboptimal) $\mathcal{O}(n \lg \lg n)$-time algorithm of Kempa and Kosolobov [11]. We also improve Move-r itself by engineering internal rank and select data structure.

We implement different variations of the r-index and Move-r and show trade-offs between index size and query performance in our experiments.

2 Preliminaries

Let $\Sigma \subseteq \mathbb{N}$ be an integer alphabet and $T \in \Sigma^n$ a *text* over Σ of length n. In this work, we are interested in pattern matching queries asking for occurrences of a given pattern $P \in \Sigma^m$ of length m in T. Particularly, we are interested in the queries (a) *count*, asking for the number occ of occurrences of P in T, and (2) *locate*, asking for an enumeration of the starting positions of the occurrences. For some $i \in [1, n]$, we denote by $T[i]$ the i-th character in T. Given additionally $j \in [i, n]$, we denote by $T[i \mathbin{..} j]$ the substring $T[i]\, T[i+1] \cdots T[j-1]\, T[j]$, juxtaposition meaning concatenation. The aforementioned queries are formally defined as $\mathrm{locate}(T, P) = \{i \in [1, n-m+1] \mid T[i \mathbin{..} i+m-1] = P\}$ and $\mathrm{count}(T, P) = |\mathrm{locate}(T, P)|$. We argue about running times in the word RAM model, where we can do operations on words of length $\omega = \Theta(\lg n)$ bits in constant time. Unless explicitly stated otherwise, logarithms are given as base-two.

2.1   Lempel-Ziv Parsings and Random Access

Lempel-Ziv (LZ) parsings factorize a text T into $z \leq n$ *phrases* $f_1, \ldots, f_z$ such that their concatenation $f_1 \cdots f_z = T$. They form a family of dictionary compressors, with arguably the most popular representative being Lempel-Ziv 77 (LZ77) [23]. There, we define the phrase f_i (for $i \in [1, z]$) as either (1) a new symbol that does not occur in $T[1 \mathbin{..} |f_1 \cdots f_{i-1}|]$, or (2) the longest prefix of $T[|f_1 \cdots f_{i-1}| + 1 \mathbin{..} n]$ that has an occurrence in T starting at a position $\leq |f_1 \cdots f_{i-1}|$. In the second case, we can encode the phrase as a *reference* to a previous occurrence, which can potentially be stored in less bits than storing the phrase explicitly. Even if we relax the definition of referencing phrases, it is typically $z \ll n$ if T is repetitive. Thus, LZ parsings are a popular choice for compressing T and are used in myriad everyday utilities such as *gzip*.

Random Access. We are interested in efficient random access on T in its compressed form, i.e., we wish to extract a substring $T[x \mathbin{..} x + \ell]$ for some $x \in [1, n]$ and $\ell \geq 0$ without decoding substantial portions of T. Regarding just the character $T[x]$, it is contained in the phrase f_i for $i = \min\{i \in [1, z] \mid |f_1 \cdots f_i| \geq x\}$. We say that phrase f_i *covers* position x. We can store the set $E = \{|f_1 \cdots f_j| \mid j \in [1, z]\}$ (the end positions of the phrases) in $z\lceil \lg n \rceil$ bits (i.e., in space $\mathcal{O}(z)$) and compute i in time $\mathcal{O}(\lg z)$ using binary search, or use a static successor data structure to compute i in time $\mathcal{O}(\lg \lg(n/z))$ [20].[1] An inherent disadvantage of the classic LZ77 scheme defined above, albeit achieving very good compression in practice, is that random access cannot be done efficiently. Phrases may refer to arbitrary prior positions in T, and thus to decode $T[x]$, we may have to decode all phrases $f_1, \ldots, f_i$ up to (a prefix of) the phrase that covers x. We now look at two variants that resolve this issue at the cost of worse compression.

LZ-End. Kreft and Navarro introduced the scheme *LZ-End* [13]. Here, each phrase f_i is represented as a triple (j, ℓ, α), where $j < i$ is the *source phrase*, $\ell \geq 0$ is the *copy length* and $\alpha \in \Sigma$ is a character such that $f_i = T[|f_1 \cdots f_j| - \ell + 1 \mathbin{..} |f_1 \cdots f_j|]\,\alpha$ for maximal possible ℓ and there is no $k < i$ such that f_i is a suffix of $T[1 \mathbin{..} |f_1 \cdots f_k|]$. We allow $f_0 := \epsilon$ as a valid source phrase such that the above is well-defined. Intuitively, f_i extends the length-ℓ suffix of $T[1 \mathbin{..} |f_1 \cdots f_j|]$ by a new character α and j is picked greedily such that ℓ is maximized.

---

[1] Alternatively, we can build the characteristic bit vector B_E of n bits where the j-th bit is set iff $j \in E$. Since exactly z bits are set in B_E, we can build a data structure of size $\lceil \lg \binom{z}{m} \rceil + o(n) + \mathcal{O}(\lg \lg z)$ bits that supports constant-time rank and select queries on B_E [21]. With this, we can then also compute i in constant time. In this work, however, our aim is to focus on compressed space $\mathcal{O}(z)$.

Since each phrase adds exactly one character to a previously occurring substring, the end position of which is explicitly stated in the encoding triple, we can decode $T[x .. x + \ell]$ in time $\mathcal{O}(h + \ell)$ once we know the phrase that covers position $x + \ell$. Here, h is the length of the longest phrase. This gives us total random access time $\mathcal{O}(\lg \lg(n/z_{\text{end}}) + h + \ell)$, where $z_{\text{end}} = \mathcal{O}(z \lg^2 n)$ is the number of LZ-End phrases of T [12].

When computing the parsing, we can artificially constrain h to obtain parameterized random access time at the cost of introducing only at most $\mathcal{O}((n \lg n)/h)$ additional phrases [12] ($\mathcal{O}(n/h)$ of which come naturally from a phrase length restriction of h).[2]

Relative Lempel-Ziv. Kuruppu et al. proposed a variant of Lempel-Ziv parsings where we do not refer to earlier parts of T itself, but instead to a given reference $R \in \Sigma^*$ [14]. This is useful especially in scenarios where we want to store a collection of texts that are highly similar (e.g., genomic sequences from the same species). Formally, the phrase f_i is the longest prefix of $T[|f_1 \cdots f_{i-1}| + 1 .. n]$ that occurs in R, or a single character that does not occur in R. This scheme is referred to as *Relative Lempel-Ziv* (RLZ).

To decode $T[x .. \ell]$, we can directly access the substring in R that the phrase covering x refers to, which can be done in total time $\mathcal{O}(\lg \lg(n/z_R) + \ell)$, where z_R is the number of RLZ phrases of T for reference R. The compression depends on how well R represents T, and R must be stored alongside the compressed form of T in order to be able to decode T.

2.2   Suffix Arrays, Burrows-Wheeler Transform and Compression

In the *suffix array* A of T, we store the starting positions of the suffixes of T in their lexicographical order [15]. This ordering causes suffixes that begin with equal prefixes to be grouped in consecutive intervals. A text book algorithm to answer count queries in time $\mathcal{O}(m \lg n)$ finds the interval $[b, e] \subseteq [1, n]$ of A that contains all (and only the) suffixes of T beginning with P, the query time stemming from binary searches for b and e, respectively. To answer locate, we simply need to enumerate $A[b .. e]$. We can store A in $n\lceil \lg n \rceil$ bits of space and construct it in time $\mathcal{O}(n)$ [17].

The *Burrows-Wheeler transform* (BWT) of T is a reversible transform of T defined as $L[i] := T[A[i] - 1]$ (or $L[i] := T[n]$ if $A[i] = 1$) [3]. The BWT of repetitive texts tends to contain long equal-letter runs, which can be exploited by run-length compression. We denote by r the number of these runs.

Compressed Differential Suffix Arrays. In practice, storing A plainly is prohibitive for large T. Even though it is a permutation over $[1, n]$ and thus not inherently compressible, different ways to compress A have been shown (we refer

---

[2] We thank the anonymous reviewer for the insightful comments on the behaviour of LZ-End under phrase length restrictions.

to [10] for an overview). In this work, we focus on compressing the *differential suffix array* $A^d \in \mathbb{Z}^n$, where $A^d[1] := A[1]$ and $A^d[i] := A[i] - A[i-1]$ for $i \in [2, n]$.

Gonzet al. first exploited the interesting property that the number of distinct values in A^d is bounded by the number r of BWT runs and repetitiveness in T implies repetitiveness in A^d [9]. In this work, we are interested in the approach by Puglisi and Zhukova [19], who instead considered RLZ to compress A^d. They describe a strategy to extract R from A^d by selecting segments based on the frequencies of representative substrings, and show that this outperforms using a random sample of A^d (for which bounds on the expected compression have been shown [8]). We denote by $\hat{z}_R$ the number of RLZ phrases of A^d computed this way.

For random access on A, we want to avoid having to compute $A[x] = \sum_{i=1}^x A^d[i]$ for some $x \in [1, n]$ in time $\mathcal{O}(n)$. Rather, we create a sample A' that contains a subsequence of A. Let y be the greatest sampled position $\leq x$, then we can compute $A[x] = A'[y] + \sum_{i=y+1}^x A^d[i]$ in time $\mathcal{O}(\delta)$, where δ is the maximum distance between any position and the previous sample.

For example, in [19], we take a sample of A for every RLZ phrase. This gives us $\delta < h$ and using a static successor data structure of size $\mathcal{O}(\hat{z}_R)$ (since $|A'| = \hat{z}_R$), random access is possible in time $\mathcal{O}(\lg \lg(n/\hat{z}_R) + h)$, where h is the length of the longest RLZ phrase.

2.3 (Move-)r-Index

The r-index is a recent advancement in compressed data structures for pattern matching [6] that is also highly relevant in practice. It is a self-index that encodes the BWT of T and auxiliary data structures in $\mathcal{O}(r)$ space. Using the move data structure of [16] and assuming $|\Sigma| = \mathcal{O}(\text{polylog } n)$, we obtain optimal $\mathcal{O}(m)$ time for count and optimal additional time $\mathcal{O}(\text{occ})$ for locate queries, where occ is the number of occurrences of the search pattern.

3 LZ-End Compression of Suffix Arrays

Following the idea of [19] to apply LZ compression on the differential suffix array, we explore its compression using LZ-End. To give an intuition as to why this may be fruitful, LZ-End (1) allows for efficient random access on the compressed input and (2) achieves competitive compression in practice. We first show the following for compressing any integer sequence A.

Theorem 1. *Let $A \in [1, n]^n$ be an integer sequence. In time and space $\mathcal{O}(n)$, we can construct a data structure of size $\mathcal{O}(\hat{z}_{end})$ such that for $x \in [1, n]$ and $\ell \geq 0$, we can reconstruct $A[x \mathrel{..} x + \ell]$ in time $\mathcal{O}(\lg \lg(n/\hat{z}_{end}) + h + \ell)$, where $\hat{z}_{end}$ is the number of LZ-End phrases of the differential representation A^d of A and h the length of the longest phrase.*

Proof. The differential representation A^d of A can be computed in time and space $\mathcal{O}(n)$ and by [11], the same holds for the LZ-End parsing of A^d. We represent the triples defining the parsing as three arrays:

1. the array src, where the i-th entry contains the number $\in [1, i-1]$ of the source phrase that f_i refers to,
2. the array end, where the i-th entry contains the position $|f_1 \cdots f_i| \in [1, n]$ at which phrase f_i ends in A^d, and
3. the array ext, where the i-th entry contains the value $\in [-n, n]$ from A^d that extends the suffix of $A^d[1 .. |f_1 \cdots f_{src[i]}|]$.

Each array can be stored in space $\mathcal{O}(\hat{z}_{end})$. We also build a static successor data structure over end that allows for successor queries in time $\mathcal{O}(\lg \lg(n/\hat{z}_{end}))$, which we can do in time and space $\mathcal{O}(\hat{z}_{end})$. Finally, in time at most $\mathcal{O}(\hat{z}_{end})$, we sample the $\hat{z}_{end}$ values from A at the positions stored in end in a new array A' and store them also in space $\mathcal{O}(\hat{z}_{end})$.

Given $x \in [1, n]$ and $\ell \geq 0$, we decode $A[x .. x+\ell]$ as follows: we first extract the range $A^d[x .. x+\ell]$ from the LZ-End parsing in time $\mathcal{O}(\lg \lg(n/\hat{z}_{end}) + h + \ell)$ using the extraction algorithm from [13] (the length of a phrase f_i can trivially be computed in constant time as $|f_i| = end[i] - end[i-1]$). Using the successor data structure, we can find in time $\mathcal{O}(\lg \lg(n/\hat{z}_{end}))$ the position of a relevant sample of A that is stored in A'. Then, in time at most $\mathcal{O}(h + \ell)$, we accumulate the relevant differential values from $A^d[x .. x+\ell]$ to reconstruct $A[x .. x+\ell]$.

Corollary 1. *Let $T \in \Sigma^n$ be a string of length n. In time and space $\mathcal{O}(n)$, we can construct a data structure of size $\mathcal{O}(\hat{z}_{end})$ such that for $x \in [1, n]$ and $\ell \geq 0$, we can compute the suffix array interval $A[x .. x+\ell]$ in time $\mathcal{O}(\lg \lg(n/\hat{z}_{end}) + h + \ell)$, where $\hat{z}_{end}$ is the number of LZ-End phrases of the differential representation A^d of the suffix array A of T and h is the length of the longest phrase.*

3.1 Practical LZ-End Parsing

To implement the computation of LZ-end parsings, we adopt and modify the algorithm by Kempa and Kosolobov [11] that does so in time $\mathcal{O}(n \lg \lg n)$ in a left-to-right scan of the text T of length n (a linear-time algorithm exists [11], but we conjecture it to be hardly practical). At its core lies a dynamic predecessor/successor data structure M that marks the lexicographic ranks of suffixes of the reverse input $\overleftarrow{T}$ at which already computed phrases end. In the following, we briefly describe our modifications and refer to the full version [5] for details.

First, we make M associative, so that at each marked suffix, we also store the number of the phrase that ends at the suffix. This removes a level of indirection and even allows us to completely discard the suffix array after initialization.

Second, say that $f_1 \cdots f_x$ is the LZ-End parsing computed thus far for $x < z_{end}$ and we now read the next character $\alpha \in \Sigma$. One of the possible cases for the next LZ-end phrase is merging the phrases f_{x-1} and f_x to a new phrase $f_{x-1}f_x\alpha$. In this case, clearly, we cannot use phrase $x-1$ as a source phrase to copy from. In the original algorithm, we temporarily unmark phrase $x - 1$ in M to guarantee that it is not reported as a possible a source phrase, and then afterwards mark it back in M unless it was merged with phrase x. We reduce the overall number of updates on M by performing an additional predecessor or successor query in case $x - 1$ is reported, and only ever unmark a phrase in M when it is actually removed.

Third, the parsing is computed processing T left to right, but suffixes of the reverse $\overleftarrow{T}$ are considered. We save arithmetic computations by using a variant $A^{\leftarrow-1}$ of the inverse suffix array of T that is defined as $A^{\leftarrow-1}[A[n - i - 1]] := i$. Then, it is $A^{\leftarrow-1}[i] = A^{-1}[n - i]$ and we read $A^{\leftarrow-1}$ left to right as we process T.

Finally, our implementation of the algorithm is written in a way that Σ may be an arbitrary integer alphabet such that, e.g., we can compute the parsing for a differential suffix array.

We set the maximum phrase length to $h := 2^{13}$, giving us the best balance in preliminary experiments: choosing lower h notably increased the number of phrases, whereas higher—or unbounded—h had an increasingly negative impact on access performance. Furthermore, we store the array end of end positions plainly using $z\lceil \lg n \rceil$ bits and use a simple $\mathcal{O}(\lg z)$-time binary search with no auxiliary data structure to find the phrase covering a position in question. Preliminary experiments have shown that despite its simplicity, this approach is the fastest given the relatively low number z.

In the full version of this paper [5], we present results of experiments showing that our implementation of LZ-End is competitive with the *lz-end-toolkit* and faster on general (non-highly repetitive) inputs.

4   Improved RLZ Compression of Suffix Arrays

Puglisi and Zhukova [19] considered compressing the differential suffix array using Relative Lempel-Ziv (henceforth referred to as RLZSA). However, their source code remains closed. With the aim of reproducing their results for further research, we reimplemented RLZSA as described there and in Zhukova's doctoral thesis [22] to the best of our capabilities. In this process, we found several clues as to how to improve upon their work. We summarize our improvements here and refer to the full version [5] for an in-depth description of the individual steps.

First, we shrink the overall representation by separating data pertaining to literal phrases (encoding explicitly a character $\alpha \in \Sigma$) or copying RLZ phrases (encoding the source position in R and the number of characters to copy). We store a bit vector of length z_R with support for rank/select that allows us to identify the type of each phrase and use rank and select queries to access its data in separate arrays. This allows us to drop the requirement that each copying phrase needs to be preceded by a literal phrase, which also improves RLZ compression. Using the new representation, we can reduce the time for randomly accessing a suffix array interval $A[b .. e]$ from $\mathcal{O}(|e - b| + h + \lg(z_R/a) + a)$ to $\mathcal{O}(|e - b| + \lg(na/z_R) + a)$, where $a \geq 1$ is an integer sampling parameter (for sampling phrase starting positions) and h is the length of the longest RLZ phrase.

We then proceed to improve upon the construction of RLZSA. By using Big-BWT [2], we can make the construction of A^d semi-external, reducing the memory usage from $\mathcal{O}(n)$ to $\mathcal{O}(|\mathsf{PFP}|)$, where PFP is a prefix free parsing of T. By allowing the selection of arbitrary segments from A^d (instead of partitioning A^d and only allowing aligned segments) and by setting the considered k-mer length to $k := 1$, we can reduce the time and space required for reference construction from $\mathcal{O}(n)$ to $\mathcal{O}(r^{1-\epsilon}n^\epsilon)$ and $\mathcal{O}(r)$, respectively, where $\epsilon \in [0,1]$ is a parameter. The new segment selection strategy also improves the reference quality, leading to better compression (as shown later in Table 1).

Finally, we speed up and reduce the memory usage for computing the RLZ parsing of A^d for the computed reference R by replacing the FM-index by Move-r over $\overleftarrow{R}$ using an optimized rank/select data structure for large alphabets.

We set the size of the RLZ reference $|R| := \min(5.2r, n/3)$, which gave us the best results overall in preliminary experiments. RLZ phrases are limited to maximum length $h := 2^{16}$, which allows storing their length in 16-bit integers.

5   Applications to the (Move)-r-Index

The main bottleneck when answering locate queries using the r-index in practice are the applications of the function Φ required to enumerate occurrences.

There have been at least two different approaches to resolve this, both of which have been shown to be relevant in practice: Puglisi and Zhukova store the RLZ-compressed differential suffix array next to the r-index, which allows for up to two orders of magnitude faster locate queries (with many occurrences) at the cost of using 2–13 times as much memory [19]. Bertram et al., on the other hand, implement the move data structure by [16], speeding up queries (both count and locate) by an order of magnitude while only doubling the required space [1].

We propose variations and combinations of the above and evaluate them in our experiments (Sect. 6). Namely, we explore storing an LZ-End-compressed differential suffix array next to the r-index to find out whether we can obtain a trade-off similar to [19]. Furthermore, albeit much faster than in the original r-index, the Φ steps for enumerating occurrences remain a bottleneck for the locate queries of [1], each causing cache misses. We thus also consider storing either compressed differential suffix array next to Move-r. The space requirement for the whole index increases by at most a polylogarithmic factor, which follows from known bounds on LZ-End and differential suffix arrays [7,12][3].

The r-index (as well as Move-r) already maintains a sampling A'_r of the suffix array at the boundary of every BWT run, i.e., $|A'_r| = r$. This creates redundancy regarding the sampling A' of suffix array values at LZ phrase end positions proposed by [19] and in the proof of Theorem 1. For reconstructing a suffix array value using A_d, we can as well use A'_r. This worsens the worst-case access time to $\mathcal{O}(n)$, because there is no general bound on the length of a BWT

---

[3] We again thank the anonymous reviewer for their notes regarding the additional required space.

run. In practice, however, the average length of a BWT run is reasonably short even for repetitive inputs (see, e.g., column $\lfloor n/r \rfloor$ in Table 1).

Alternatively, one could consider replacing the sampling A'_r by A' in the r-index or Move-r. However, then, to retrieve the suffix array value at the end of a run, we must spend up to $\mathcal{O}\left(\lg\lg(n/\hat{z}) + h\right)$ extra time time for random access, which would worsen the performance of locate queries, conflicting with our motivations. It would also increase the index size in practice as empirically, it holds that $\hat{z} > r$. Therefore, we do not further consider this sampling method.

6   Experiments

In our experiments, we evaluate the construction and locate query performance of the following variations of the r-index and Move-r:

- r-index – the original r-index of [6],
- r-rlz – the r-index plus the RLZ-compressed differential suffix array,
- r-lzend – the r-index plus the LZ-End-compressed differential suffix array,
- move-r – the Move-r index of [1] (with improved internal rank/select),
- move-r-rlz – Move-r plus the RLZ-compressed differential suffix array, and
- move-r-lzend – Move-r plus the LZ-End-compressed differential suffix array.

Note that only move-r-rlz contains the improved RLZSA construction that we described in Sect. 4, whereas r-rlz is based on a reimplementation of [19] described in the full version. We do this to better argue about our improvements. However, we use Big-BWT [2] for all variants to compute suffix arrays. As mentioned in the list above, we also applied improvements to Move-r itself by engineering a new rank/select data structure tailored specifically for its internal queries. We refer to the full version for details.

We implemented all index variants in C++20 and make the source code publicly available[4]. We compiled using the GCC 13.3.0 compiler with flags set for highest optimization (-march=native -DNDEBUG -Ofast).

Table 1 lists the input texts that we considered in our experiments alongside relevant statistics. The texts einstein and english are part of the Pizza&Chili Corpus[5], whereas dewiki is a highly repetitive text manually constructed from German Wikipedia entries. From the *National Center for Biotechnology Information*[6] (NCBI) database, we constructed chr19, consisting of concatenated human chromosome 19 haplotypes, and sars2, a collection of Sars-Cov-2 genomes. From all text files, we erased all zero bytes.

For each text, we generated two sets of query patterns (hence two lines per file in Table 1) using our tool move-r-patterns (also included in our source code repository). The sets differ in the pattern length m, as well as the average number $\overline{occ}$ of occurrences in the respective text. We chose the patterns in the

---

[4] Our source code: https://github.com/LukasNalbach/Move-r.
[5] Pizza&Chili Corpus: https://pizzachili.dcc.uchile.cl/.
[6] NCBI: https://www.ncbi.nlm.nih.gov/.

first set such that $\overline{occ} \approx m$. This implies that when locating those patterns, we measure a blend of backward-search and suffix array extraction. The performance of counting queries was measured against this set. The patterns in the second set were chosen such that $\overline{occ} \approx 10^5 m$. When locating these, we measure mostly suffix array extraction, which is a particularly relevant measure for our experiments.

We note that we do not consider $m \gg$ occ, i.e., where we have long patterns with few occurrences. Measurements in this realm would essentially measure the performance of count queries, which has already been done for both the r-index as well as Move-r [1].

All experiments were done on a Ubuntu 24.04 system with two AMD EPYC 7452 CPUs ($32/64 \times 2.35$–3.35GHz, $2/16/128$MB L1/2/3 cache) and 1TB of RAM (3200 MT/s DDR4).

Table 1. The input files for our experiments. For each input, we give the size n, the size $|\Sigma|$ of the alphabet and the compression ratios n/r, $n/\hat{z}_R$ and $n/\hat{z}_{\text{end}}$ (higher values mean more repetitive). Here, $\hat{z}_R$ is the number of RLZ phrases of A^d following the construction of [19], whereas $\hat{z}_{R'}$ refers to our improved construction from Sect. 4. As in Sect. 3.1, $\hat{z}_{\text{end}}$ denotes the number of LZ-End phrases of A^d. By N, we denote the number of queried patterns, by m the pattern length and by $\overline{occ}$ the average number of occurrences of the patterns. Per input, the first line indicates N, m and $\overline{occ}$ for $m \approx \overline{occ}$, the second line for $m \ll \overline{occ}$.

text	n [GB]	$|\Sigma|$	$\lfloor n/r \rfloor$	$\lfloor n/\hat{z}_R \rfloor$	$\lfloor n/\hat{z}_{R'} \rfloor$	$\lfloor n/\hat{z}_{\text{end}} \rfloor$	N	m	$\overline{occ}$
einstein	0.47	140	1,611	118	183	1,081	100,000	800	736
							10,000	7	72,644
sars2	10.00	80	548	60	61	336	3,000	2,700	2,745
							100	24	178,948
dewiki	10.00	207	377	122	146	306	100,000	300	323
							1,000	9	76,372
chr19	10.00	53	46	12	25	34	1,000	25,000	19,531
							1,000	100	107,991
english	2.21	240	3	2	4	3	500,000	35	37
							300	7	91,964

6.1 Construction Performance

We first look at the construction of the competing indexes. Figure 1 shows the construction throughput as well as the peak memory usage during construction.

To no surprise, compressing the differential suffix array dominates the time and space needed for construction (comparing r-index and move-r to the variants storing a compressed suffix array). Regarding the two different compression schemes, we see that LZ-End (move-r-lzend and r-lzend) is relatively slow to compute overall, but competitive with r-rlz regarding both time and space.

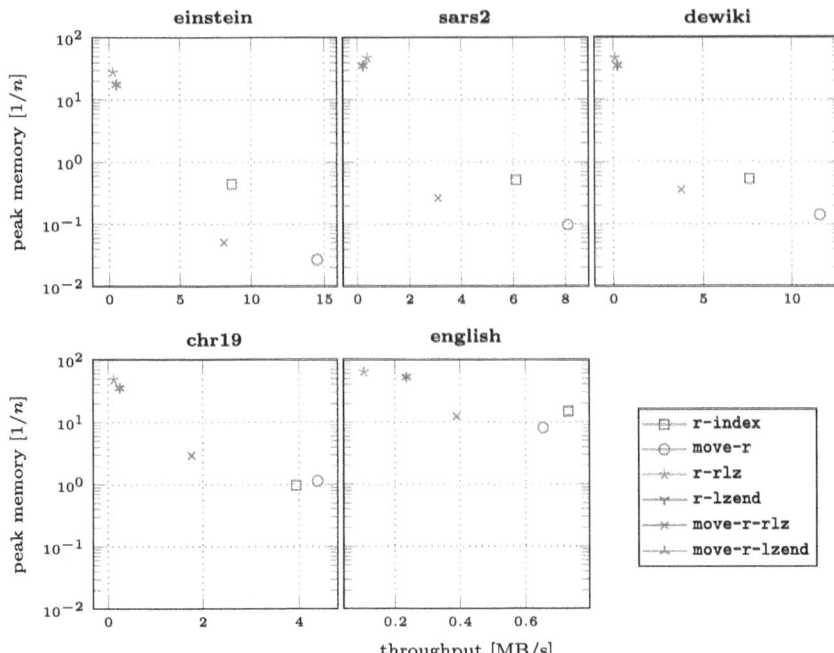

Fig. 1. Construction time versus peak memory usage (in bytes per input character) of our implemented index data structures for the given inputs. Memory usage is given on a logarithmic scale in order to highlight the marginal differences between `r-index`, `move-r` and `move-r-rlz`. Data points for `r-lzend` and `move-r-lzend` do, in fact, overlap nearly precisely.

Our improved RLZSA construction from Sect. 4 (`move-r-rlz`), however, clearly outperforms the other variants that compress the suffix array: it is faster by a factor of up to ten (einstein) and the required space is sometimes even lower than for just computing the r-index. It also clearly outperforms the construction of our reimplementation of RLZSA (`r-rlz`).

6.2   Locate Query Performance

We now look at locate queries for the two pattern sets described above (one query per pattern). Figure 2 shows the query throughput as well as the size of the considered indexes. For reference, we also give the throughput of count queries, which does not involve any compressed suffix arrays (because we only report the size of the corresponding suffix array interval, not its contents).

We can assert that the performance of `move-r` is somewhat improved over [1] (the experiments there were done on the same machine). The trade-off compared to `r-index` remains the same: we require roughly twice the amount of space, but queries are considerably faster overall.

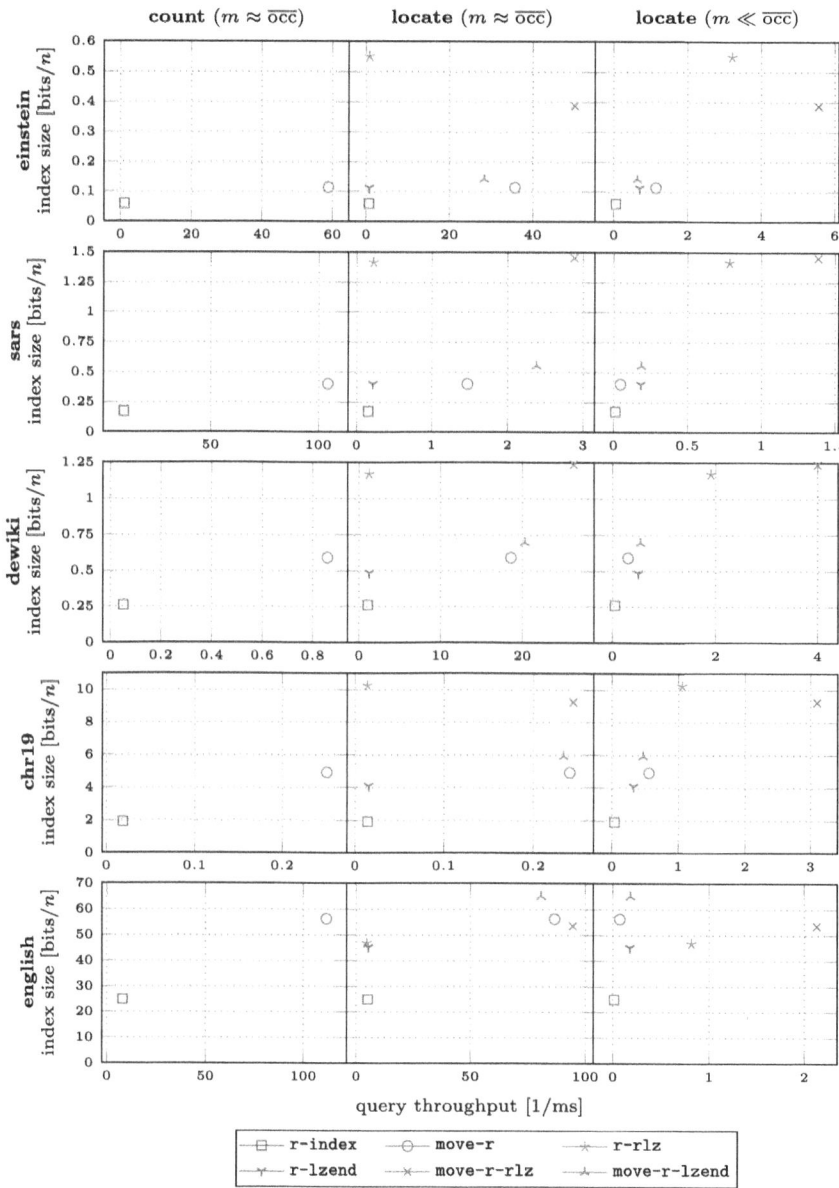

Fig. 2. Size (in bits per input character) versus locate query throughput (queries per millisecond) of our implemented index data structures given medium ($m \approx \overline{occ}$) or short ($m \ll \overline{occ}$)patterns for the given inputs. For reference, we also give the count query throughput of the base index data structures, `r-index` and `move-r` for medium patterns.

As expected, enhancing the r-index by compressed suffix arrays (`r-rlz` and `r-lzend`) considerably improves the performance of locate queries for patterns with many occurrences. This confirms the results of [19]. We see how `r-rlz` achieves overall higher throughputs than `r-lzend` (by a factor of 4 for $m \ll \overline{occ}$). This is expected, as random access on RLZ-compressed data incurs only one cache miss per phrase, as opposed to up to h cache misses for LZ-End. However, we see that LZ-End achieves better compression, which is also confirmed in Table 1 when comparing columns $\lfloor n/\hat{z}_R \rfloor$ and $\lfloor n/\hat{z}_{end} \rfloor$, making it a trade-off.

When enhancing Move-r with compressed suffix arrays (`move-r-rlz` and `move-r-lzend`), the picture differs a bit. Here, using LZ-End (`move-r-lzend`) can sometimes even slow down locate queries (e.g., on einstein and chr19). Using RLZ (`rlzsa`), on the other hand, improves query performance by a great deal particularly for frequent patterns $m \ll \overline{occ}$ (e.g., by a factor of over 16 for sars). Again, however, LZ-End yields much better compression than RLZ in most cases (now comparing $\lfloor n/\hat{z}_{R'} \rfloor$ and $\lfloor n/\hat{z}_{end} \rfloor$ in Table 1). Interestingly however, on english, the improved RLZSA construction (`move-r-rlz`) achieves better compression than LZ-End (`move-r-lzend`), which is a topic for further research.

Overall, our improved RLZSA (`move-r-rlz`) achieves better compression than that of [19] (`r-rlz`). This is particularly evident for einstein and chr19, where `move-r-rlz` is smaller than `r-rlz` despite storing more information (e.g., compare `move-r` against `r-index`).

7   Conclusions and Future Work

We enhanced the recent r-index as well as Move-r by compressed suffix arrays with efficient random access to speed up locate queries. For this, we explored two different compression schemes: Relative Lempel-Ziv and LZ-End. The experiments show that the idea works, confirming and expanding upon the results of [19]. We can achieve different trade-offs regarding construction performance, index size and query performance by choosing different combinations of index and compressed suffix arrays. For both compression schemes, we gave new strategies and algorithms that improve upon their predecessors.

In future research, enhancing the subsampled r-index by Cobas et al. [4] may be considered. We also saw that reference construction for Relative Lempel-Ziv is still an interesting topic of research beyond [8,19]. By improving upon the segment selection strategy of [19], we were able to improve the quality of the reference and thus compression.

References

1. Bertram, N., Fischer, J., Nalbach, L.: Move-R: optimizing the R-index. In: 22nd International Symposium on Experimental Algorithms (SEA). LIPIcs, vol. 301, pp. 1:1–1:19. Dagstuhl (2024). https://doi.org/10.4230/LIPICS.SEA.2024.1
2. Boucher, C., Gagie, T., Kuhnle, A., Langmead, B., Manzini, G., Mun, T.: Prefix-free parsing for building big BWTs. Algorithms Mol. Biol. **14**(1), 13:1–13:15 (2019). https://doi.org/10.1186/S13015-019-0148-5

3. Burrows, M., Wheeler, D.: A block-sorting lossless data compression algorithm. Technical report, 124, Digital Equipment Corporation (1994)

4. Cobas, D., Gagie, T., Navarro, G.: A fast and small subsampled R-index. In: 32nd Annual Symposium on Combinatorial Pattern Matching (CPM). LIPIcs, vol. 191, pp. 13:1–13:16. Dagstuhl (2021). https://doi.org/10.4230/LIPICS.CPM.2021.13

5. Dinklage, P., Fischer, J., Nalbach, L., Zumbrink, J.: RLZ-R and LZ-End-R: enhancing move-R (full version). CoRR abs/2507.17300 (2025). http://arxiv.org/abs/2507.17300

6. Gagie, T., Navarro, G., Prezza, N.: Optimal-time text indexing in BWT-runs bounded space. In: 29th Annual ACM-SIAM Symposium on Discrete Algorithms (SODA), pp. 1459–1477. SIAM (2018). https://doi.org/10.1137/1.9781611975031.96

7. Gagie, T., Navarro, G., Prezza, N.: Optimal-time text indexing in BWT-runs bounded space (extended version). CoRR abs/1705.10382 (2025). http://arxiv.org/abs/1705.10382

8. Gagie, T., Puglisi, S.J., Valenzuela, D.: Analyzing relative Lempel-Ziv reference construction. In: Inenaga, S., Sadakane, K., Sakai, T. (eds.) SPIRE 2016. LNCS, vol. 9954, pp. 160–165. Springer, Cham (2016). https://doi.org/10.1007/978-3-319-46049-9_16

9. González, R., Navarro, G., Ferrada, H.: Locally compressed suffix arrays. ACM J. Exp. Algorithmics **19**(1) (2014). https://doi.org/10.1145/2594408

10. Grossi, R.: A quick tour on suffix arrays and compressed suffix arrays. Theor. Comput. Sci. **412**(27), 2964–2973 (2011). https://doi.org/10.1016/J.TCS.2010.12.036

11. Kempa, D., Kosolobov, D.: LZ-end parsing in linear time. In: 25th European Symposium on Algorithms (ESA). LIPIcs, vol. 87, pp. 53:1–53:14. Dagstuhl (2017). https://doi.org/10.4230/LIPIcs.ESA.2017.53

12. Kempa, D., Saha, B.: An upper bound and linear-space queries on the LZ-end parsing. In: Naor, J.S., Buchbinder, N. (eds.) 33rd Annual ACM-SIAM Symposium on Discrete Algorithms (SODA), pp. 2847–2866. SIAM (2022). https://doi.org/10.1137/1.9781611977073.111

13. Kreft, S., Navarro, G.: LZ77-like compression with fast random access. In: 2010 Data Compression Conference (DCC), pp. 239–248. IEEE (2010). https://doi.org/10.1109/DCC.2010.29

14. Kuruppu, S., Puglisi, S.J., Zobel, J.: Relative Lempel-Ziv compression of genomes for large-scale storage and retrieval. In: Chavez, E., Lonardi, S. (eds.) SPIRE 2010. LNCS, vol. 6393, pp. 201–206. Springer, Heidelberg (2010). https://doi.org/10.1007/978-3-642-16321-0_20

15. Manber, U., Myers, E.W.: Suffix arrays: a new method for on-line string searches. SIAM J. Comput. **22**(5), 935–948 (1993)

16. Nishimoto, T., Tabei, Y.: Optimal-time queries on BWT-runs compressed indexes. In: 48th International Colloquium on Automata, Languages, and Programming (ICALP). LIPIcs, vol. 198, pp. 101:1–101:15. Dagstuhl (2021). https://doi.org/10.4230/LIPICS.ICALP.2021.101

17. Nong, G., Zhang, S., Chan, W.H.: Two efficient algorithms for linear time suffix array construction. IEEE Trans. Comput. **60**(10), 1471–1484 (2011). https://doi.org/10.1109/TC.2010.188

18. Puglisi, S.J., Zhukova, B.: Relative Lempel-Ziv compression of suffix arrays. In: Boucher, C., Thankachan, S.V. (eds.) SPIRE 2020. LNCS, vol. 12303, pp. 89–96. Springer, Cham (2020). https://doi.org/10.1007/978-3-030-59212-7_7

19. Puglisi, S.J., Zhukova, B.: Smaller RLZ-compressed suffix arrays. In: 2021 Data Compression Conference (DCC), pp. 213–222. IEEE (2021). https://doi.org/10.1109/DCC50243.2021.00029
20. Pătraşcu, M., Thorup, M.: Time-space trade-offs for predecessor search. In: 31st Annual ACM Symposium on Theory of Computing (STOC), pp. 232–240. ACM (2006). https://doi.org/10.1145/1132516.1132551
21. Raman, R., Raman, V., Rao, S.S.: Succinct indexable dictionaries with applications to encoding K-ary trees and multisets. In: 13th Annual ACM-SIAM Symposium on Discrete Algorithms (SODA), pp. 233–242. SIAM (2002)
22. Zhukova, B.: New space-time trade-offs for pattern matching with compressed indexes. Ph.D. thesis, University of Helsinki, Finland (2024). http://hdl.handle.net/10138/570140
23. Ziv, J., Lempel, A.: A universal algorithm for sequential data compression. IEEE Trans. Inform. Theory **23**(3), 337–343 (1977). https://doi.org/10.1109/TIT.1977.1055714

Massively Parallel Computation
of Matching Statistics

Anastasia C. Diseth$^{(\boxtimes)}$, Keijo Heljanko, and Simon J. Puglisi

University of Helsinki, Helsinki, Finland
{anastasia.diseth,keijo.heljanko,simon.puglisi}@helsinki.fi

Abstract. The *matching statistics* of a string S relative to another string R is a sequence of $|S|$ integer pairs, (p_i, ℓ_i), one for each position in S, such that $S[i..i + \ell_i] = R[p_i..p_i + \ell_i]$ and ℓ_i is the length of the longest substring starting at position i in S that also occurs in R. Matching statistics have a variety of applications in sequence processing, such as approximate pattern matching, suffix sorting, compressed indexing, data compression, and multiple sequence alignment. Often the strings involved are large, making efficient computation paramount. In this paper we describe algorithms for matching statistics computation on massively parallel architectures, namely, graphics processing units (GPUs). We show that when implemented carefully these methods are much faster than CPU-based algorithms.

Keywords: matching statistics · suffix array · parallel algorithms · GPU

1 Introduction

Given two strings R and S, the *matching statistics* of S relative to R, denoted $M_{S|R}$, is a sequence of $|S|$ integer pairs, (p_i, ℓ_i), such that $S[i..i + \ell_i] = R[p_i..p_i + \ell_i]$ is the longest substring starting at position i in S that also occurs somewhere in R. For example, given $R = acaatca$ and $S = tcaaacaa$ then $M_{S|R} = (4, 3), (1, 3), (2, 2), (2, 2), (0, 4), (0, 3), (0, 2), (0, 1)$.

Computation of matching statistics has a number of important applications in sequence processing, including approximate pattern matching [4], suffix sorting [16,20], data compression (in particular, RLZ parsing [2,10,14]), indexing [7, 8,22], and computing anchors for multiple sequence alignment [9,17,19,21].

In many applications the so-called *reference sequence*, R, is fixed and the matching statistics must be computed relative to it for many different S strings. Often the strings involved are large and highly similar to each other, such as genome sequences of individuals from the same species. Efficient computation of matching statistics in this setting will be our main concern throughout.

In this paper we describe algorithms for matching statistics computation on massively parallel architectures, in particular, graphics processing units (GPUs).

This work is supported in part by the Research Council of Finland via grant 369068.

G. Badkobeh et al. (Eds.): SPIRE 2025, LNCS 16073, pp. 79–94, 2026.
https://doi.org/10.1007/978-3-032-05228-5_8

We show that when implemented carefully these algorithms are much faster than CPU-based algorithms. Our main contributions can be summarized as follows.

– We describe a simple algorithm for matching statistics computation that is amenable to GPU implementation, which we refer to as BRACKETS. Given a string R and a collection of m strings $S_1, S_2, \ldots, S_m$, BRACKETS computes the suffix array of the string $RS_1 S_2 \ldots S_m$ formed by concatenating the input strings. For every suffix $S_i[j..]$ of every S_i string, it then finds the lexicographically preceding and succeeding suffix from R, from which the jth value in the matching statistics can be easily computed.
– We compare BRACKETS to the current fastest CPU-based method, the LMP algorithm [15,16], as well as a GPU version of LMP that we have implemented. Our results indicate BRACKETS is 20% faster than the GPU version of LMP, which is in turn 10x faster than the CPU implementation.
– The bottleneck in BRACKETS is suffix array construction for the concatenated string collection. We observe that because the order of non-reference suffixes relative to each other is irrelevant, it is possible to compute the suffix array of S_i suffixes individually and merge these with the suffix array of R to find lexicographic predecessors and successors, saving both time and memory. Our MERGE algorithm makes use of the LCP array of R and each S_i and we provide the first GPU implementation of the LCP array construction algorithm by Kärkkäinen et al. [11], which is of independent interest.
– We show that a losslessly compressed representation—the so-called *heads* [16]—of the matching statistics can be computed efficiently in parallel, reducing the data transferred from the GPU to the CPU, further saving time.

We also release CUDA implementations of all the methods we describe[1].

This paper is structured as follows. In the next section we set notation and define basic concepts and structures used throughout. Section 3 reviews related work. Section 4 presents a fast GPU-based method for LCP array construction. Section 5 presents and compares three novel GPU-based matching statistics algorithms. Conclusions and reflections are then offered.

2   Background and Notation

Let Σ be an ordered alphabet of size σ. A string $T[0..n]$ over Σ is a finite sequence of $n+1$ characters from Σ, except the last character of T is the *sentinel character*, denoted \$, which does not occur elsewhere in the string and is assumed to be smaller than all characters in Σ. The ith character of T is denoted $T[i]$, its length is $|T| = n$, and $T[i..j]$ denotes the substring $T[i] \cdots T[j]$. If $i > j$, then $T[i..j]$ is the empty string ϵ. The suffix $T[i..] = T[i..n]$ is referred to as the ith suffix. When T is clear from context, we sometimes write "suffix i" for $T[i..n]$.

---

[1] https://github.com/anadis504/SPIRE_MS/tree/main/MS_GPU.

2.1   Matching Statistics

Let R and S be two strings, and let us denote the sentinel character of R as $\#$ and of S as \$, where $\# < \$$. The *matching statistics* [4] $M_{S|R}$ of S w.r.t. R (or just M, when clear from context) is an array of length $|S|$, whose entries are integer pairs defined as follows. For $1 \leq i \leq |S|$, let U_i be the longest prefix of suffix $S[i..|S|]$ which occurs as a substring in R; we refer to U_i as the *matching factor* at position i. Then $M[i] = (p_i, \ell_i)$, where $\ell_i = |U_i|$, and p_i is an occurrence of U_i in R if $U_i \neq \epsilon$, and $p_i = -1$ otherwise.

2.2   Suffix Arrays and Related Structures

The *suffix array SA* of a string T is a permutation of $[1, n]$ such that $SA[i] = j$ iff $T[j..|T|]$ is the ith in lexicographic order among all suffixes. For a substring Y of T, all suffixes prefixed by Y appear contiguously in SA; the interval $[s, e]$ of the SA containing all occurrences of Y is called *Y-interval* or *SA-range of Y*.

The *inverse suffix array ISA* is the inverse permutation of SA, namely, for all $1 \leq i \leq n$, $ISA[SA[i]] = i$. Another permutation, denoted Φ, stores the lexicographical predecessor of each suffix, i.e., for $i > 0$, $\Phi[i] = SA[i-1]$. The *longest common prefix* of two strings T and S is the longest string U which is a prefix of both T and S. We denote by $lcp(i, j)$ the length of the longest common prefix of suffixes i and j. The *longest-common-prefix array LCP* is another array closely related to the SA. It is given by: $LCP[1] = 0$, and for $i > 1$, $LCP[i] = lcp(SA[i], SA[i-1])$, i.e. the length of the longest common prefix of lexicographically consecutive suffixes $SA[i-1]$ and $SA[i]$. The LCP-array can be computed in linear time [13] from the SA.

The Burrows-Wheeler Transform L [3] is a reversible permutation of the input text T. It is defined as $L[i] = \$$ if $SA[i] = 1$, and $L[i] = T[SA[i] - 1]$ otherwise.

2.3   GPU Computing

GPUs are massively parallel processors with hundreds of thousands of concurrent active threads. The programming model is called the single instruction multiple thread (SIMT) model, where the programs are written for individual threads, each thread working on a different data item.

In the underlying hardware implementations of the programming model, threads are grouped into single-instruction-multiple-data (SIMD) instructions with a typical width 32 threads per instruction—a *warp*—and each warp executes all the threads of the warp in parallel. As a result, any *if-then-else* statements (that cause threads of a warp to execute different instructions) are executed in a fashion where first only the threads taking the *then* branch execute their code, followed by only the threads of the warp executing in the *else* branch executing their code. This can cause wasted execution resources and the effect is called *branch divergence*. There are also mechanisms available with which threads inside a warp can execute arbitrarily complex control flows but the dangers of branch divergence and its effect on performance need to be kept in mind.

A group of threads (multiple warps) is called a *thread block*. Thread blocks are scheduled to be run on streaming multiprocessors (SMs) on the GPU. The NVIDIA A100 has in total 108 SMs. The memory hierarchy of GPUs is organized into a large global memory accessible by all threads within the device (e.g., 80 GB on an A100 GPU), 40 MB of L2 cache shared across all SMs, smaller but faster configurable L1 cache/shared memories for each SM (192 KB per SM of L1 cache, out of which 164 KB can be allocated for shared memory on an A100), and local registers for each thread in the thread block (256 KB per SM on an A100).

If all the threads inside a warp access global memory locations within the same cache line (128 byte cache lines are typical), memory operations of all 32 threads can do a *coalesced access* in a single cache line operation, while if they all access memory locations in completely separate cache lines, 32 cache line operations are needed. Thus memory access optimizations are crucial to performance. NVIDIA GPUs support a set of warp-wide instructions (e.g., shuffles and ballots) that allow threads within a warp to communicate with each other.

Another important factor in the use of GPUs is latency to transfer data between CPU and GPU memory, which can add significantly to overall runtime. An A100 GPU has a transfer speed of up 64 GB/s, but can be much less.

2.4   Data and Experimental Setup

We measure and report on experiments with our new methods throughout the paper and so define our experimental machine and test data now.

Data. In all experiments reported here we use a subset of a collection of 3682 assembled E.coli genomes[2], in particular, a set of 400 of these genome sequences, concatenated into a single string with an ASCII $ symbol separating each. To assess the effect of input size on various algorithms, we use increasing prefixes of this sequence. Table 1 shows some statistics of this data.

Test Environment. For CPU experiments we used an Intel(R) Xeon(R) Gold 6248 running AlmaLinux 8.4. For GPU experiments we used an NVIDIA A100 with 80 GB of GPU memory. Both machines were otherwise idle.

3   Related Work

We now review work on matching statistics computation on the CPU and as well as GPU methods for construction of suffix arrays and related data structures.

---

[2] Available at https://zenodo.org/records/6577997.

Table 1. Statistics of data used in experiments. Each row corresponds to a string that is a prefix of the string of the next row. The 400 E.coli genomes are samples from the collection of 3862 genomes available at https://zenodo.org/records/6577997. As can be see from the table, each genome is around 5 million letters long. For purposes of matching statistics computation, throughout our experiments a single genome, randomly sampled from the 400, is used as string R. The column "# heads" refers to the number of matching statistics entries $M[i] = (p_i, \ell_i)$ for which $S[i-1] \neq R[p_i-1]$ (such entries, called *heads* in [16] imply all other entries and so are a compressed representation).

# genomes	Size (bytes)	Avg. LCP	Max. LCP	# heads
50	258,840,545	2,518	125,877	95,117,978
100	522,486,408	5,094	227,332	195,938,122
150	779,482,542	7,020	329,756	294,041,823
200	1,035,209,449	11,235	2,662,396	390,309,245
250	1,293,933,509	11,520	2,662,396	486,998,379
300	1,553,127,212	11,835	2,662,396	583,331,111
350	1,803,444,213	11,493	2,662,396	677,113,351
400	2,064,004,895	12,212	2,662,396	776,236,276

3.1 CPU Algorithms for Matching Statistics

Most algorithms for computing $M_{S|R}$ first construct an index data structure for R and then use this index to determine the match lengths for each position in S. Chang and Lawler [4] index R with a suffix tree. Ohlebusch et al. explore the use of BWT-based compressed indexes [23]. Rossi et al. [26] use a run-length compressed BWT index. Cunial et al. also use a compressed index to compute a compact representation of the matching statistics lengths, in which the difference between the ith and $(i+1)$st length is stored as a unary code, taking overall $2|S|$ bits [5]. The current fastest algorithm in practice is due to Lipták et al. [16]. We describe this algorithm in more detail in Sect. 5.1. We use it as a baseline in our experiments and also implement a version of it for the GPU. We refer the reader to [15] for a detailed performance comparison of different approaches, including CPU-parallel implementations.

3.2 GPU-Based SA Construction Algorithms

The majority of published research on string algorithms on GPUs concerns suffix array construction. The first papers are due to Osipov [24], who adapts the doubling algorithm of Manber and Myers [18], and Deo and Keely [6], who adapt the skew algorithms of Kärkkäinen et al. [12]. Neither paper has source code available. Wang, Baxter, Owens [28,29], improve the skew implementation of Deo and Keeley and present a novel hybrid skew/prefix-doubling method, making heavy use of parallel primitives like radix sort and segmented sort. Their source code is not available either [25], but a tuned implementation of their approach due to Grebnov is[3]. We use this implementation in our GPU experiments.

3.3 GPU-Based ISA and BWT Construction

Computing ISA from the SA is straightforward on the GPU using a parallel scan. SA is divided up into equal sized segments, with each segment being given to a different warp. The warp assigned the segment $SA[is..(i+1)s-1]$ writes the values $ISA[SA[is+j]] = is+j$ for $i \in 0..s-1$. A similar approach can be used to obtain the BWT sequence from SA and T. To give a sense of the kind of speed up achieved by a GPU for this simple transformation, computing the ISA as just described with a *single CPU thread* that assigns $ISA[SA[i]] = i$ for every $i \in 0..n$ takes 3252 ms on a 150 MB prefix of the *E.Coli* dataset. Using 40 CPU threads reduces the time to 319 ms. On a GPU the computation takes just 24 ms—a speed up of over two orders of magnitude.

```
         —  Store lex. ranks in ISA                    —  Store lex. predecessors in Φ
 1:      for i ← 0 to n − 1 do             1:      for i ← 0 to n − 1 do
 2:            ISA[SA[i]] ← i              2:            Φ[SA[i]] ← SA[i−1]
         —  Compute LCP array                      —  Turn Φ into PLCP
 3:      ℓ ← 0                             3:      ℓ ← 0
 4:      for i ← 0 to n − 1 do             4:      for i ← 0 to n − 1 do
 5:            if ISA[i] > 0 then          5:            j ← Φ[i]
 6:                  j = SA[ISA[i] − 1]    6:            while t[i+ℓ] = t[j+ℓ] do
 7:                  while t[i+ℓ] = t[j+ℓ] do  7:                  ℓ ← ℓ+1
 8:                        ℓ ← ℓ+1         8:            PLCP[i] ← ℓ
 9:                  LCP[ISA[i]] = ℓ       9:            ℓ ← max(ℓ − 1, 0)
10:            ℓ ← max(ℓ − 1, 0)                   —  Permute PLCP into LCP
                                          10:      for i ← 0 to n − 1 do
                                          11:            LCP[i] ← PLCP[SA[i]]
```

Fig. 1. LCP array construction algorithms. The algorithm listed on the left is due to Kasai et al. [13]. On the right is the PLCP algorithm of Kärkkäinen et al. [11].

4 GPU-Based LCP Array Construction

This section contains our first contribution, which is an implementation of fast LCP array construction algorithms for the GPU. We construct LCP arrays in one of our GPU-based matching statistics algorithms later in the paper.

The LCP array is a vital adjunct to the suffix array in many applications (see e.g., [1]). The first linear time algorithm for suffix array construction is described by Kasai et al. [13]. Its main trick is to exploit the observation that if $h = lcp(SA[i-1], SA[i])$ then $LCP[j]$, where $SA[j] = i+1$ must be at least $h-1$.

---

[3] https://github.com/IlyaGrebnov/libcubwt.

Another linear time algorithm is the so-called PLCP algorithm of Kärkkäinen et al. [11], which essentially reorders the computations of Kasai's algorithm to be more cache friendly, making it the preferred sequential algorithm in practice. Pseudocode for these two algorithms is listed in Fig. 1.

Deo and Keeley [6] describe an adaption of the Kasai et al. [13] algorithm for the GPU. They observe that there is a dependency in Kasai's algorithm whenever the LCP value computed for suffix i is used as a lower bound when computing the LCP value for suffix $i+1$ (allowing $\ell-1$ characters to be skipped). They observe this dependency is broken whenever ℓ becomes 0 in line 10, which happens for entries where $T[SA[i]] \neq T[SA[i-1]]$ (in which case $LCP[i] = 0$). Determining these suffixes is done in a parallel scan of SA, and allows work on the whole LCP array computation to be divided into (at most) $|\Sigma|$ independent parts. This idea can be generalized by considering prefixes longer than one letter. Unfortunately, Deo and Keeley's implementations for LCP array construction are not available.

Shun [27] describes multicore parallel (i.e. not GPU) methods for LCP construction adapting sequential approaches by Kasai et al. [13] and the PLCP algorithm [11]. Shun's implementation of Kasai et al. is similar to Deo and Keeley's, but uses off-the-shelf load balancing tools.

We have implemented the Kasai and PLCP LCP array construction algorithms for the GPU. Parallelism is achieved by conceptually dividing text positions $[1, n]$ into contiguous segments, each of size b. Segments are processed in parallel, with the LCP values for positions in a given segment computed by the same thread, which, essentially, executes the loop starting at line 4 of the relevant algorithm in Fig. 1 on its segment. This means the LCP value corresponding to the first suffix of each segment must be computed from scratch, but this extra work is easily paid for by the parallelism achieved.

In implementations of both PLCP and Kasai, LCP values are computed in text order. In the Kasai approach, LCP values are written to the LCP array directly in SA order, while in the PLCP algorithm they are first written to the Φ array (overwriting the values there) in text order, before being permuted into SA order in parallel, similar to SA inversion as described in the previous section.

We implemented multicore versions of Kasai and PLCP algorithms, similar to those of Shun [27], as a performance baseline. An experimental comparison of methods is shown in Fig. 2. We observe that the GPU-based methods are clearly faster than the parallel CPU methods, which are, in turn, much faster than those using a single thread. A significant fraction of the time taken by the GPU methods is in transferring data between the CPU and the GPU (i.e., the input text and SA, and the output LCP). For example, on the biggest test instance in Fig. 2, a file of over 2GB, the multicore implementation using 40 threads requires 5862 ms. The GPU implementation takes 3038ms, of which 1903ms is transfer time. Thus, in applications where the SA already resides on the GPU (e.g., because it is constructed there) or if the LCP array used on the GPU and then discarded, the advantage of the GPU methods increases.

5   Computing Matching Statistics on a GPU

This section describes algorithms for computing matching statistics on a GPU. We start by reviewing the current fastest sequential algorithm for the problem—the LMP algorithm—and examine its suitability for parallelization. We then describe a completely new approach that is more amenable to GPU implementation, an algorithm we call BRACKETS. We then derive a kind of hybrid of LMP and BRACKETS. We then show how the matching statistics can be compressed in parallel to reduce data transferred between GPU and CPU memory.

5.1   The Algorithm of Lipták et al.

The current state-of-the-art CPU algorithm for computing matching statistics for collections of long similar strings in which we are interested is due to Lipták et al. [15,16]. We refer to this algorithm as LMP. Because we use LMP as a baseline in our experiments, and implement a GPU version of it, we now describe it briefly.

Algorithm LMP computes matching statistics via a series of extendRight and contractLeft operations on the suffix array of R. Given an interval of $SASA[s, e]$ containing all occurrences of string U', extendRight(s,e,c) returns the interval $SA[s', e']$ containing all occurrences of string Uc, if it exists in R, or an empty range otherwise. The operation contractLeft(s,e) returns the range $SA[s', e']$ that contains all occurrences of string $U[1..|U| - 1]$ (i.e., U with the first symbol removed).

Pseudocode is listed in Fig. 3. At a generic step in the algorithm we have computed entry $M[i] = (p_i, \ell_i)$ and know the SA_R-range $[s_i, e_i]$ of U_i. To compute the next entry $M[i + 1]$ we first perform a contractLeft operation, obtaining the interval containing all occurrences of U', where $U_i = S[i]U'$. U' is a prefix of the matching factor U_{i+1}. We then perform at most $\ell_i + 1$ extendRight operations, at which point an empty interval is returned and we know $M[i]$.

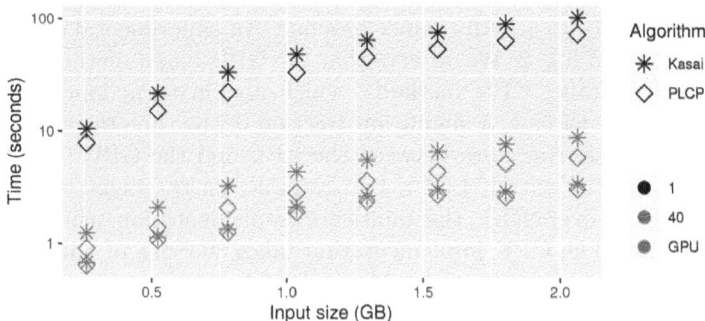

Fig. 2. Comparison of CPU and GPU LCP array construction using PLCP and Kasai algorithms. For CPU implementations, times are shown for 1 thread and 40 threads. The GPU implementation of PLCP uses 60000 as the segment size. Note log scale.

To implement extendRight and contractLeft, the LMP algorithm uses the arrays SA_R, ISA_R and LCP_R over the reference string R and data structures answering *extended PSV/NSV* queries on LCP_R[4]. Having computed entry $M[i]$ and the SA_R-range $[s_i, e_i]$ of U_i, to compute the next entry $M[i+1]$ we:

1. Contract left: i.e., compute the SA_R-range $[s', e']$ of U', where $U_i = S[i]U'$. We have $s' = PSV(LCP_R, ISA_R[SA_R[s_i] + 1]], |U'|)$ and $e' = NSV(LCP_R, ISA_R[SA_R[e_i]+1]], |U'|) - 1$.
2. Extend the factor U' to the right with character $c = S[i+1+|U'|]$ as long as $S[i+1..i+1+|U'|]$ occurs in R. This is done by searching for the longest common prefix of $S[i+1..]$ that occurs in R using two binary searches on SA_R to obtain the $U'c$-interval $[s, e]$ if it exists.

```
1:      ℓ ← 0
2:      [s, e] ← [0, |R|]
3:      for i ← 0 to |S| do
            — Extend match to the right until mismatch
4:          [s', e'] ← extendRight([s, e], ℓ, S[i + ℓ])
5:          while [s', e'] not empty do
6:              [s, e] ← [s', e']
7:              ℓ ← ℓ + 1
8:              [s', e'] ← extendRight([s, e], ℓ, S[i + ℓ])
            — Assign matching statistic for current position
9:          M[i] = (p_i, ℓ_i) ← (SA[s], ℓ)
10:         [s, e] ← contractLeft([s, e], ℓ)
            — SA[s, e] now contains all occurrences of S[p_i + 1..p_i + ℓ_i]
```

Fig. 3. A textbook solution for matching statistics (see, e.g., Gusfield [9], Sect. 7.8.1) that is the basis of Algorithm LMP of Lipták et al. [15,16].

The authors of [15] utilize heuristics that allow one to often skip the second step above and greatly improve running time in practice. The most significant heuristic is the *longest repeating factors*-array (LRF) of the string R, and is given by $LRF_R[i] = max\{LCP_R[i], LCP_R[i + 1]\}$. Given the entry $MS[i] = (p_i, ℓ_i)$, if $ℓ_i - 1 > LRF_R[p_i + 1]$ then $MS[i + 1] = (p_i + 1, ℓ_i - 1)$, because there are no other occurrences of factor $R[p_i + 1..p_i + ℓ_i]$ in R that need to be checked. We enable this and other heuristics described in [15] in our experiments.

The straightforward way to parallelize the LMP algorithm is to divide string S up into equal size segments and apply the loop starting at line 3 in Fig. 3 in parallel to the positions in each segment. Doing this potentially increases the work done at the start of each segment, as the matching statistic there must be

---

[4] For array $A[0, n - 1]$ and index i, *extended PSV* and *extended NSV*, are defined as follows: $PSV(A, i, x) = max\{i' < i : A[i'] < x\}$ and $NSV(A, i, x) = min\{i' > i : A[i'] < x\}$, the previous (respectively next) smaller values *with respect to* x.

computed from scratch, instead of being possibly inferred from the values of the previous position (i.e., at line 10 in Fig. 3).

Figure 4 shows the performance of LMP for various segments sizes on two data sets. Somewhat surprisingly, the best performance is achieved when the segment size is one, meaning that each entry $M[i]$ is computed from scratch at each position, with no information reused between positions. That is, as if we remove line 10 from the algorithm in Fig. 3 and move lines 1 and 2 inside the for loop just after line 3. The reason is thread divergence: the contractLeft operation involves so much internal branching that threads are too often idle.

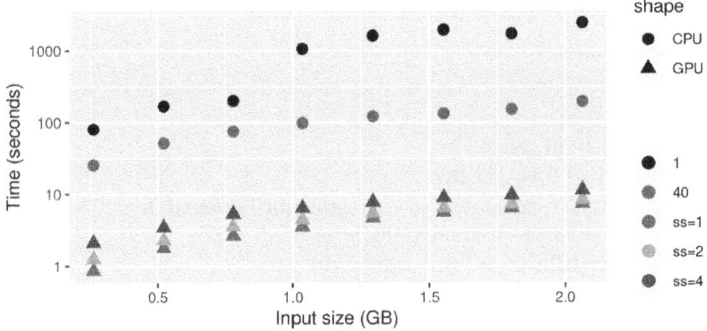

Fig. 4. Effect of segment size in the GPU-based LMP algorithm. CPU methods use 1 and 40 threads. For the GPU methods, ss is the segment size (see text). Note log scale.

5.2 Matching Statistics via Lexicographic Brackets

Let $R\#S\$$ be the concatenation of two strings R and S, where $\#$ and $\$$ are two sentinels not appearing in S or R, and let SA denote the suffix array of $R\#S\$$. For every suffix $SA[i] > |R|$ (i.e., suffix in S), we find the pair (l_i, r_i), $l_i < i < r_i$ such that suffix $SA[l_i]$ (resp. $SA[r_i]$) is the lexicographically largest (resp. smallest) suffix from R that is lexicographically smaller (larger) than suffix $SA[i]$, or -1 if no such suffix exists. In other words, $SA[l_i]$ is the lexicographic predecessor of $SA[i]$ and $SA[r_i]$ its lexicographic successor among the suffixes of R. We call l_i (resp. r_i) the *left bracket* (resp. *right bracket*) of $SA[i]$ and we call $SA[l_i]$ the *left-bracket suffix*. In the matching statistics, the suffix of $SA[l_i]$ and $SA[r_i]$ that has longer common prefix with $SA[i]$ will be a valid value for $p_{SA[i]}$.

Parallel computation of brackets and matching statistics for every $SA[i] > |R|$ proceeds as follows. The total work done is linear in $|R| + |S|$.

1. Let $N = |R\#S\$|$. In a parallel scan of SA, we mark the positions of the left brackets in a bit array $F_l[0..N-1]$, so that $F_l[j] = 1$ iff suffix $SA[j] < |R|+1$ and suffix $SA[j+1] > |R|$, for $j < N-1$. In the same scan we mark the right brackets in a bit array $F_r[0..N-1]$, so that $F_r[j] = 1$ iff $SA[j] < |R|+1$ and $SA[j-1] > |R|$, for $j > 0$.

2. Using a parallel prefix sum routine[5], we compute prefix sums over the bit arrays F_l and F_r, yielding the arrays $Rank_l$ and $Rank_r$.

3. In a parallel scan of F_l and $Rank_l$ (resp. F_r and $Rank_r$), we copy the left-bracket suffix (resp. right-bracket suffix) into a new array B_l (resp. B_r) of length $Rank_l[N-1] \leq |R|$. In particular, for $i \in \{i \mid F_l[i] = 1\}$ $B_l[Rank_l[i]] = SA[i]$ (and for $i \in \{i \mid F_r[i] = 1\}B_r[Rank_r[i]] = SA[i]$). To ensure corner cases where a suffix from S does not have a lexicographically smaller (resp. larger) suffix in R, we assign $B_l[0]$ (and $B_r[Rank_r[N-1]+1]$) to -1.

4. In a parallel scan of B_l and B_r, we fill arrays $P_l[0..|S|]$ and $P_r[0..|S|]$ with the left- and right-bracket suffixes of S suffixes in text order. That is, if $SA[i] > |R|$, $P_l[SA[i] - |R|] = B_l[Rank_l[i]]$ and $P_r[SA[i] - |R|] = B_l[Rank_r[i] + 1]$.

5. From those arrays we compute arrays L_l and L_r, where $L_l[i]$ gives the lcp value between suffixes $S[i..]$ and $R[P_l[i]..]$. S is divided into equal segments, which are worked on in parallel. The first $lcp(\cdot, \cdot)$ value in each segment is computed from scratch. The subsequent $L_l[i]$ entries are computed in the same manner as the $PLCP$ values in the loop starting at line 4 in Fig. 1, but using array P_l pointing into R instead of Φ. L_r is defined and computed similarly, but using P_r instead of P_l. This ensures linear comparisons.

The ith matching statistics entry is given by $\ell_i = \max\{L_l[i], L_r[i]\}$ and p_i is the suffix from the corresponding P array, depending on which L value is greater.

A heuristic we found improved the performance of step 5 above was to test at an iteration i whether $P[i] = P[i-1]+1$. In the positive case, the $L[i]$th value can then be written directly as $L[i-1]-1$ avoiding a memory access into R.

5.3 Matching Statistics via Merging of Suffix Arrays

A nice aspect of the LMP algorithm is that the suffix array for R is computed only once and computation of the SA for $R\#S\$$, which in some applications will be a highly repetitive string, is avoided. We now describe an algorithm that can be viewed as a hybrid of these two approaches. The idea is to compute the suffix arrays of R and S separately and then merge them with the help of their LCP arrays. The merged suffix array is never stored, but the merging process locates the brackets for each S suffix, allowing the matching statistics to be computed (Fig. 5).

Pseudocode for merging SA_R and SA_S is listed below. At a generic step in the algorithm we want to determine the relative order of some suffix $s_{i,R} = SA_R[i_R]$ from R and another suffix $s_{i,S} = SA_S[i_S]$ from S. The algorithm uses information about LCPs of adjacent suffixes in SA_R and SA_S to determine the longest common prefix ℓ of suffixes $s_{i,R}$ and $s_{i,S}$, allowing the order to be deduced from the symbols $R[s_{i,R} + \ell]$ and $S[s_{i,S} + \ell]$.

The algorithm is easily adapted to compute matching factors, not just lexicographic predecessors and successors. The time is linear in $|R| + |S|$ apart from

---

[5] In particular, `InclusiveSum` in NVidia's CUB library.

—*Given strings R, S with suffix arrays SA_R, SA_S.*
—*Assume symbols $R[|R| - 1] < S[|S| - 1]$ are unique and smaller than all other symbols.*
1:   $i_R \leftarrow i_S \leftarrow 0$
2:   $p \leftarrow -1$
3:   $\ell \leftarrow lcp(SA_R[i_R], SA_S[i_S])$
     —*suffixes of S not having a lex. pred. in R will be assigned a lex. pred. of -1*
4:   **while** $i_R < |R|$ **and** $i_S < |S|$ **do**
     —*the next loop moves i_S until suffix $SA_S[i_s]$ is the lex. succ. in S of suffix $SA_R[i_R]$*
5:       **while** $R[SA_R[i_R] + \ell] > S[SA_S[i_S] + \ell]$ **do**
     —*assign p as the index in SA_R of the lex. pred. of $SA[i_S]$*
6:           $P[SA[i_S]] \leftarrow p$
7:           $i_S \leftarrow i_S + 1$
8:           **if** $i_S = |S|$ **then break**
9:           $\ell \leftarrow \min(\ell, LCP_S[i_S])$
10:          **while** $R[SA_R[i_R] + \ell] = S[SA_S[i_S] + \ell]$ **do** $\ell \leftarrow \ell + 1$
11:      **if** $i_S \neq |S|$ **then**
     —*the next loop moves i_R until suffix $SA_R[i_R]$ is the lex. succ. in R of suffix $SA_S[i_S]$*
12:          **while** $R[SA_R[i_R] + \ell] < S[SA_S[i_S] + \ell]$ **do**
13:              $i_R \leftarrow i_R + 1$
14:              **if** $i_R = |R|$ **then break**
15:              $\ell \leftarrow \min(\ell, LCP_R[i_R])$
16:              **while** $R[SA_R[i_R] + \ell] = S[SA_S[i_S] + \ell]$ **do** $\ell \leftarrow \ell + 1$
     —*update p to be the lex. pred. in R of suffix $SA_S[i_S]$*
17:          $p \leftarrow i_R - 1$
     —*assign p as the index in SA_R of the lex. pred. for any suffixes remaining in SA_S*
18: **while** $i_S < |S|$ **do**
19:     $P[SA[i_S]] \leftarrow p$
20:     $i_S \leftarrow i_S + 1$

Fig. 5. Given two strings and their suffix arrays and LCP arrays, an algorithm for computing, for each suffix of S, its lexicographic predecessor suffix in R. Predecessors are stored in array P.

lines 10 and 16, where symbols from each string are compared to determine $lcp(s_{i,R}, s_{i,S})$. The number of comparisons in those lines is non-constant only in the case that $\ell = LCP_S[i_S]$ at line 9 (and $\ell = LCP_R[i_R]$ in line 15). The number of symbol comparisons made this way can be quadratic in $\min(|R|, |S|)$, for example, when $R = S$ and every letter in the strings is distinct.

To parallelize this approach, we select every bth suffix from SA_S (i.e., $SA_S[0]$, $SA_S[b]$, $SA_S[2b]$, ...) and then find the rank of the lexicographic predecessor of each in SA_R. Treating each suffix as a pattern we can find these ranks by binary search in SA_R, doing $O(|S|^2 \log |R|/b)$ total work. Segments of SA_S between samples (each of size b) can then be assigned to different threads and the rank of the first suffix in the segment used as the starting point for i_R above.

Run time (not including CPU-GPU transfer time or time to output to disk) of the three GPU-based methods we have described—the GPU implementation of the LMP algorithm, the BRACKETS algorithm, and the MERGE algorithm—is shown in Fig. 6. MERGE and BRACKETS are consistently about 30% faster than the GPU implementation of LMP and over an order of magnitude faster than the 40-thread CPU-parallel implementation of LMP.

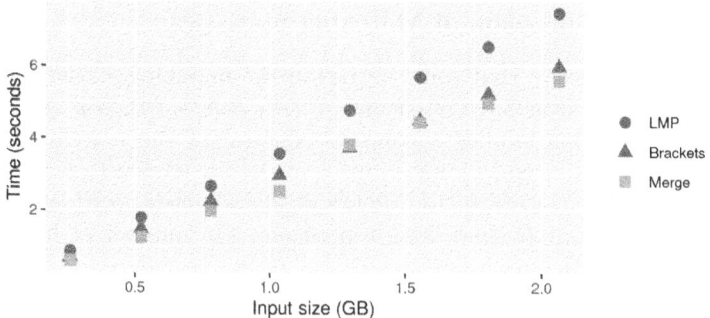

Fig. 6. Performance comparison of GPU-based matching statistics algorithms LMP, BRACKETS, and MERGE. LMP uses a segment size of 1.

In both BRACKETS and MERGE, more than half of the time is spent on computing the suffix array of the concatenation of a collection of strings. A heuristic we found significantly reduced the running time was based on the observation that while the lexicographical ordering between suffixes of the reference string and another given string from the collection is essential to the algorithm, an ordering of suffixes among strings in the collection is irrelevant.

In the BRACKETS algorithm, we computed suffix arrays of concatenations of R and smaller subsets of strings from the collection. The tradeoff in this approach is that computing suffix arrays of small subsets of strings is faster, but R must be included in each subset. Our preliminary experiments showed that concatenating about 13 strings at a time was the fastest ($R\#S_i\$\ldots S_{i+11}\$$). Experiments with the MERGE algorithm showed that computing the suffix arrays for single strings separately was the fastest. Figure 7 shows the time taken for SA construction and matching statistics computation in BRACKETS and MERGE.

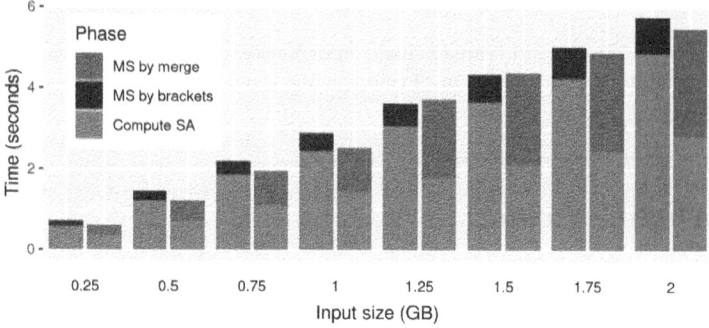

Fig. 7. Time taken for suffix array construction versus matching statistics computation in the BRACKETS and MERGE algorithms. Time spent on suffix array construction varies between the BRACKETS and MERGE algorithms because SA construction is performed on different subsets of the collection in each (see text).

5.4   Parallel Compression of Matching Statistics

Lipták et al. [16] observe that some entries in the matching statistics sequence can be inferred from others. In particular, if $\ell_{i+1} = \ell_i - 1$ then $p_i + 1$ is a valid position for p_{i+1}. This implies a compressed form of the matching statistics where, instead of pairs (p_i, ℓ_i) for every i, we store triples (i, p_i, ℓ_i) only for positions i for which $\ell_i \neq \ell_{i-1} - 1$. Lipták et al. call such positions *heads*, and show that when S and R share long repetitions, the number of heads can be much smaller than $|S|$.

Given transfers between GPU and CPU memory are costly, transferring heads instead of full matching statistics can reduce end-to-end running time if heads can be extracted fast. The heads can be extracted in parallel from the full matching statistics, M, as follows. In a parallel scan of M, whenever $R[p_i - 1] \neq S[i - 1]$ we set $B[i] = 1$ in a bitvector B of length $|S|$. We will output the heads to an array $H[0..h - 1]$, where $h = |\{i | B[i] = 1\}|$. We then perform a parallel prefix sum on B, storing the result in P. Finally, we perform a parallel scan of B and whenever $B[i] = 1$ we copy $M[i]$ to position $P[i]$ in the output buffer H.

Figure 8 shows that computing and transferring the heads to the CPU takes roughly half the time needed to transfer the full matching statistics.

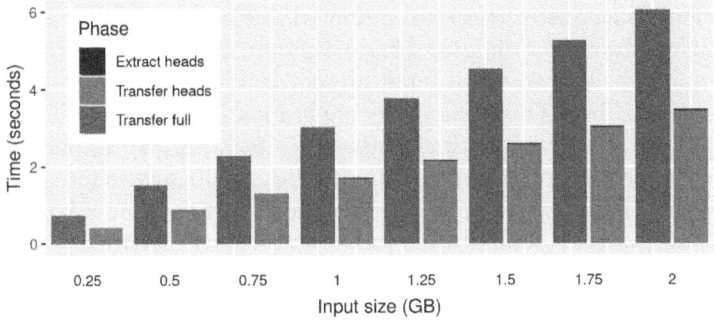

Fig. 8. Time to compute and transfer head information from the GPU vs. transferring full matching statistics information. Time for the extract heads phase in very small and so is barely visible in the plots.

6   Concluding Remarks

We have described GPU-based algorithms for matching statistics and LCP array computation, and have shown that by careful use of this specialized hardware, handsome performance gains can be achieved in practice compared to sequential and multicore algorithms. Given the massive parallelism available on GPUs, it is perhaps not particularly surprising that our algorithms are faster, but our experiments also show that certain approaches exploit this parallelism much more effectively than others.

References

1. Abouelhoda, M.I., Kurtz, S., Ohlebusch, E.: Replacing suffix trees with enhanced suffix arrays. J. Discr. Algorithms **2**(1), 53–86 (2004)
2. Bille, P., Gørtz, I.L., Puglisi, S.J., Tarnow, S.R.: Hierarchical relative Lempel-ZIV compression. In: Proceedings of the 21st International Symposium on Experimental Algorithms (SEA), pp. 18:1–18:16. LIPIcs 265, Schloss Dagstuhl - Leibniz-Zentrum für Informatik (2023)
3. Burrows, M., Wheeler, D.J.: A block-sorting lossless data compression algorithm. Tech. rep, DIGITAL System Research Center (1994)
4. Chang, W.I., Lawler, E.L.: Sublinear approximate string matching and biological applications. Algorithmica **12**(4/5), 327–344 (1994)
5. Cunial, F., Denas, O., Belazzougui, D.: Fast and compact matching statistics analytics. Bioinform. **38**(7), 1838–1845 (2022)
6. Deo, M., Keely, S.: Parallel suffix array and least common prefix for the GPU. SIGPLAN Not. **48**(8), 197–206 (2013). https://doi.org/10.1145/2517327.2442536
7. Do, H.H., Jansson, J., Sadakane, K., Sung, W.: Fast relative Lempel-ZIV self-index for similar sequences. Theor. Comput. Sci. **532**, 14–30 (2014)
8. Gagie, T., Gawrychowski, P., Kärkkäinen, J., Nekrich, Y., Puglisi, S.J.: A faster grammar-based self-index. In: Dediu, A.-H., Martín-Vide, C. (eds.) LATA 2012. LNCS, vol. 7183, pp. 240–251. Springer, Heidelberg (2012). https://doi.org/10.1007/978-3-642-28332-1_21
9. Gusfield, D.: Algorithms on Strings, Trees, and Sequences. Cambridge University Press (1997)
10. Hoobin, C., Puglisi, S.J., Zobel, J.: Relative Lempel-ZIV factorization for efficient storage and retrieval of web collections. Proc. VLDB Endow. **5**(3), 265–273 (2011)
11. Kärkkäinen, J., Manzini, G., Puglisi, S.J.: Permuted longest-common-prefix array. In: Kucherov, G., Ukkonen, E. (eds.) CPM 2009. LNCS, vol. 5577, pp. 181–192. Springer, Heidelberg (2009). https://doi.org/10.1007/978-3-642-02441-2_17
12. Kärkkäinen, J., Sanders, P., Burkhardt, S.: Linear work suffix array construction. J. ACM **53**(6), 918–936 (2006)
13. Kasai, T., Lee, G., Arimura, H., Arikawa, S., Park, K.: Linear-time longest-common-prefix computation in suffix arrays and its applications. In: Amir, A. (ed.) CPM 2001. LNCS, vol. 2089, pp. 181–192. Springer, Heidelberg (2001). https://doi.org/10.1007/3-540-48194-X_17
14. Kuruppu, S., Puglisi, S.J., Zobel, J.: Relative Lempel-ZIV compression of genomes for large-scale storage and retrieval. In: Chavez, E., Lonardi, S. (eds.) SPIRE 2010. LNCS, vol. 6393, pp. 201–206. Springer, Heidelberg (2010). https://doi.org/10.1007/978-3-642-16321-0_20
15. Lipták, Z., Luca, M., Masillo, F., Puglisi, S.J.: Fast matching statistics for sets of long similar strings. In: Prague Stringology Conference (PSC), pp. 3–15. Czech Technical University (2024)
16. Lipták, Z., Masillo, F., Puglisi, S.J.: Suffix sorting via matching statistics. Algorithms Mol. Biol. **19**(1), 11 (2024). https://doi.org/10.1186/S13015-023-00245-Z
17. Mäklin, T., Alanko, J.N., Biagi, E., Puglisi, S.J.: Sequence alignment with k-bounded matching statistics. bioRxiv (2025). https://doi.org/10.1101/2025.05.19.654936
18. Manber, U., Myers, E.W.: Suffix arrays: a new method for on-line string searches. SIAM J. Comput. **22**(5), 935–948 (1993)

19. Marçais, G., Delcher, A.L., Phillippy, A.M., Coston, R., Salzberg, S.L., Zimin, A.: Mummer4: a fast and versatile genome alignment system. PLOS Comput. Biol. **14**(1), 1–14 (2018)
20. Masillo, F.: Matching statistics speed up BWT construction. In: Proceedings of the 31st Annual European Symposium on Algorithms (ESA), pp. 83:1–83:15. LIPIcs 274, Schloss Dagstuhl - Leibniz-Zentrum für Informatik (2023). https://doi.org/10.4230/LIPICS.ESA.2023.83
21. Mäkinen, V., Belazzougui, D., Cunial, F., Tomescu, A.: Genome Scale Algorithm Design, 2nd edn. Cambridge University Press (2023)
22. Navarro, G., Sepúlveda, V.: Practical indexing of repetitive collections using Relative Lempel-ZIV. In: Proceedings of the 29th Data Compression Conference (DCC), pp. 201–210 (2019)
23. Ohlebusch, E., Gog, S., Kügel, A.: Computing matching statistics and maximal exact matches on compressed full-text indexes. In: Chavez, E., Lonardi, S. (eds.) SPIRE 2010. LNCS, vol. 6393, pp. 347–358. Springer, Heidelberg (2010). https://doi.org/10.1007/978-3-642-16321-0_36
24. Osipov, V.: Parallel suffix array construction for shared memory architectures. In: Calderón-Benavides, L., González-Caro, C., Chávez, E., Ziviani, N. (eds.) SPIRE 2012. LNCS, vol. 7608, pp. 379–384. Springer, Heidelberg (2012). https://doi.org/10.1007/978-3-642-34109-0_40
25. Owens, J.D.: Personal communication
26. Rossi, M., Oliva, M., Langmead, B., Gagie, T., Boucher, C.: MONI: a pangenomic index for finding maximal exact matches. J. Comput. Biol. **29**(2), 169–187 (2022)
27. Shun, J.: Fast parallel computation of longest common prefixes. In: International Conference for High Performance Computing, Networking, Storage and Analysis, SC 2014, New Orleans, LA, USA, 16-21 November 2014, pp. 387–398. IEEE Computer Society (2014). https://doi.org/10.1109/SC.2014.37
28. Wang, L., Baxter, S., Owens, J.D.: Fast parallel suffix array on the GPU. In: Träff, J.L., Hunold, S., Versaci, F. (eds.) Euro-Par 2015. LNCS, vol. 9233, pp. 573–587. Springer, Heidelberg (2015). https://doi.org/10.1007/978-3-662-48096-0_44
29. Wang, L., Baxter, S., Owens, J.D.: Fast parallel skew and prefix-doubling suffix array construction on the GPU. Concurr. Comput. Pract. Exp. **28**(12), 3466–3484 (2016). https://doi.org/10.1002/CPE.3867

Cache-Friendly Compressed Boolean Matrices

Antonio Fariña[1], Adrián Gómez-Brandón[1,3(✉)],
Asunción Gómez-Colomer[2,3], and Gonzalo Navarro[2,3]

[1] Universidade da Coruña, CITIC, A Coruña, Spain
{fari,adrian.gbrandon}@udc.es
[2] Department of Computer Science, University of Chile, Santiago, Chile
asuncion.gomez@ug.uchile.cl, gnavarro@uchile.cl
[3] Millennium Institute for Foundational Research on Data (IMFD), Santiago, Chile

Abstract. We introduce a new compressed representation of sparse Boolean matrices that enjoys reference locality properties. We build on an existing representation based on LOUDS-deployed cardinal trees, and design one based instead on DFUDS. While this brings various complications, we show that the resulting matrix representation is considerably faster to carry out sums and multiplications, with speedups of up to 60%.

Keywords: Compact Data Structures · Algebra · Binary Matrices · Cache-Friendly

1 Introduction

Sparse binary matrices arise in a number of applications, including the representation of labeled graphs (e.g., in graph databases, social networks, etc.) and Machine Learning [7]. Recent work [1,2], aiming at solving regular path queries on graph databases, showed that a representation based on k^2-trees [4], which exploits sparsity and clustering of the 1 s in the matrix, efficiently supported a general Boolean matrix algebra including transposition, sums and other pointwise operations, multiplication, and transitive closure. In this paper we focus on better representations to support such matrix algebra.

The k^2-tree is a cardinal tree of arity k^2 with a bounded height. Its original representation [4] is analogous to the so-called LOUDS deployment [10], which represents *ordinal* trees in levelwise node order, using 2 bits per node. The *cardinal* LOUDS version, instead, uses k^2 bits per *internal* node to store a *signature* telling which children exist. The sparse matrices referenced above used $k = 2$, so the representation uses 4 bits per internal node. While efficient for navigation, because it relies on the most basic primitives for bitvectors, *rank* and *select* [5,11],[1] this levelwise representation suffers from poor reference locality.

---

[1] This adds $o(1)$ bits per bit of the representation, which we are disregarding to streamline the discussion.

© The Author(s), under exclusive license to Springer Nature Switzerland AG 2026
G. Badkobeh et al. (Eds.): SPIRE 2025, LNCS 16073, pp. 95–108, 2026.
https://doi.org/10.1007/978-3-032-05228-5_9

A way to improve reference locality is to replace LOUDS by a variant of the so-called DFUDS deployment of *ordinal* trees [3], which also uses 2 bits per node but deploys the nodes in depth-first order. The existing *cardinal* DFUDS-based representations [3] [12, Sec. 8.3.1], however, would use $k^2 + 2$ bits per (internal and leaf) node, 6 in our case.[2] This is much more than the space used by LOUDS.

The cardinal DFUDS representation consists of the ordinal DFUDS plus the LOUDS signatures in depth-first order. The ordinal DFUDS is used to navigate the tree, whereas the LOUDS signatures mark which children exist. A further burden of cardinal DFUDS when used on k^2-trees is that the ordinal DFUDS sequence needs to explicitly represent all leaves, even those at the maximum possible depth, and thus we must also include their LOUDS signatures (k^2 zeros). The LOUDS levelwise representation, instead, can ignore those empty signatures because they would appear contiguously at the end of the sequence. This sums to $k^2 + 2$ bits per node, instead of LOUDS' k^2 bits per internal node.

In this paper we note and exploit the fact that *the ordinal DFUDS representation of a node is a function of its LOUDS signature*, so as to navigate the tree using, essentially, the node signatures deployed in DFUDS' depth-first order, and getting rid of the ordinal DFUDS sequence. Further, we manage to remove the empty signatures of the existing cardinal DFUDS representation, at the price of extending the signatures of internal nodes by one bit. As a consequence, our representation uses only $k^2 + 1$ bits per *internal* node. Since, in this representation, subtrees are deployed compactly in the sequence, tree traversals are more cache-friendly. We dub our representation *kache-trees*.

Our kache-trees use the signatures of (sequences of) nodes and transform them on the fly to their ordinal DFUDS representations, then resorting to the classic DFUDS navigation implemented over rmM-trees [13]. Note that this is in the line of the theoretical proposal of Farzan and Munro [8], yet our representation is specialized to k^2-trees, practical and simple to implement.

In order to evaluate the impact of the improved locality of reference, we implemented our kache-trees and used them to solve sums and multiplications on Boolean matrices. Our kache-trees turn out to be considerably faster—offering speedups of up to 60%—on the denser matrices, which are the most time-consuming to multiply. This indeed turns out to be a consequence of cache misses, which are reduced by about a half. On sums, where both representations traverse the trees sequentially with few cache misses, our representation performs similarly on the denser matrices and speeds up by up to 60% on the sparser ones, because there are more cases of copying whole submatrices to the output, and kache-trees deploy those submatrices contiguously.

2   LOUDS, DFUDS, and k^2-Trees

Rooted trees can be classified into two main categories: ordinal and cardinal. Ordinal trees distinguish only the order among children, whereas cardinal trees

---

[2] They [3] [12, Sec. 8.4] introduce representations using $O(\log k)$ bits per node, but those are likely to be practical only for large enough k values.

have a fixed set of possible children, each of which can be present or not. LOUDS and DFUDS are two popular succinct representations of ordinal trees, which use $2n + o(n)$ bits of space for an ordinal tree of n nodes and support various navigation operations, like going to a node's parent or child, in constant time.

2.1 LOUDS Representation of Ordinal Trees

LOUDS (Level-Order Unary Degree Sequence) [10] represents an ordinal tree of n nodes using a single bitvector B, where the nodes are deployed in level order. This representation encodes each tree node having c children ($c = 0$ for leaves) with its so-called *description*, 1^c0. The identifier of a node is the position where its description starts in B. Bitvector B is prepended with 10 to avoid certain special cases. It is easy to show that the length of B is $2n + 1$.

LOUDS supports basic operations, such as navigating to the parent and children of a node v, in a constant amount of basic *rank* and *select* primitives, which run themselves in $O(1)$ time if one spends $o(n)$ further space [5,11]. For example, the t-th child of v is $\texttt{child}(v,t) = select_0(B, rank_1(B, v - 1 + t)) + 1$.

2.2 DFUDS Representation of Ordinal Trees

DFUDS (Depth-First Unary Degree Sequence) [3] deploys the same node descriptions of LOUDS, yet in depth-first order. It regards the bits as parentheses; the description of a node with c children is '$(^c)$'. The bit sequence is now prepended with '$(\ )$' to prevent special cases, and B is of length $2n + 2$.

An advantage of this representation is that all nodes within a subtree are deployed contiguously in the bitvector. Together with the use of parentheses, this enables many more operations not supported in LOUDS. The DFUDS operations build mostly on two basic primitives apart from *rank* and *select*:

$$\texttt{fwd\_search}(B, i, d) = \min\{j > i, \texttt{excess}(B, j) = \texttt{excess}(B, i) + d\} \cup \{n + 1\}$$
$$\texttt{bwd\_search}(B, i, d) = \max\{j < i, \texttt{excess}(B, j) = \texttt{excess}(B, i) + d\} \cup \{0\},$$

where $\texttt{excess}(B, j)$ is the number of opening minus closing parentheses in $B[1..j]$. For example, $\texttt{child}(v, t)$ is calculated as $\texttt{fwd\_search}(B, \texttt{succ}_0(B, v) - t, -1) + 1$, where $\texttt{succ}_0(B, v) = select_0(rank_0(B, v - 1) + 1)$.

Those primitives are more complex to implement than *rank* and *select*. A popular implementation uses $o(n)$ further bits to store a so-called rmM-tree [13]. This is a perfect binary tree whose leaves cut B into blocks of b parentheses, storing (at least) the minimum and the total excess of the block. Internal rmM-tree nodes store the same summary information on the subsequence spanned by its descendant leaves. It is easy to build the rmM-tree bottom-up in linear time.

To compute $\texttt{fwd\_search}(B, i, d)$ using the rmM-tree, we begin by scanning the block of B that contains the i-th parenthesis. Within this block, we search for the smallest position $j > i$ where the excess is $\texttt{excess}(B, i) + d$. If we do not find it, we use the rmM-tree to move upwards and locate the nearest ancestor whose subtree contains the excess sought, based on its stored minimum and total

Matrix

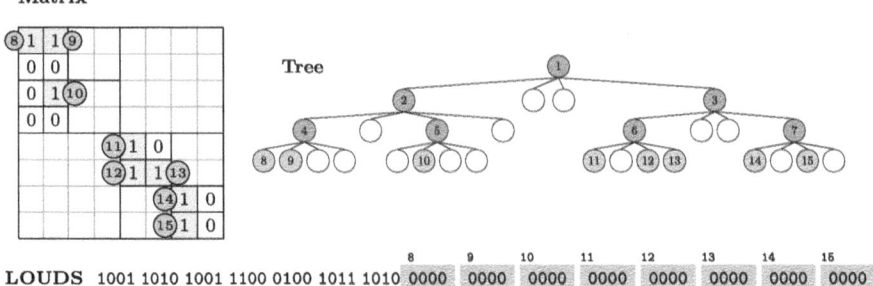

Fig. 1. An 8×8 matrix (left, above), its corresponding k^2-tree (right, above), and its cardinal LOUDS representation (below); the highlighted leaves 0000 are not represented.

excess. Then, we move downward towards the child that contains the position we are looking for, until reaching the rmM-tree leaf that contains the answer. This process ends by scanning that leaf, where j is finally found.

2.3 K^2-Trees and Cardinal LOUDS

The k^2-tree [4] is a compact representation of binary $l \times l$ matrices by recursively dividing them into k^2 quadrants of equal area. Non-empty quadrants are further subdivided recursively, up to height $h = \lceil \log_k l \rceil$, where the submatrices represent individual cells (if l is not a power of k, the matrix is padded with 0s up to side k^h). The k^2-tree excels in representing sparse and clustered matrices.

The k^2-tree is a cardinal tree of arity k^2, and is represented, as in LOUDS, by deploying the nodes levelwise. This time, however, each node is represented with a *signature* of k^2 bits that indicates with 1s which children of the node exist (i.e., which of its submatrices are nonempty). The nodes are identified by the position of their signature (counting number of signatures, not of bits), and the navigation operations are even simpler than for ordinal LOUDS. For example, the formula to descend to a(n existing) child is $\text{child}(v, t) = rank_1(B, k^2(v - 1)) + t + 1$.

The k^2-tree is a particular case of the cardinal LOUDS representation in that, since the height is fixed to h, there is no need to store the last level of leaves, as they are always empty. Therefore, the representation of a k^2-tree uses k^2 bits per *internal* node, without using any bits for the last-level leaves.

Figure 1 illustrates a matrix with eight 1 s along with its corresponding k^2-tree representation for $k = 2$. We also show the cardinal LOUDS representation of the tree, highlighting the final leaves that are omitted from the representation.

2.4   DFUDS Representation of Cardinal Trees

Cardinal trees can also be represented with a DFUDS-inspired representation [3] [12, Sec. 8.3.1]. This stores the original DFUDS for ordinal trees (using the rmM-tree we described above) along with an additional bitvector, S, that contains the

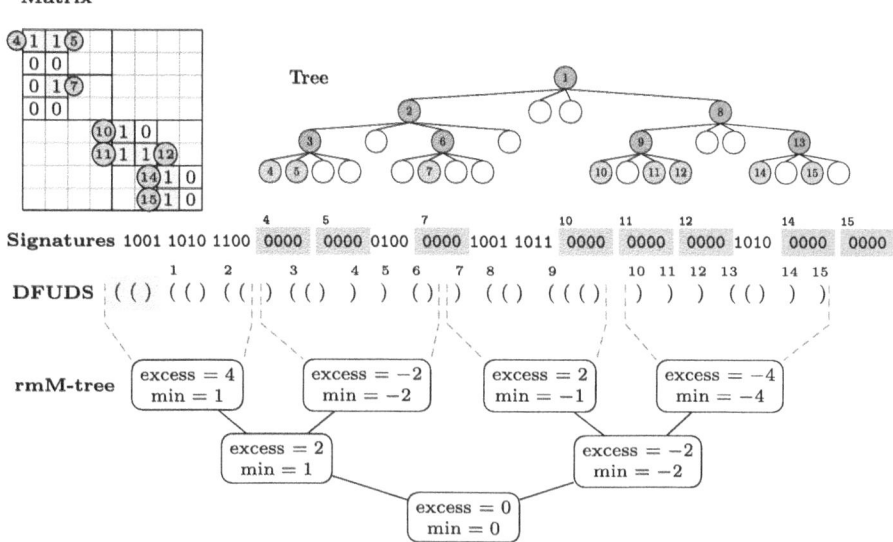

Fig. 2. Cardinal DFUDS representation of the same k^2-tree of Fig. 1, using block size $b = 8$ for the rmM-tree.

k^2-bit signatures of the nodes in depth-first order. The DFUDS sequence provides the navigation, while the signatures indicate which children of the current node exist. The DFUDS node v corresponds to the $rank_0(B, v)$-th signature.

Figure 2 illustrates this representation on our running example. Note that leaf signatures 0000 are now spread along the sequence and are not easily discarded.

3   LOUDS-Based Matrix Algebra

A sparse Boolean matrix representation [1,2] builds on the cardinal LOUDS described in Sect. 2.3. The authors implement an algebra where matrices are input to and returned from operators in this format. We focus on two of the most popular operations: the Boolean sum and product.

To compute the sum $A + B$, they sequentially traverse the representation of both k^2-trees, in levelwise order. The algorithm uses a queue of tasks of types: *copy* tasks, which just copy nodes from either A or B to the output and enqueue the tasks of copying their children, and *merge* tasks, which output the bitwise-or of both signatures and enqueue the tasks of copying or merging the corresponding children of A and B (depending on whether only A, or only B, or both, have each child). While this one-pass procedure is very efficient, we notice that copying a whole subtree to the output when the other operand's subtree is empty, is done signature by signature, as the nonempty subtree is not deployed in contiguous form in the LOUDS format.

For the multiplication $A \times B$, they use the standard recursive algorithm:

$$\begin{pmatrix} A_0 & A_1 \\ A_2 & A_3 \end{pmatrix} \times \begin{pmatrix} B_0 & B_1 \\ B_2 & B_3 \end{pmatrix} = \begin{pmatrix} A_0 \times B_0 + A_1 \times B_2 & A_0 \times B_1 + A_1 \times B_3 \\ A_2 \times B_0 + A_3 \times B_2 & A_2 \times B_1 + A_3 \times B_3 \end{pmatrix}.$$

Zero-bit children in k^2-trees indicate empty submatrices, in which case the multiplication is avoided. The sums are implemented as explained; sums with empty submatrices are also converted into copies. Those copies are done level by level because, again, subtrees are not deployed contiguously in LOUDS.

4   Our Cardinal DFUDS Representation for k^2-Trees

We introduce a new k^2-tree representation, named *kache-tree*, based on cardinal DFUDS, improving the one seen in Sect. 2.4. An advantage of a DFUDS-based representation, compared to the standard LOUDS-based one, is that the signatures are concatenated in depth-first order, and thus every subtree is deployed in a contiguous sequence and its traversal becomes more cache-friendly. A disadvantage of the structure of Sect. 2.4 is that, in addition to the n node signatures, it adds the $2n$-bits ordinal DFUDS sequence of the same tree in order to support navigation.

4.1   The Key Idea

Our new representation gets rid of those $2n$ bits of the ordinal DFUDS representation, together with the empty signatures of last-level leaves. It only retains the sequence of signatures of internal nodes in depth-first order. Unlike in cardinal LOUDS, it is not immediate how to navigate those depth-first-ordered signatures without the help of the ordinal DFUDS sequence.

A first insight to solving this challenge is given in the following lemma.

Lemma 1. *In a k^2-tree, the ordinal DFUDS description of a node is a function of its signature and of the depth of the node.*

Proof. Let the signature of node v have c 1 s; therefore v has c children. If the depth of v is not that of the last level of the internal nodes, its DFUDS description is '$(^c)$' (i.e., c opening parentheses followed by a closing one). Otherwise, v is the parent of c last-level leaves and its DFUDS description is '$(^c)$$)^c$', that is, the same as an internal node with c children followed by the c closing parentheses of the leaves. ☐

An example of those cases can be seen in Fig. 3. The signature of node 1 is '1010' and, since the node is not at the last level of internal nodes, its DFUDS representation is '(( )'. The signature of node 3 is '1100' and, since it is at the last level of internal nodes, its DFUDs representation is '(( )))'.

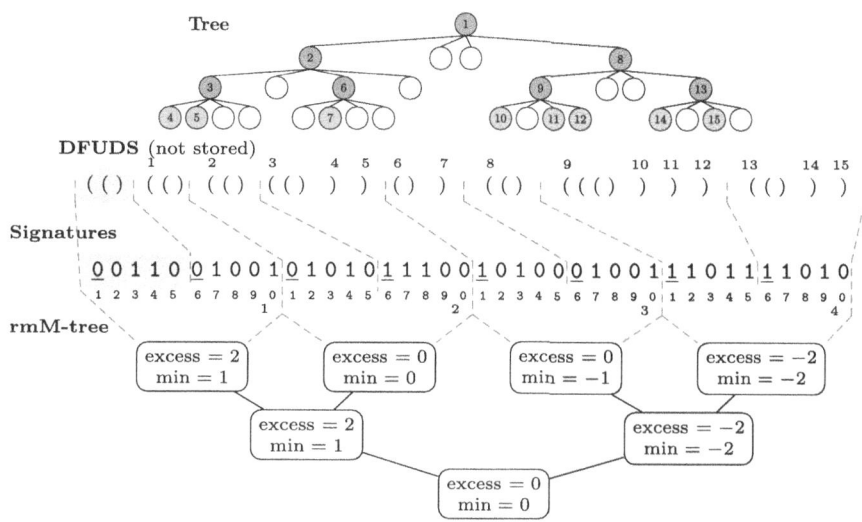

Fig. 3. The kache-tree representation of our running example, using blocks of $b = 2$ signatures for the rmM-tree leaves. Note that those leaves cover an irregular number of DFUDS parentheses (those parentheses are not stored in the kache-tree).

4.2   The Data Structure

In order to distinguish both cases, we will prepend the node signatures with an additional bit that is 1 when the node is at the last depth of internal nodes, and 0 otherwise. With those enhanced signatures, we can obtain the DFUDS representation of a node *on the fly*, without having to spend *two* bits per *node*. Instead, we spend *one* bit per *internal node*. In addition, we do not represent the empty signatures of last-level leaves.

In our example above, the enhanced signature of node 1 would be '01010' and that of node 3 would be '11100'. The cardinal LOUDS representation of this tree uses 28 bits (7 4-bit signatures, see Fig. 1), the original cardinal DFUDS representation uses 92 bits (15 4-bit signatures plus 32 from ordinal DFUDS, see Fig. 2), and our new representation uses 40 bits (8 5-bit signatures, see Fig. 3).

As we presented in Sect. 2.2, operations such as *fwd_search* on DFUDS need to scan a subsequence of parentheses to compute its *excess* and *minimum excess*. To speed up that computation in our sequence of signatures and remain cache-friendly, we precompute a small universal table $T[0, 2^{15} - 1]$ that is independent of the sequence itself. Each entry $T[x]$ contains the field e (excess) and m (minimum excess) for x, where x is the number that corresponds to a sequence of 3 signatures of 5 bits. Hence, reading 15 bits, we can compute the *excess* and *minimum excess* of three consecutive nodes simultaneously. The total space of T is 64 KB, which allows it fit in cache. A smaller table $t[0, 2^5 - 1]$ handles individual signatures, storing the same fields e and m, a field c with the number of 1 s in the signature, and a field p with the actual parentheses sequence represented by the signature.

Table T allows us simulate the efficient scanning of parentheses performed by *fwd_search*. As seen in Sect. 2.2, that operation is sped up by an rmM-tree built on the DFUDS sequence. In our approach, each rmM-tree leaf covers a fixed number b of nodes, that is, a variable number of parentheses. Apart from that, our rmM-tree works just as the original one.

The construction of rmM-tree, assuming that b is a multiple of 3, scans consecutive subsequences of $\frac{b}{3}$ chunks of 15 bits. From the $T[\cdot]$ fields of the chunks, the excess and minimum excess of each leaf is computed. The fields of the internal nodes are then completed as for the original rmM-tree; see Sect. 2.2.

4.3   Forward Search

In the DFUDS format, operation $\texttt{fwd\_search}(i, d)$ requires finding relative excess d forwards from parenthesis position i. In our representation, this may imply starting the search from the middle of a signature. Consequently, we redefine the operation as $\texttt{fwd\_search}(i, d, \delta)$, meaning that the scan should start from the δth parenthesis inside the ith signature, and that we must return the signature that contains the corresponding parenthesis position. We then proceed as follows:

1. We read the ith signature, x, obtain its DFUDS representation from $t[x].p$, and scan it from the δth parenthesis. We return the answer i directly if the difference d is found within $t[x].p$; otherwise we update d by subtracting the excess of the traversed parentheses.
2. We scan the rest of the corresponding rmM-tree leaf, from the $(i + 1)$th signature onwards, by chunks of 3 signatures, skipping chunks y where $d < T[y].m$ and in that case updating d to $d - T[y].e$.
3. If $d \geq T[y].m$ for some chunk y, we scan y signature-wise, using $t[x].m$ and $t[x].e$, until finding the one containing the answer, and return it.
4. If we scan the whole rmM-tree leaf without finding the answer, we continue the navigation on the rmM-tree in the standard way.
5. The rmM-tree sends us to another leaf where we resume the scanning as in points 2 and 3. This time the answer should be found.

To compute $\texttt{child}(i, t)$, the tth child of node i (with $0 \leq t < 4$), we find with $t[\cdot].c$ the number of children of i, and compute $j = \texttt{fwd\_search}(i, -1, c-t-1)$, so we know that the parenthesis preceding the tth child of i is at the jth signature. This implies that the signature of the tth child of i is the $(j + 1)$th signature.

5   Solving Matrix Sums and Multiplications

In this section, we address the algebraic operations of sum and multiplication.

5.1   Boolean Sum

Given two kache-trees A and B representing matrices of the same size (so the trees have the same maximum depth), we can compute the Boolean sum $S = A + B$ with a sequential scan of the signatures of both kache-trees. We start with two read pointers a and b at the beginning of the signatures of A and B, and a write pointer s at the beginning of the signatures of the resulting kache-tree S. Then, we proceed as follows:

1. The bitwise-or of the next 5 bits starting at a and b is written into the result at position s. The three pointers are then advanced to the next signature. Note that the bit that marks last-level internal nodes must be the same in both signatures because we are summing two matrices of the same size.
2. If the obtained signature is not in the last-level of internal nodes, we check if each child exists in A and B.
   - In case the child appears in both, we continue recursively generating the signatures of S.
   - If the child occurs only in A, the next m signatures of the subtree of a are copied to S. The last signature of that subtree is computed as $j = \mathtt{fwd\_search}(a, -1, 0)$, so $m = j - i + 1$. Finally, a and s increase by m.
   - Otherwise, the child occurs only in B, so we append to S the corresponding subtree of B and increase the pointers b and s as above.

Note that every time a subtree is copied to the final result, we just need to copy a consecutive subsequence of bits. This is the main difference from the k^2-tree, where the subtree representation is broken into different portions along the levels. Therefore, the k^2-tree requires a more resource consuming level-order traversal to copy each subtree [2, Sec. 5].

5.2   Boolean Multiplication

Regarding Boolean product, $M = A \times B$, we use the recursive divide-conquer algorithm of Sect. 3. In each step, we need to solve eight multiplications of the form $M_k = A_i \times B_j$, where A_i and B_j are submatrices of A and B, respectively. In the kache-tree, since both A and B support $\mathtt{child}(i, t)$ operations, these multiplications can be computed recursively. Once we obtain those partial results, we have to combine them according to Sect. 3, by applying the sum operation.

For efficiency, we do not build the rmM-trees of the partial results M_k. As a consequence, during the sum of partial results, we cannot use rmM-trees to compute $\mathtt{fwd\_search}(i, -1, 0)$ when copying a whole subtree. Instead, we compute it with a sequential scan of the sequence, that is, the same procedure that was used to scan the leaves of the rmM-tree (see Sect. 4.3). This does not increase the times much because we will traverse the sequence anyway in order to copy it. We only build the final rmM-tree of the result, as described in Sect. 4.2, so that the output of the operation is a fully-functional kache-tree.

Table 1. Experiments on the synthetic scenario. Average times for sum and multiplication are shown in msec; the average space of the source matrices is in KB.

| | | k^2-tree | | | | kc-tree (b=1024) | | |
| | $d=10^{-1}$ | $d=10^{-2}$ | $d=10^{-3}$ | $d=10^{-4}$ | $d=10^{-1}$ | $d=10^{-2}$ | $d=10^{-3}$ | $d=10^{-4}$ |
v								
Sum 1,000	5.10	1.07	0.14	0.02	4.99	0.97	0.12	0.01
4,000	77.94	14.76	2.16	0.28	77.67	13.84	1.69	0.19
8,000	320.86	58.57	8.33	1.19	310.63	55.08	6.49	0.74
16,000	1,284.38	237.13	33.75	4.57	1,244.32	219.63	25.72	2.85
Mult 1,000	3,355.56	126.61	3.61	0.12	2,105.61	100.31	3.11	0.10
4,000	213,325.00	8,337.57	234.56	6.71	134,149.00	6,568.05	199.20	6.24
8,000	1,709,220.00	67,233.60	1,902.69	53.75	1,073,840.00	52,758.70	1,607.03	49.48
16,000	13,863,200.00	537,481.00	15,331.70	432.80	8,588,380.00	423,232.00	12,953.30	396.39
Space 1,000	82.33	16.59	2.70	0.54	97.21	19.45	3.02	0.47
4,000	1,311.54	261.71	40.08	5.59	1,551.01	309.34	47.21	6.42
8,000	5,243.45	1,045.94	159.51	21.69	6,201.42	1,236.88	188.47	25.46
16,000	20,969.54	4,182.88	637.10	86.05	24,801.23	4,947.04	753.32	101.57

6   Experimental Results

We now compare our proposal (kc-tree, with $b = 1024$) with the previous solution, k^2-tree [2], evaluating their performance at performing sums and multiplications of Boolean matrices. The implementations are in C/C++, and the g++ compiler was used with options -std=c++11 -O3. All our experiments ran on an Intel(R) i7-8700k @4.70 GHz, which has 6 cores (12 siblings), with 64 GB of RAM. Each core has 32 KB + 32 KB of L1- plus 300 KB of L2-cache. The last-level cache has 12 MB. We ran experiments under two different scenarios.

- The first one operates over synthetic matrices. We created 16 sets of $v \times v$ matrices, for $v \in \{1000, 4000, 8000, 16000\}$, with 1 s uniformly distributed with densities $d \in \{10^{-4}, 10^{-3}, 10^{-2}, 10^{-1}\}$. For the sets with $v \in \{1000, 4000, 8000\}$ we created 19 matrices and for those with $v = 16000$ we generated 9 matrices. Within each set, we performed both sum and multiplication of every pair of consecutive matrices, $M_i + M_{i-1}$ and $M_i \times M_{i-1}$.
- In the second scenario, we took four real matrices of varying size from the web graphs at https://law.di.unimi.it/datasets.php, namely cnr-2000, eu-2005, uk-2014-host, and enwiki-2023. To account for typical operations of interest we computed, for each matrix M, $M + M^T$ (which connects the nodes regardless of the direction), $M \times M$ (which detects paths of length 2, i.e., node pairs (x, y) such that $x \rightarrow z \rightarrow y$ occurs in the graph for some node z), and $M \times M^T$ (which detects nodes pointing to the same node, i.e., node pairs (x, y) such that $x \rightarrow z \leftarrow y$ occurs in the graph for some node z).

Table 2. Experiments on real scenario. Times for sum and multiplication are shown in msec and sec respectively; the space of the source matrices is in MB. Ratio indicates the value k^2-tree/kc-tree. The product $M \times M^T$ did not finish for enwiki-2023 and uk-2014-host for lack of main memory space.

		cnr-2000	enwiki-2023	eu-2005	uk-2014-host
	nodes (v)	325,557	6,625,370	862,664	4,769,354
	edges (#1s)	3,216,152	165,206,104	19,235,140	50,829,923
k^2-tree	Space $\|M\|$	1.45	360.89	10.04	59.74
	Time $M + M^T$	56.94	17,082.80	402.18	2,849.48
	Time $M \times M$	6.13	48,406.20	115.20	5,811.81
	Time $M \times M^T$	300.67	–	2,524.89	–
kc-tree	Space $\|M\|$	1.71	426.84	11.87	70.66
	Time $M + M^T$	49.89	13,416.60	356.57	2,132.31
	Time $M \times M$	4.14	48,219.10	89.58	5,077.20
	Time $M \times M^T$	102.69	–	1,118.66	–
Ratio	Space $\|M\|$	0.85	0.85	0.85	0.85
	Time $M + M^T$	1.14	1.27	1.13	1.34
	Time $M \times M$	1.48	1.00	1.29	1.14
	Time $M \times M^T$	2.93	–	2.26	–

Synthetic Scenario. Table 1 shows the average elapsed times needed to perform each sum or multiplication, in milliseconds. We also show the average space required, in KB, to represent the source matrices from each set.

The k^2-tree space usage is almost exactly as expected from dv^2 points distributed uniformly on a $v \times v$ matrix, $2\log_2(1/d)$ bits per 1 in the matrix.[3] The kc-tree uses 12%-18% more space than the k^2-tree, which is below the 25% we would expect from using 5-bit signatures instead of 4-bit ones. This is a consequence of using a relatively large rmM-tree leaf size $b = 1024$,[4] yielding an extra space under 1% on the kc-tree, compared to the 6% extra space of the *rank/select* data structures used by the k^2-tree.

On the other hand, the kc-tree is considerably faster than the k^2-tree in almost all cases. In sums, the difference is larger as the matrices are sparser, because as the density decreases there are more opportunities for copying whole submatrices, which as explained the kc-tree can do more efficiently. For example, on the largest matrices, the kc-tree is from 3% to 60% faster than the k^2-tree (i.e., k^2-tree/kc-tree = 1.03 to 1.60) as the density decreases.

Multiplications are orders of magnitude more expensive than sums, taking hours on the largest matrices. On the denser matrices, which are the most expensive to multiply, the kc-tree is around 60% faster than the k^2-tree. This

---

[3] The matrix is essentially full up to level $\log_{k^2}(dv^2)$ and, from there to the last level, $\log_{k^2} v^2$, each 1 basically induces a distinct path. Thus, each 1 induces $\log_{k^2}(1/d) = \frac{1}{2}\log_k(1/d)$ k^2-bit signatures, which is the dominant term in the space.

[4] To tune b, we tried values $b \in \{256, 512, 1024, 2048, 4096\}$ in the synthetic scenario: $b = 1024$ offered the best space/time tradeoff in practice.

speedup decreases with the density, reaching a point where the `kc-tree` is just around 8% faster. The decrease can be attributed to our need, when skipping over a subtree that is multiplied with a zero submatrix, to scan subtrees in order to find where they finish, which is not an issue in the k^2-`tree` representation.

Real Scenario. Table 2 shows results on the real-life matrices. Those are very large and have very low density, from $d \approx 2 \cdot 10^{-6}$ to $3 \cdot 10^{-5}$. The space usage of the k^2-`tree` is much lower than $2\log_2(1/d)$, which was an excellent predictor on uniformly distributed matrices; this shows that the k^2-tree representation is very efficient in exploiting clustering of the matrix. The space of our `kc-tree` is always around 18% over that of the k^2-`tree`.

While the advantage of the `kc-tree` over the k^2-`tree` seems to decrease on non-uniform matrices, it is still significant: the `kc-tree` is 13%–34% faster than the k^2-`tree` for sums, and 0%–48% faster on multiplications of the form $M \times M$. Interestingly, the difference becomes much larger on multiplications of the form $M \times M^T$, where the `kc-tree` is 2 to 3 times faster than the k^2-`tree`. This suggests that `kc-tree` outperforms the k^2-`tree` more sharply when the output is larger: $M \times M^T$ tends to produce bigger outputs than $M \times M$; indeed, for lack of main memory space our experiments ran only on the two smaller matrices.

The speedups are largely attributable to the reduced amount of cache misses: `perf` shows that the k^2-`tree` generated 31%–42% and 7%–225% more cache-misses than the `kc-tree` on sums and multiplications, respectively. The only exception is the sum in the smallest dataset (which fits in less than 2 MB) where the k^2-`tree` produces 40% less cache misses than the `kc-tree`.

7   Conclusions and Future Work

We have introduced *kache-trees*, a new representation of k^2-trees that, instead of the original level-wise deployment of nodes, traverses the nodes in depth-first order. This poses various new challenges, like efficiently distinguishing the last-level k^2-tree nodes, which we manage by slightly increasing the space—by less than 20% in practice. Navigating the tree also requires more operations, because we must build on parentheses sequences instead of just *rank/select* on bitvectors. In reward, the depth-first order yields the important advantage of making navigation operations more cache-friendly. We demonstrate this aspect by experimentally comparing our new representation with standard k^2-trees to represent sparse Boolean matrices, and obtaining speedups of up to 60%, even when our traversal operations require more operations.

Not only k^2-trees have many more applications than representing Boolean matrices, and our new kache-trees are likely to outperform them in all those, but our technique can also be used to represent general cardinal trees (of low arity) in cache-friendly form. In this case, leaves may need to be explicitly represented with k zeros instead of extending the signatures of internal nodes by one bit.

There are several lines of future work. One is to complete the implementation of the algebra. This includes, most prominently, to implement the transitive closure operation, where we expect kache-trees to excel because the resulting matrices tend to be considerably denser than the original one.

In the longer term, we aim to produce a fully cache-oblivious implementation of Boolean matrices. This is currently not true because the rmM-tree is deployed in heap format, which (even if it is much smaller than the sequence of signatures) makes subtrees not represented compactly in memory. We plan to change its representation to a depth-first order. This will imply that, whenever a submatrix has $O(B)$ points, it will be represented within a contiguous memory area of $O(B)$ rmM-tree nodes and another one of $O(B)$ signatures. Therefore, summing two matrices of n and m 1 s stored in external memory with a(n unknown) block size of B, will require $O((n + m)/B)$ I/Os, whereas multiplying them will require $O((n + m + r)/B)$ I/Os, where r is the number of 1 s in the product. This matches current results on plain representations of sparse matrices [6,9].

Finally, we aim to explore the possibility of stopping the recursive decomposition into submatrices at sizes below $r \times r$, for some moderate r, and represent those using Huffman codes. This reported important space reductions on various k^2-tree applications [4], but poses some challenges when implementing a matrix algebra, where new matrices are generated on the fly. It may also be less cache-friendly for not so small r, due to the Huffman tables. Encodings that are inferior to Huffman, but are local and easily computed on the fly, like differentially encoding the positions of the 1 s, can be a better alternative. Special procedures to operate pairs of encoded submatrices must also be developed.

Acknowledgments. This work is supported by ANID Millennium Science Initiative Program Code ICN17_002, and Fondecyt Grant 1-230755; CITIC is funded by the Xunta de Galicia and co-financed by the EU through FEDER Galicia [ED431G 2023/01]; MCIN/AEI and EU/ERDF A way of making Europe Grants [PID2022-141027NB-C21 (EarthDL), PID2021-122554OB-C33 (Oassis)]; MCIN/AEI and NextGenerationEU/PRTR Grants [PID2020-114635RB-I00 (EXTRACOMPACT), TED2021-129245B-C21 (PLAGEMIS)].

References

1. Arroyuelo, D., Gómez-Brandón, A., Navarro, G.: Evaluating regular path queries on compressed adjacency matrices. In: Proceedings of the 30th International Symposium on String Processing and Information Retrieval (SPIRE), pp. 35–48 (2023)
2. Arroyuelo, D., Gómez-Brandón, A., Navarro, G.: Evaluating regular path queries on compressed adjacency matrices. Very Large Databases J. **34**, 2 (2025)
3. Benoit, D., Demaine, E.D., Munro, J.I., Raman, R., Raman, V., Rao, S.S.: Representing trees of higher degree. Algorithmica **43**(4), 275–292 (2005)
4. Brisaboa, N.R., Ladra, S., Navarro, G.: Compact representation of Web graphs with extended functionality. Inf. Syst. **39**(1), 152–174 (2014)
5. Clark, D.R.: Compact PAT Trees. PhD thesis, University of Waterloo, Canada (1996)
6. Dusefante, M., Jacob, R.: Cache oblivious sparse matrix multiplication. In: Proceedings of the 13th Latin American Theoretical Informatics Symposium (LATIN), pp. 437–447 (2018)

7. Elgohary, A., Boehm, M., Haas, P.J., Reiss, F.R., Reinwald, B.: Compressed linear algebra for declarative large-scale machine learning. Commun. ACM **62**(524), 83–91 (2019)
8. Farzan, A., Munro, J.I.: A uniform paradigm to succinctly encode various families of trees. Algorithmica **68**(1), 16–40 (2012)
9. Gleinig, N., Besta, M., Hoefler, T.: I/O-optimal cache-oblivious sparse matrix-sparse matrix multiplication. In: Proceedings of the IEEE International Parallel and Distributed Processing Symposium (IPDPS), pp. 36–46 (2022)
10. Jacobson, G.: Space-efficient static trees and graphs. In: Proceedings of the 30th IEEE Symposium on Foundations of Computer Science (FOCS), pp. 549–554 (1989)
11. Munro, J.I.: Tables. In: Proceedings of the 16th Conference on Foundations of Software Technology and Theoretical Computer Science (FSTTCS), pp. 37–42 (1996)
12. Navarro, G.: Compact Data Structures – A practical approach. Cambridge University Press (2016)
13. Navarro, G., Sadakane, K.: Fully-functional static and dynamic succinct trees. ACM Trans. Algorithms **10**(3), 16 (2014)

Tight Additive Sensitivity on LZ-Style Compressors and String Attractors

Yuto Fujie[1], Hiroki Shibata[1] , Yuto Nakashima[2] ,
and Shunsuke Inenaga[2(✉)]

[1] Joint Graduate School of Mathematics for Innovation, Kyushu University,
Fukuoka, Japan
{fujie.yuto.104,shibata.hiroki.753}@s.kyushu-u.ac.jp
[2] Department of Informatics, Kyushu University, Fukuoka, Japan
{nakashima.yuto.003,inenaga.shunsuke.380}@m.kyushu-u.ac.jp

Abstract. The *worst-case additive sensitivity* of a string repetitiveness measure c is defined to be the largest difference between $c(w)$ and $c(w')$, where w is a string of length n and w' is a string that can be obtained by performing a single-character edit operation on w. We present $O(\sqrt{n})$ upper bounds for the worst-case additive sensitivity of the smallest string attractor size γ and the smallest bidirectional scheme size b, which match the known lower bounds $\Omega(\sqrt{n})$ for γ and b [Akagi et al. 2023]. Further, we present matching upper and lower bounds for the worst-case additive sensitivity of the Lempel-Ziv family - $\Theta(n^{\frac{2}{3}})$ for LZSS and LZ-End, and $\Theta(n)$ for LZ78.

Keywords: data compression · sensitivity · Lempel-Ziv family · bidirectional macro schemes · string attractors

1 Introduction

Measuring the *repetitiveness* of strings is one of the most fundamental studies on strings which attract recent attention. Examples of string repetitiveness measures are the *substring complexity* δ [8], the smallest *string attractor* size γ [6], the number r of runs in the *Burrows-Wheeler transform* [3], the size b of the smallest *bidirectional macro scheme* [12], the size z of the Lempel-Ziv factorization [13] (and its variants [7,9,12]), and the smallest grammar size g [4,11]. We refer readers to the survey [10] for comparisons of these measures and others.

Akagi et al. [1] introduced the notion of *sensitivity* of string repetitiveness measures, which evaluates the perturbation of the measures after an edit operation (insertion, deletion, or substitution of a character) is performed on the input string. Sensitivity is a mathematical model for robustness of string compressors/repetitiveness measures against dynamic changes and/or errors. Of two versions of sensitivity, the work in [1] mostly focuses on the worst-case *multiplicative sensitivity* $\mathsf{MS}_{\mathrm{edit}}(c,n)$ with edit operation $\mathrm{edit} \in \{\mathrm{ins},\mathrm{del},\mathrm{sub}\}$ that is defined to be the maximum of $c(T')/c(T)$, where c is the measure, T is any string of length n, and T' is any string that can be obtained by a single-character edit operation from T. The other alternative is the worst-case *additive sensitivity* $\mathsf{AS}_{\mathrm{edit}}(c, n)$, which is defined to be the maximum of $c(T') - c(T)$. We note that most of the results on the additive sensitivity in [1] are byproducts of their multiplicative sensitivity, and they are evaluated in terms of the measure c. For instance, the multiplicative sensitivity 3 of the overlapping LZSS z immediately leads to a $2z$ additive sensitivity [1]. The behavior of $\mathsf{AS}_{\mathrm{edit}}(c, n)$ in terms of n is not well understood for a majority of measures, except for the trivial bound for δ [1] and the $\Omega(\sqrt{n})$ lower bounds for γ, b and r [1,5].

In this paper, we show tight upper and lower bounds of $\mathsf{AS}_{\mathrm{edit}}(c, n)$ in terms of n for the following measures c: the smallest string attractor γ, the smallest bidirectional macro scheme b, the Lempel-Ziv family including the overlapping/non-overlapping LZSS z and z_{no} [12], the optimal LZ-End z_{end} [7], and LZ78 z_{78} [14]. We present $O(\sqrt{n})$ upper bounds for the worst-case additive sensitivity of γ and b, which match the known lower bounds $\Omega(\sqrt{n})$ for γ and b [1]. Further, we present matching upper and lower bounds $\Theta(n^{\frac{2}{3}})$ for the worst-case additive sensitivity of z, z_{no}, and z_{end}. We also present an $\Omega(n)$ lower bound instance for the worst-case additive sensitivity of z_{78}, which matches a naïve $O(n)$ upper bound. Our results and previous results are summarized in Table 1.

Independently to our work, Blocki et al. [2] presented an $O(n^{\frac{2}{3}})$ upper bound and an $\Omega(n^{\frac{2}{3}}/\log^{\frac{1}{3}} n)$ lower bound for the worst-case additive sensitivity of the non-overlapping LZ77 factorizations [13]. While they provide a specific constant factor for the upper bound, their lower bound is slightly loose (up to a logarithmic

Table 1. The worst-case additive sensitivity of the string compressors/repetitiveness measures studied in this paper, evaluated in terms of the length n of the string. All the bounds hold for any type of edit operations (substitutions, insertions, and deletions). All the bounds except those from [1,5] are our new results.

string compressor/repetitiveness measure	worst-case additive sensitivity	
	upper bound	lower bound
substring complexity δ	1 [1]	1 [1]
smallest string attractor γ	$O(\sqrt{n})$	$\Omega(\sqrt{n})$ [1]
smallest bidirectional macro scheme b	$O(\sqrt{n})$	$\Omega(\sqrt{n})$ [1]
non-overlapping LZSS z_{no}	$O(n^{\frac{2}{3}})$	$\Omega(n^{\frac{2}{3}})$
overlapping LZSS z	$O(n^{\frac{2}{3}})$	$\Omega(n^{\frac{2}{3}})$
non-overlapping LZ77 $z_{77\mathrm{no}}$	$O(n^{\frac{2}{3}})$ [2]	$\Omega(n^{\frac{2}{3}})$
overlapping LZ77 z_{77}	$O(n^{\frac{2}{3}})$	$\Omega(n^{\frac{2}{3}})$
optimal LZ-End z_{end}	$O(n^{\frac{2}{3}})$	$\Omega(n^{\frac{2}{3}})$
greedy LZ-End z_{e}	-	$\Omega(n^{\frac{2}{3}})$
LZ78 z_{78}	$O(n)$	$\Omega(n)$
run-length BWT r	-	$\Omega(\sqrt{n})$ [5]

factor). We remark that our tight $\Theta(n^{\frac{2}{3}})$ sensitivity bounds for (non)overlapping LZSS can readily extend to (non)overlapping LZ77.

2   Preliminaries

2.1   Strings

Let Σ be an *alphabet*. An element of Σ^* is called a *string*. The length of a string T is denoted by $|T|$. The empty string ε is the string of length 0. Let Σ^n denote the set of strings of length n. For a string $T = xyz$, x, y and z are called a *prefix*, *substring*, and *suffix* of T, respectively. Let $\mathsf{Substr}(T)$ denote the set of substrings of s. The i-th symbol of a string T is denoted by $T[i]$ for $1 \le i \le |T|$. Let $T[i..j]$ denote the substring of T that begins at position i and ends at position j for $1 \le i \le j \le |T|$. For convenience, let $T[i..j] = \varepsilon$ when $i > j$. For a sequence $s_1, \ldots, s_k$ of strings, let $\prod_{i=1}^{k} s_k = s_1 \cdots s_k$ denote their concatenations.

2.2   String Attractors

Let T be a non-empty string of length n. A set $\Gamma = \{p_1, \ldots, p_\ell\}$ of ℓ positions in T is said to be a *string attractor* [6] of T if any substring $s \in \mathsf{Substr}(T)$ has an occurrence $[i, j]$ such that $s = T[i..j]$ and $i \le p_k \le j$ for some $p_k \in \Gamma$. Since the set $\{1, \ldots, n\}$ of all positions in T clearly satisfies the definition, any string has a string attractor. The size of a string attractor is the number of positions in it. Let $\gamma(T)$ denote the size of the smallest string attractor(s) for T. For instance, the set $\{5, 7\}$ of positions is a string attractor of string $T = \texttt{baaaabbaaa}$ (see also Fig. 1), and it is the smallest since T contains two distinct characters $\texttt{a}, \texttt{b}$.

2.3   Bidirectional Macro Scheme

A sequence $B = f_1, \ldots, f_b$ of non-empty substrings of a string T is said to be a *bidirectional macro scheme* of T if (1) $T = f_1 \cdots f_b$ and (2) each *phrase* $f_k = T[p_k, p_k + |f_k| - 1]$ that occurs at position $p_k = |f_1 \cdots f_{k-1}| + 1$ is either a single character or a copy of a *source* substring $T[q_k, q_k + |f_k| - 1]$ that occurs to the left or to the right of f_k (i.e. $q_k \ne p_k$). If $|f_k| = 1$, then f_k is called a *ground phrase*. A bidirectional macro scheme $B = f_1, \ldots, f_b$ for T with $p_k = |f_1 \cdots f_{k-1}| + 1$ induces a function $F_B : [1, n] \cup \{0\} \to [1, n] \cup \{0\}$ such that

$$F_B(0) = 0$$
$$F_B(p_k) = 0 \qquad \text{if } f_k \text{ is a grand phrase,}$$
$$F_B(p_k + j) = q_k + j \qquad \text{if } |f_k| > 1 \text{ and } 0 \le j < |f_k|.$$

Let $F_B^0(p_k) = p_k$ and $F_B^m(p_k) = F(F^{m-1}(p_k))$ for $m \le 1$. A bidirectional macro scheme B is called *valid* if there exists an $m \ge 1$ such that $F_B^m(i) = 0$ for every $i \in [1, n]$. The string T can be reconstructed from the bidirectional macro scheme if and only if it is valid. The size of a bidirectional macro scheme B is the number of phrases of B. Let $b(T)$ denote the size of the smallest valid bidirectional macro scheme(s) for T.

2.4 LZ-Style Compressors

A sequence $f_1, \ldots, f_z$ of non-empty substrings of a string T is said to be an *LZ-style factorization* of T if $T = f_1 \cdots f_z$ and each f_k is either a fresh character not occurring to its left, or a copy of a previous substring occurring to its left.

The *overlapping LZSS* factorization of a string T is the greedy LZ-style factorization of T such that each f_k is taken as long as possible. Hence, each copied phrase f_k in the overlapping LZSS factorization is the longest prefix of $T[|f_1 \cdots f_{k-1}| + 1..n]$ that occurs at least twice in $T[1..|f_1 \cdots f_k|]$. Let $\mathsf{LZSS}(T)$ denote the overlapping LZSS factorization of a string T and $z(T)$ the number of phrases in $\mathsf{LZSS}(T)$.

In the *non-overlapping LZSS* factorization, each copied phrase f_k is the longest prefix of $T[|f_1 \cdots f_{k-1}| + 1..n]$ that occurs at least once in $T[1..|f_1 \cdots f_{k-1}|]$. Let $\mathsf{LZSS}_{no}(T)$ denote the non-overlapping LZSS factorization of a string T and $z_{no}(T)$ the number of phrases in $\mathsf{LZSS}_{no}(T)$.

An LZ-style factorization $f_1, \ldots, f_z$ of a string T is said to be an *LZ-End factorization* of T if each copied phrase f_k has a previous occurrence $[i..j]$ in $T[1..|f_1 \cdots f_{k-1}|]$ such that j is the ending position of a previous phrase, namely $j = |f_1 \cdots f_h|$ for some $1 \le h < k$. The *optimal LZ-End factorization* is the LZ-End factorization whose size is the smallest. Let z_{end} denote the size of the optimal LZ-End factorization of string T. The *greedy LZ-End factorization* is the LZ-End factorization such that each copied phrase f_k is taken as long as possible, namely, f_k is the longest prefix of $T[|f_1 \cdots f_{k-1}|+1..n]$ that is a suffix of $T[1..|f_1 \cdots f_h|]$ for some $1 \le h < k$. Let $\mathsf{LZEnd}(T)$ denote the greedy LZ-End factorization of a string T, and $z_e(T)$ the number of phrases in $\mathsf{LZEnd}(T)$.

An LZ-style factorization $f_1, \ldots, f_z$ of a string T is said to be an *LZ78 factorization* of T if each copied phrase f_k is the shortest prefix of $T[|f_1 \cdots f_{k-1}| + 1..n]$ such that $f_k[1..|f_k| - 1] = f_h$ for some $1 \le h < k$ and $f_k \ne f_\ell$ for any $1 \le \ell < k$. Let $\mathsf{LZ78}(T)$ denote the LZ78 factorization of a string T, and $z_{78}(T)$ the number of phrases in $\mathsf{LZ78}(T)$.

Figure 1 shows concrete examples of the aforementioned string compressors and repetitiveness measures.

non-overlapping LZSS	b	a	a	a	a	b	b	a	a	a		6
overlapping LZSS	b	a	a	a	a	b	b	a	a	a		5
greedy LZ-End	b	a	a	a	a	b	b	a	a	a		7
LZ78	b	a	a	a	a	b	b	a	a	a		6
Smallest Bidirectinal macro scheme	b	a	a	a	a	b	b	a	a	a		5
Smallest stirng attractor	b	a	a	a	$\dot{a}$	b	$\dot{b}$	a	a	a		2

Fig. 1. Examples of string compressors/repetitiveness measures for string $T = $ baaabbaaa.

2.5 Additive Sensitivity of String Repetitiveness Measures

For two strings S, T, let $\mathsf{ed}(S, T)$ denote the edit distance between S and T.

Let $c(T)$ denote the size of the output of compressor (such as LZSS) or a repetitiveness measure (such as the smallest string attractor) for a string T of length n. The worst-case *additive sensitivity,multiplicative sensitivity* of c is defined by

$$\mathsf{AS}_{\mathsf{edit}}(c, n) = \max_{T \in \Sigma^n, T' \in \Sigma^m} \{ c(T') - c(T) \mid \mathsf{ed}(T') = 1 \},$$
$$\mathsf{MS}_{\mathsf{edit}}(c, n) = \max_{T \in \Sigma^n, T' \in \Sigma^m} \{ c(T')/c(T) \mid \mathsf{ed}(T') = 1 \}$$

where $\mathsf{edit} \in \{\mathsf{sub}, \mathsf{ins}, \mathsf{del}\}$ represents the type of the edit operation (substitution, insertion, deletion), and $m = n$ for $\mathsf{edit} = \mathsf{sub}$, $m = n + 1$ for $\mathsf{edit} = \mathsf{ins}$, and $m = n - 1$ for $\mathsf{edit} = \mathsf{del}$.

3 Additive Sensitivity of String Attractors

In this section, we consider the additive sensitivity of the smallest string attractor size γ. The following lower bound is known:

Theorem 1 ([1]). $\mathsf{AS}_{\mathsf{edit}}(\gamma, n) = \Omega(\sqrt{n})$ *holds for* $\mathsf{edit} \in \{\mathsf{sub}, \mathsf{ins}, \mathsf{del}\}$.

Together with Theorem 2, we obtain a tight bound $\Theta(\sqrt{n})$ for $\mathsf{AS}_{\mathsf{edit}}(\gamma, n)$.

Theorem 2. $\mathsf{AS}_{\mathsf{edit}}(\gamma, n) = O(\sqrt{n})$ *holds for* $\mathsf{edit} \in \{\mathsf{sub}, \mathsf{ins}, \mathsf{del}\}$.

For ease of discussion, we consider the case of substitution where $\mathsf{edit} = \mathsf{sub}$. Let T be the original string of length n and T' be the string obtained by replacing the i-th character $T[i] = a$ with a distinct character b ($\neq a$). Namely, $T' = T[1..i-1] \cdot b \cdot T[i+1..n]$. We consider three subsets S_1, S_2, S_3 of $\mathsf{Substr}(T')$ as follows:

$S_1 = \{ s \in \mathsf{Substr}(T') \mid s \text{ has an occurrence in } T \text{ that contains position } i \}$,
$S_2 = \{ s \in \mathsf{Substr}(T') \mid s \text{ has an occurrence in } T' \text{ that contains position } i \}$,
$S_3 = \mathsf{Substr}(T') - (S_1 \cup S_2)$.

Lemma 1. *There exists an integer set $A \subseteq [1, n]$ such that $|A| = O(\sqrt{n})$ holds and every $s \in S_1$ has an occurrence in T' that contains a position in A.*

Proof. We consider two cases w.r.t. the length of $s \in S_1$.

1. Assume that $|s| \geq 1 + \lceil \sqrt{n} \rceil$. Let

$$A_1 = \left\{ \lfloor \sqrt{n} \rfloor, 2\lfloor \sqrt{n} \rfloor, \ldots, \lfloor \sqrt{n} \rfloor * \lfloor \sqrt{n} \rfloor \right\} \cup \left\{ \left\lfloor \frac{\lfloor \sqrt{n} \rfloor^2 + n}{2} \right\rfloor \right\}.$$

We can see that $n - \left\lfloor \frac{\lfloor \sqrt{n} \rfloor^2 + n}{2} \right\rfloor \leq 1 + \sqrt{n}$, $\left\lfloor \frac{\lfloor \sqrt{n} \rfloor^2 + n}{2} \right\rfloor - \lfloor \sqrt{n} \rfloor^2 \leq 1 + \sqrt{n}$, and $k\lfloor \sqrt{n} \rfloor - (k-1)\lfloor \sqrt{n} \rfloor = \lfloor \sqrt{n} \rfloor \leq \sqrt{n}$ for every $k \in [1, \lfloor \sqrt{n} \rfloor]$. These facts imply that every s has an occurrence in T that contains a position in A_1. It is clear that $|A_1| \leq \lfloor \sqrt{n} \rfloor + 1 = O(\sqrt{n})$ holds.

2. Assume that $|s| \leq \lceil \sqrt{n} \rceil$. We consider a subset $S_1' \subseteq S_1$ such that $|s| \leq \lceil \sqrt{n} \rceil$ holds and the occurrence (as an interval) of s is not nested by an occurrence of any other elements in S_1'. From such structures, we can see that $|S_1'| \leq \lceil \sqrt{n} \rceil$. Let $s \in S_1'$ and d be the position in s that corresponds to the edit position. We choose an occurrence j of s in T' and also consider a position $j + d - 1$ (see Fig. 2). We can see that this position stabs s and also shorter substrings $s' \notin S_1'$. Thus a set A_2 of such positions for every $s \in S_1'$ covers an occurrence of each substring in this case. It is clear that $|A_2| \leq \lceil \sqrt{n} \rceil = O(\sqrt{n})$ holds.

From the above discussion, $A = A_1 \cup A_2$ satisfies the statement.   □

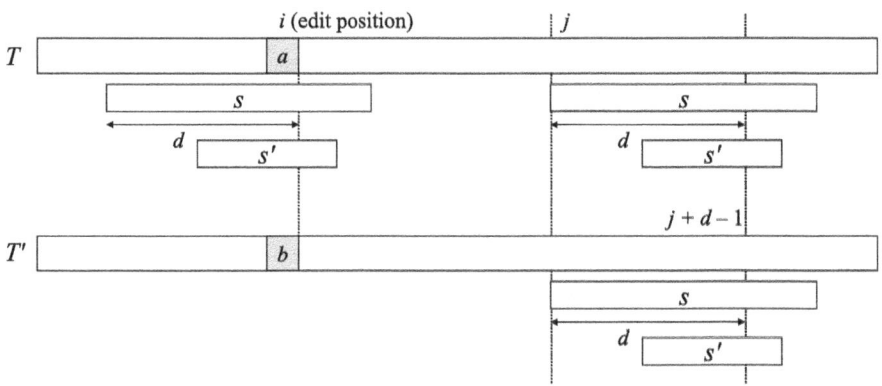

Fig. 2. This figure illustrates the case that the i-th character in T is substituted by b. A position $j + d - 1$ stabs both s and s' in T'.

We show the following lemma.

Lemma 2. *For any strings T of length n, there exists a string attractor of T' of size $\gamma(T) + O(\sqrt{n})$.*

Proof. Let Γ be a string attractor of T. We show that there exists a string attractor Γ' of T' whose size is $|\Gamma| + O(\sqrt{n})$. By Lemma 1, positions in A that is defined in Lemma 1 cover substrings in S_1. It is easy to see that position i stabs all substrings in S_2. For any substring in S_3, it has an occurrence in both strings that contain a position in Γ. Thus Γ covers substrings in S_3. Overall, there exists a string attractor $\Gamma' = \Gamma \cup A \cup \{i\}$ of size $|\Gamma| + O(\sqrt{n})$.   □

Lemma 2 immediately derives Theorem 2.

4   Additive Sensitivity of Bidirectional Macro Scheme

In this section, we consider the additive sensitivity of the smallest bidirectional macro scheme size b. Akagi et al. [1] showed the following lower bounds for each edit operation.

Theorem 3 ([1]). $\mathsf{AS}_{\mathrm{edit}}(b,n) = \Omega(\sqrt{n})$ *holds for* edit $\in \{\mathrm{sub}, \mathrm{ins}, \mathrm{del}\}$.

Together with Theorem 4, we obtain a tight bound $\Theta(\sqrt{n})$ for $\mathsf{AS}_{\mathrm{edit}}(b,n)$.

Theorem 4. $\mathsf{AS}_{\mathrm{edit}}(b,n) = O(\sqrt{n})$ *holds for* edit $\in \{\mathrm{sub}, \mathrm{ins}, \mathrm{del}\}$.

To prove Theorem 4, we use the proof of the multiplicative sensitivity of bidirectional macro scheme by Akagi et al. [1]. Let T be a string and T' the string obtained by editing the i-th character of T, and $f_1, \ldots, f_b$ be a valid bidirectional macro scheme of T. They showed how to construct a valid bidirectional macro scheme of T' from $f_1, \ldots, f_b$ and the increasing number of phrases in the operations. We denote the beginning position of phrase f_j by p_j (i.e., $f_j = T[p_j..p_j + |f_j| - 1]$) and its beginning position of source by q_j (i.e., $T[q_j..q_j + |f_j| - 1]$). Akagi et al. considered the three types of operations.

Proposition 1 (Theorem 11 of [1]). *A valid bidirectional macro scheme of* T' *can be obtained by applying the following operations to* $f_1, \ldots, f_b$.

(1) *If* $i \in [p_j, p_j + |f_j| - 1]$, *we divide the phrase* f_j *into at most five phrases.*
(2) *If* $i \notin [p_j, p_j + |f_j| - 1]$ *and* $i \notin [q_j, q_j + |f_j| - 1]$, *we don't need any changes.*
(3) *If* $i \notin [p_j, p_j + |f_j| - 1]$ *and* $i \in [q_j, q_j + |f_j| - 1]$, *we divide the phrase* f_j *at most three phrases.*

In case (1), the number of phrases increases by at most four, and in case (2), the number of factor remains unchanged. Here, we give an upper bound of increasing number of phrases in case (3) w.r.t. n. First, we consider the number of phrases in case (3) whose length is at least $\sqrt{n}$. Then, such phrases can occur at most $O(\sqrt{n})$ times in T, since $|T| = n$. Hence the total increasing number of phrases from such phrases is $O(\sqrt{n})$, since each phrase is divided into at most three phrases by Proposition 1. Next, we consider the number of phrases whose length is smaller than $\sqrt{n}$. Let f_k be a phrase in case (3) such that there exists a phrase $f_{k'}$ that is a superstring of f_k ($k' \neq k$). For such phrase f_k, a corresponding substring of $f_{k'}$ can be a new source of f_k even if the original source of f_k is lost by editing. Thus, in this case, it is unnecessary to divide f_k. On the other hand, the maximum number of phrases in case (3) whose sources do not nest each other is $O(\sqrt{n})$ since the sources contain the edited position i. For such $O(\sqrt{n})$ phrases, we divide them into at most three phrases by Proposition 1. Hence, the total increasing number of phrases from such phrases is also $O(\sqrt{n})$.

Overall, the total increasing number of phrases is $O(\sqrt{n})$, and we obtain $\mathsf{AS}_{\mathrm{edit}}(b,n) = O(\sqrt{n})$.

5   Additive Sensitivity of LZSS and LZ-End

In this section, we consider the additive sensitivity of the size of LZ-style factorization. First, we describe LZ-style factorization, excluding LZ78. Specifically, they are overlapping LZSS, non-overlapping LZSS, optimal LZ-End factorization, and greedy LZ-End factorization.

5.1   Lower Bounds

For any function $c : \Sigma^n \to \{1, \dots, n\}$, the following theorem holds:

Theorem 5. *If $z(T) \le c(T) \le z_e(T)$ for any string $T \in \Sigma^+$, then* $\mathsf{AS}_{\mathrm{edit}}(c, n)$ $\in \Omega(n^{\frac{2}{3}})$ *holds for* edit $\in \{\mathrm{sub}, \mathrm{ins}, \mathrm{del}\}$.

Proof. Let T be any string of length n, and T' any string such that $\mathrm{ed}(T, T') = 1$. By the assumption, $c(T') - c(T) \ge z(T') - z_e(T)$ holds. Therefore, to prove $\mathsf{AS}(z, n) \in \Omega(n^{\frac{2}{3}})$ for measure c, it is sufficient to show that there exist strings T, T' that satisfy $z(T') - z_e(T) \in \Omega(n^{\frac{2}{3}})$.

To demonstrate this, we define the following sequence: Let $p \ge 2$ be an integer. We define the sequence $\mathsf{Pair}(p)$ of pairs (l_i, r_i) such that

$$
(l_i, r_i) = \begin{cases} (1, 1) & \text{if } i = 1, \\ (l_{i-1} - 1, r_{i-1} + 1) & \text{if } l_{i-1} \ne 1 \text{ and } r_{i-1} \ne p, \\ (l_{i-1} + r_{i-1} + 1, 1) & \text{otherwise}, \end{cases}
$$

and $l_i + r_i \le 2p$ for any i. Intuitively, $\mathsf{Pair}(p)$ enumerates all pairs $(l_i, r_i) \in [1, p]^2$ such that $l_i + r_i = k$ in increasing order of $k \in \{2, 3, \dots, 2p\}$, and for each fixed k in increasing order of l_i. For instance, $\mathsf{Pair}(2) = (1, 1), (1, 2), (2, 1), (2, 2)$. The number of elements in $\mathsf{Pair}(p)$ is p^2 and thus $\sum_{i=1}^{p^2}(l_i + r_i) = \Theta(p^3)$ holds.

Let $\Sigma = \{a_1, \dots, a_p, b_1, \dots, b_p, x, y, \#_1, \dots, \#_{p^2}\}$, where $p \ge 2$. Furthermore, for each $i \in [1, p]$, we define the strings $A(i) = a_1 \cdots a_i$ and $B(i) = b_i \cdots b_1$. Then, we consider the following string T.

$$
T = b_p \cdots b_1 x a_1 \cdots a_p \#_1 b_1 x a_1 \#_2 b_2 b_1 x a_1 \#_3 \cdots \#_i B(l_i) x A(r_i) \#_{i+1} \cdots \#_{p^2}
$$

$$
= B(p) x A(p) \prod_{i=1}^{p^2} \left(\#_i B(l_i) x A(r_i) \right).
$$

Observe that $|T| = \Theta(p^3)$.

The greedy LZ-End factorization of T is as follows.

$$
\mathsf{LZEnd}(T) = b_p | b_{p-1} | \cdots | b_1 | x | a_1 | a_2 | \cdots | a_p | \prod_{i=1}^{p^2} \left(|\#_i| B(l_i) x A(r_i)| \right).
$$

Since each character in the interval $[1, 2p + 1]$ is new, they become a phrase of length 1 each. Next, for $i \in [1, p^2]$, $\#_i B(l_i) x A(r_i)$ is partitioned as $|\#_i| B(l_i) x A(r_i)|$. This is because $\#_i$ is a new character, $B(l_i) x A(r_i)$ has a previous occurrence in $T[p+1-l_i..p+1+r_i]$, and $T[p+1+r_i] = a_{r_i}$ is a single-character phrase, and the character $\#_{i+1}$ immediately following $B(l_i) x A(r_i)$ is new. Thus, we have $z_e(T) = 2p^2 + 2p + 1$.

As for substitutions, consider the string

$$
T' = b_p \cdots b_1 \underline{y} a_1 \cdots a_p \#_1 b_1 x a_1 \#_2 b_2 b_1 x a_1 \#_3 \cdots \#_i B(l_i) x A(r_i) \#_{i+1} \cdots \#_{p^2}
$$

$$
= B(p) \underline{y} A(p) \prod_{i=1}^{p^2} \left(\#_i B(l_i) x A(r_i) \right)
$$

that is obtained by substituting the first occurrence of $x = T[p+1]$ with a character y (underlined) which does not occur in T.

Let us analyze the structure of overlapping LZSS of T'. Since each character in the interval $T'[1..2p+1]$ is new, they become a phrase of length 1 each.

For $i = 1$, $\#_1 a_1 x b_1$ is divided into $|\#_1|a_1|x|b_1|$ because $\#_1$ is a new character, neither $a_1 x$ nor $x b_1$ has previous occurrences, and $\#_2$ which immediately follows b_1 is a new character.

For each $i \in [2, p^2]$, the substring $\#_i B(l_i) x A(r_i)$ is divided into 3 phrases: Since $\#_i$ is a new character, it becomes a single-character phrase. Hence $B(l_i) x A(r_i)$ is divided into 2 phrases, and the second phrase ends immediately before $\#_{i+1}$. We have two cases:

– When $r_i \neq 1$, $B(l_i) x A(r_{i-1})$ is the longest prefix of $B(l_i) x A(r_i)$ that has a previous occurrence. There are two reasons for this. The first is that the occurrence $B(l_i) x A(r_i) = T[p+1-l_i..p+1+r_i]$ is lost in T' due to the substitution of x with y. The second is that there exists $j \in [1, i-1]$ such that $l_j = l_i$ and $r_j = r_i - 1$. In particular, if $l_i + r_i = k$, then $l_j + r_j = k - 1 < k$. Consequently, by the ordering property of $\mathsf{Pair}(p)$, the pair (l_j, r_j) is guaranteed to appear before (l_i, r_i). Consequently, $B(l_i) x A(r_i)$ is divided into $|B(l_i) x A(r_{i-1})|a_{r_i}|$.
– When $r_i = 1$, then $A(r_i) = A(1) = a_1$. The previous occurrence of $B(l_i) x a_1 = T[p+1-l_i..p+2]$ is also lost in T' by the substitution. Then, $B(l_i)$ is the longest prefix of $B(l_i) x a_1$ that has a previous occurrence. $B(l_i)$ occurs in $T[p+1-l_i, p]$, but $B(l_i) x$ does not. Consequently, $B(l_i) x a_1$ is divided into $|B(l_i)|x a_1|$.

Now it follows that for each $i \in [2, p^2]$, the substring $\#_i B(l_i) x A(r_i)$ is divided into 3 phrases. Hence, $z(T) = 3p^2 + 2p + 2$ holds. Recall $|T| = n = \Theta(p^3)$. As a result, we have $z(T') - z_e(T) \geq 3p^2 + 2p + 2 - (2p^2 + 2p + 1) = p^2 \in \Theta(n^{\frac{2}{3}})$.

For insertions and deletions, by considering T'' and T''' obtained from T by inserting y between $x = T[p+1]$ and $b_1 = T[p+2]$ and deleting $x = T[p+1]$ respectively, we can get similar decompositions in the case of substitutions, which lead to $z(T'') = 3p^2 + O(p)$ and $z(T''') = 3p^2 + O(p)$ holds. This gives us $z(T'') - z_e(T) \in \Omega(n^{\frac{2}{3}})$ and $z(T''') - z_e(T) \in \Omega(n^{\frac{2}{3}})$.

Overall, $\mathsf{AS}_{\mathrm{sub}}(c, n), \mathsf{AS}_{\mathrm{ins}}(c, n), \mathsf{AS}_{\mathrm{del}}(c, n) \in \Omega(n^{\frac{2}{3}})$ holds. $\square$

Since $z(T) \leq z_{\mathrm{no}}(T) \leq z_{\mathrm{end}}(T) \leq z_e(T)$ holds [7,9], the worst-case additive sensitivity for $z(T)$, $z_{\mathrm{no}}(T)$, $z_{\mathrm{end}}(T)$, and $z_e(T)$ are all $\Omega(n^{\frac{2}{3}})$ by Theorem 5.

5.2   Upper Bounds

Overlapping/Non-Overlapping LZSS Factorization. Next, we describe the upper bound on the additive sensitivities of LZ-style factorizations. Specifically, we consider overlapping LZSS, non-overlapping LZSS, and optimal LZ-End.

Theorem 6. $\mathsf{AS}_{\mathrm{edit}}(z, n) = O(n^{\frac{2}{3}})$ and $\mathsf{AS}_{\mathrm{edit}}(z_{\mathrm{no}}, n) = O(n^{\frac{2}{3}})$ hold for $\mathrm{edit} \in \{\mathrm{sub}, \mathrm{ins}, \mathrm{del}\}$.

First, we consider the non-overlapping LZSS factorization. To prove Theorem 6, we use the proof of the multiplicative sensitivity of non-overlapping LZSS by Akagi et al. [1]. For simplicity, we consider the case of substitution (the results for other cases can be shown in a similar way). Let T be a string and T' the string obtained by editing the i-th character of T, $\mathsf{LZSS}_{no}(T) = f_1 \cdots f_t$, and $\mathsf{LZSS}_{no}(T') = f'_1 \cdots f'_{t'}$. We denote the interval of phrase f_j by $[p_j, q_j]$ (i.e., $f_j = T[p_j..q_j]$). They showed each phrase of $\mathsf{LZSS}_{no}(T)$ can contain a constant number of beginning positions of phrases of $\mathsf{LZSS}_{no}(T')$ by considering the three cases.

Proposition 2 (Theorem 19 of [1]). *For $j \in [1, t]$, each interval $[p_j, q_j]$ satisfies as follows.*

(1) If $q_j < i$, $[p_j, q_j]$ contains a single beginning position of a phrase in $\mathsf{LZSS}_{no}(T')$ (i.e., $f'_j = f_j$).
(2) If $p_j \le i \le q_j$, $[p_j, q_j]$ contains at most constant number of beginning positions of phrases in $\mathsf{LZSS}_{no}(T')$.
(3) If $i < p_j$ and
(3-A) the source of f_j does not contain the position i, $[p_j, q_j]$ contains at most one beginning position of phrases in $\mathsf{LZSS}_{no}(T')$.
(3-B) the source of f_j contains the position i, $[p_j, q_j]$ contains at most constant number of beginning positions of phrases in $\mathsf{LZSS}_{no}(T')$.

In case (1) and (3-A), the number of the beginning positions of phrases in $\mathsf{LZSS}_{no}(T')$ that are contained in $[p_j, q_j]$ does not increase. In case (2), the number of beginning positions in the interval increases by a constant number, but such phrase is unique in $\mathsf{LZSS}_{no}(T)$. Here, we discuss an upper bound of the total increasing number of beginning positions of phrases in case (3-B) in terms of n. First, we consider phrases in case (3) of length at least $n^{\frac{1}{3}}$. Then, such phrases can occur at most $O(n^{\frac{2}{3}})$ times in T, since $|T| = n$. Hence the total increasing number of beginning positions of phrases from such phrases is $O(n^{\frac{2}{3}})$, since each phrase can contain a constant number of such positions by Proposition 2. Next, we consider the number of phrases whose length is smaller than $n^{\frac{1}{3}}$. If there exists $f_{j'}$ in case (3) such that $j' < j$ and $f_j = f_{j'}$, then, $f_{j'} = T'[p_{j'}..q_{j'}]$ can be a source of $T'[p_j..q_j]$. Thus, in this case, the number of beginning positions of phrases in the interval $[p_j, q_j]$ does not increase. On the other hand, the maximum number of distinct phrases in case (3) is $O(n^{\frac{2}{3}})$ since the number of substrings of length at most k that contain a position i is $O(k^2)$. For such $O(n^{\frac{2}{3}})$ phrases, $[p_j, q_j]$ contains at most a constant number of beginning positions of phrases in $\mathsf{LZSS}_{no}(T')$ by Proposition 2. Hence, the total increasing number of beginning positions of phrases from such phrases is also $O(n^{\frac{2}{3}})$.

Overall, the total increasing number of phrases is $O(n^{\frac{2}{3}})$, and we obtain $\mathsf{AS}_{\mathrm{edit}}(z, n) = O(n^{\frac{2}{3}})$.

Akagi et al. proved the multiplicative sensitivity of overlapping LZSS in a similar way [1]. We can also obtain $\mathsf{AS}_{\mathrm{edit}}(z_{\mathrm{no}}, n) = O(n^{\frac{2}{3}})$ in a similar way.

Optimal LZ-End Factorization. As for the sensitivity of optimal LZ-End factorization, we can obtain the following upper bounds. Together with Theorem 5, we can also obtain a tight bound $\Theta(n^{\frac{2}{3}})$ for $\mathsf{AS}_{\mathrm{edit}}(z_{\mathrm{end}}, n)$.

Theorem 7. $\mathsf{AS}_{\mathrm{edit}}(z_{\mathrm{end}}, n) = O(n^{\frac{2}{3}})$ *holds for* $\mathrm{edit} \in \{\mathrm{sub}, \mathrm{ins}, \mathrm{del}\}$.

Similarly to the case of LZSS (Theorem 7) in the previous section, we also use an idea for the multiplicative sensitivity of optimal LZ-End factorization. However, there is no result for it. Here, we first show an upper bound of the multiplicative sensitivity of the optimal LZ-End factorization by a similar idea. For ease of discussion, we consider the case of substitution. Let T be a string and T' the string obtained by substituting the i-th character $T[i] = a$ with b, and $f_1, \ldots, f_t$ an LZ-End factorization of T. We denote the interval of phrase f_j by $[p_j, q_j]$ (i.e., $f_j = T[p_j..q_j]$). Moreover, let I be an integer that satisfies $p_I \leq i \leq q_I$ (i.e., the index of a phrase that contains the edited position i). We consider the following operations to $f_1, \ldots, f_t$.

Proposition 3. *An LZ-End factorization of T' can be obtained by applying the following operations to $f_1, \ldots, f_t$.*

(1) If $q_j < i$, we don't need any changes.
(2) If $p_j \leq i \leq q_j$, we divide the phrase f_j into at most $I + 1$ phrases.
(3) If $i < p_j$ and
  (3-A) the source of f_j does not contain the position i, we don't need any changes.
  (3-B) the source of f_j contains the position i, we divide the phrases f_j into at most three phrases.

We show how to construct an LZ-End factorization of T' from $f_1, \ldots, f_t$. In case (1), for the first $I - 1$ phrases, T' has the same LZ-End factorization since the prefix is unchanged by the substitution. In case (2), let $T[p_I..q_I] = f_I = w_1 a w_2$ and $T'[p_I..q_I] = w_1 b w_2$. First, the substring b can become a phrase of LZ-End of T' since b is a single character. Next, since $w_1 a w_2$ is a phrase of an LZ-End of T, there exists an integer $k < I$ such that $w_1 a w_2$ is a suffix of $f_1 \cdots f_k$. Then, w_2 in the source of f_I can still be a source of a new phrase w_2 in T'. In the rest of this part, we show that $w_1 = T[p_I..p_I + |w_1| - 1]$ can be divided into at most I LZ-End phrases of T'. Since w_1 is a prefix of f_I, w_1 has a previous occurrence in its source. By traversing phrases to the left, we can find a phrase $f_{k'}$ such that $k' < I$ and the LZ-End boundary $q_{k'}$ contained in an occurrence of w_1. Assume that k' is the largest integer in such indices. Then w_1 at the position can be written as $w_1 = uv$, where u is a suffix of $f_1 \cdots f_{k'}$ and v is a prefix of $f_{k'+1} \cdots f_{I-1}$. From this structure, prefix u of $T'[p_I..q_I]$ can be an LZ-End phrase such that the source ends at $q_{k'}$. For the remaining suffix v of w_1, in a similar argument, we can find a phrase $f_{k''}$ such that $k'' < k' + 1$ ($< I$) and the LZ-End boundary $q_{k''}$ contained in an occurrence of v. Then v at the position can be written as $v = u'v'$, and the substring u' of $T'[p_I..q_I]$ can be an LZ-End phrase such that the source ends at $q_{k''}$. By traversing phrases and dividing w_1

recursively, we can finally obtained at most $I-1$ LZ-End phrases of T' such that the concatenation of these phrases is $T'[p_I..p_I + |w_1| - 1] = w_1$ (see Fig. 3 for an illustration). Thus we can obtain at most $I + 1$ LZ-End phrases of T' from f_I. We consider phrases in case (3-A). f_j has a source $T[p'_j..q'_j]$. Since $i \notin [p'_j, q'_j]$, this string also appear at the same position in T'. Our construction of an LZ-End factorization of T' only use divisions of phrases of the original factorization of T. This implies that there exists a phrase of an LZ-End factorization of T' that ends at q'_j. Hence, $T'[p_j..q_j]$ can be a phrase whose source is $T'[p'_j..q'_j]$. In case (3-B), let $f_j = w_3 a w_4$ such that a corresponds to the edited position in its source. We can see that w_3, a, w_4 can be LZ-End phrases in T', because w_3 occurs at $[p'_j, i-1]$ and there exists an LZ-End phrase in T' that ends at $i-1$ by case (2), a is a single character, and w_4 still occurs at $[i+1, q'_j]$ in T' (as a suffix of the original phrase in T).

Overall, we can obtain an LZ-End factorization of T' from a given LZ-End factorization of T by dividing each phrase in T into at most three parts.

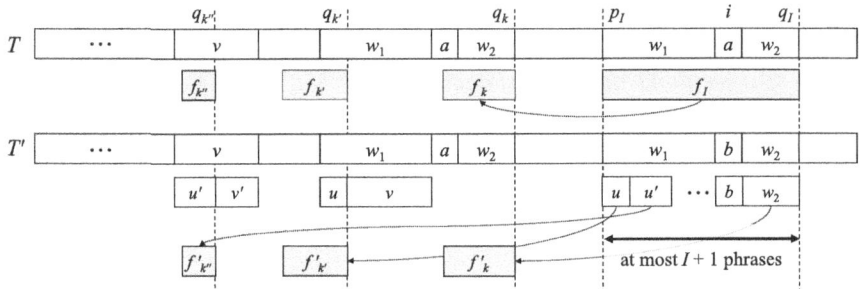

Fig. 3. Illustration of the division operation for Case (2).

Moreover, respect to deletions, a proposition that is completely the same Proposition 3 holds. As for insertions, although it is almost the same as Proposition 3, in case (3-B), f_j can be divided into at most two phrases not three. This can be proved the similar way of Akagi et al. [Theorem 18 of [1]]. Then, we get the following upper bound for the multiplicative sensitivity of z_{end}.

Corollary 1. $\mathsf{MS}_{\text{sub}}(z_{\text{end}}, n) = 3, \mathsf{MS}_{\text{del}}(z_{\text{end}}, n) = 3, \mathsf{MS}_{\text{ins}}(z_{\text{end}}, n) = 2$.

Then, we describe the upper bound of the additive sensitivity of the optimal LZ-End factorization in a similar way of Theorem 6. From Proposition 3, in case (1) or (3-A), the number of phrases remains unchanged. Here, we give an upper bound of increasing number of phrases in case (2) and (3-B) in terms of n. If $I > n^{2/3}$, then the number of phrases increases by at most $O(n^{\frac{2}{3}})$, since $|T| = n$. If $I < n^{2/3}$, then the number of phrases in the interval $[p_I, q_I]$ (case (2)) increases at most $I + 1 \in O(n^{\frac{2}{3}})$. Next, by the same way with non-overlapping LZSS, we can prove that the upper bound of increasing number of phrases is $O(n^{\frac{2}{3}})$. In any case, the total increasing number of phrases is $O(n^{\frac{2}{3}})$. and we obtain $\mathsf{AS}_{\text{edit}}(z_{\text{end}}, n) = O(n^{\frac{2}{3}})$ for edit $\in \{\text{sub}, \text{ins}, \text{del}\}$.

6   Additive Sensitivity of LZ78

In this section, we present the tight bound $\mathsf{AS}_{\mathrm{edit}}(z_{78}, n) = \Theta(n)$ for the additive sensitivity of the Lempel-Ziv 78 factorization size z_{78} for all types of edit operations $\mathrm{edit} \in \{\mathrm{sub}, \mathrm{ins}, \mathrm{del}\}$. Note that the upper bound $\mathsf{AS}_{\mathrm{edit}}(z_{78}, n) = O(n)$ trivially holds, since we have $1 \leq z_{78}(T) \leq n$ for any string T of length n.

We show the following lower bound:

Theorem 8. $\mathsf{AS}_{\mathrm{edit}}(z_{78}, n) = \Omega(n)$ *holds for* $\mathrm{edit} \in \{\mathrm{sub}, \mathrm{ins}, \mathrm{del}\}$.

Proof. Let $p \in \mathbb{N}$ and $\Sigma = \{a_1, .., a_p, b_1, .., b_p, c_1, .., c_p, \#\}$. Consider the following string

$$T = c_1 c_2 \cdots c_p a_1 a_1 b_1 a_2 a_2 b_2 \cdots a_p a_p b_p a_1 b_1 c_1 a_2 b_2 c_2 \cdots a_p b_p c_p$$
$$= \prod_{i=1}^{p} (c_i) \prod_{i=1}^{p} (a_i a_i b_i) \prod_{i=1}^{p} (a_i b_i c_i).$$

Clearly, $|T| = n = 7p$ holds. The LZ78 factorization of T is as shown below:

$$\mathsf{LZ78}(T) = |c_1|c_2| \cdots |c_p|a_1|a_1 b_1|a_2|a_2 b_2| \cdots |a_p|a_p b_p|a_1 b_1 c_1|a_2 b_2 c_2| \cdots |a_p b_p c_p|,$$

where each $|$ denotes a phrase boundary. To see why, observe that each c_i is a new character in the prefix interval $[1, p]$ in T, and thus c_i $(1 \leq i \leq p)$ is a phrase. Next, in the interval $[p+1, 4p]$, $a_i a_i b_i$ for each $1 \leq i \leq p$ is divided into $|a_i|a_i b_i|$. This is because the left a_i is a new character and it becomes the phrase of length 1. Also, the substring $a_i b_i$ has no previous occurrence as a phrase. Lastly, in the interval $[4p+1, 7p]$, the substring $a_i b_i c_i$ for each $1 \leq i \leq p$ becomes a phrase. This is because $a_i b_i$ has a previous occurrence as a phrase, but $a_i b_i c_i$ does not. Hence, we have $z_{78}(T) = 4p$.

As for substitutions, consider the following string T'

$$T' = c_1 c_2 \cdots c_p a_1 a_1 b_1 a_2 a_2 b_2 \cdots a_p a_p b_p \# b_1 c_1 a_2 b_2 c_2 \cdots a_p b_p c_p$$
$$= \prod_{i=1}^{p} (c_i) \prod_{i=1}^{p} (a_i a_i b_i) \# b_1 c_1 \prod_{i=2}^{p} (a_i b_i c_i)$$

which can obtained by substituting $T[4m+1] = a_1$ with a character $\#$ which does not occur in T. Let us analyze the structure of LZ78 factorization of T'. First, the factorization of $T'[1..4p]$ is the same as $T[1..4p]$ before the edit. Then, since $\# = T'[4p+1]$ is a new character and $b_1 = T'[4p+2]$ has no previous occurrence as a phrase, $\# b_1$ is factorized into $|\#|b_1|$. Next, the substring $c_1 a_2$ has no previous occurrence (as a phrase), but c_1 has. Therefore, $c_1 a_2$ becomes a phrase. Then, $c_{i-1} a_i b_i$ for each $2 \leq i \leq p$ is divided into $|c_{i-1} a_i|b_i|$, since c_{i-1} has a previous occurrence as a phrase, but $c_{i-1} a_i$ does not. Thus, the LZ78 factorization of T' is

$$\mathsf{LZ78}(T') = |c_1| \cdots |c_p|a_1|a_1 b_1| \cdots |a_p|a_p b_p|\#|b_1|c_1 a_2|b_2|c_2 a_3| \cdots |c_{p-1} a_p|b_p|c_p|$$

which contains $z_{78}(T') = 5p+1$ phrases. Hence, we get $\mathsf{AS}_{\mathrm{sub}}(z_{78}, n) \geq z_{78}(T') - z_{78}(T) = 5p+1 - 4p = p+1 \in \Omega(n)$ because $n = 7p$.

In the case of insertions and deletions, with the strings obtained from T by inserting $\#$ between $a_1 = T[4p+1]$ and $b_1 = T[4p+2]$ or by deleting $a_1 = T[4p+1]$ respectively, we get similar decompositions as in the case of substitutions, which give us $\mathsf{AS}_{\mathrm{ins}}(z_{78}, n), \mathsf{AS}_{\mathrm{del}}(z_{78}, n) \in \Omega(n)$. $\square$

7   Conclusions and Future Work

We studied the worst-case additive sensitivities of several repetitiveness measures and LZ-style compressors under single-character edits (substitutions, insertions, deletions). We presented $O(\sqrt{n})$ upper bounds on additive sensitivities for the smallest string attractor γ and the smallest bidirectional macro scheme b, which match the known lower bounds [1]. We also showed $\Theta(n^{\frac{2}{3}})$ tight upper and lower bounds on additive sensitivities for LZ-style compressors including the overlapping/non-overlapping LZSS and the optimal LZ-End.

An obvious future work is to close the gap between the upper and lower bounds where they are not yet tight. For example, regarding the case of the greedy LZ-End, for which we provided $\Omega(n^{\frac{2}{3}})$ lower bound in this paper, tight upper and lower bounds have not yet been established for either the additive or multiplicative sensitivity. Moreover, it is relatively easy to prove that the lower bound of the multiplicative sensitivity of the greedy LZ-End provides an asymptotic lower bound for the worst-case approximation ratio of LZ-End (i.e., the maximum of z_e/z_{end}).

Another possible open problem is to reduce the alphabet size of strings that give lower bounds (e.g., Theorem 5 uses $\Theta(n^{\frac{2}{3}})$).

Acknowledgements. This work was supported by JSPS KAKENHI Grant Numbers JP25K00136 (YN), JP23K24808, and JP23K18466 (SI).

References

1. Akagi, T., Funakoshi, M., Inenaga, S.: Sensitivity of string compressors and repetitiveness measures. Inf. Comput. **291**, 104999 (2023). https://doi.org/10.1016/J.IC.2022.104999
2. Blocki, J., Lee, S., Garcia, B.S.Y.: Differentially private compression and the sensitivity of LZ77. CoRR abs/2502.09584 (2025). https://doi.org/10.48550/ARXIV.2502.09584
3. Burrows, M., Wheeler, D.: A block-sorting lossless data compression algorithm. Tech. rep, DIGITAL SRC RESEARCH REPORT (1994)
4. Charikar, M., et al.: The smallest grammar problem. IEEE Trans. Inf. Theory **51**(7), 2554–2576 (2005). https://doi.org/10.1109/TIT.2005.850116
5. Giuliani, S., Inenaga, S., Lipták, Z., Romana, G., Sciortino, M., Urbina, C.: Bit catastrophes for the Burrows-Wheeler transform. Theory Comput. Syst. **69**(2), 19 (2025). https://doi.org/10.1007/S00224-024-10212-9

6. Kempa, D., Prezza, N.: At the roots of dictionary compression: string attractors. In: Diakonikolas, I., Kempe, D., Henzinger, M. (eds.) The 50th Annual ACM SIGACT Symposium on Theory of Computing (STOC 2018), pp. 827–840. ACM (2018). https://doi.org/10.1145/3188745.3188814

7. Kempa, D., Saha, B.: An upper bound and linear-space queries on the LZ-end parsing. In: Naor, J.S., Buchbinder, N. (eds.) The 2022 ACM-SIAM Symposium on Discrete Algorithms (SODA 2022), pp. 2847–2866. SIAM (2022). https://doi.org/10.1137/1.9781611977073.111

8. Kociumaka, T., Navarro, G., Prezza, N.: Towards a definitive measure of repetitiveness. In: Kohayakawa, Y., Miyazawa, F.K. (eds.) The 14th Latin American Symposium (LATIN 2020). LNCS, vol. 12118, pp. 207–219. Springer (2020). https://doi.org/10.1007/978-3-030-61792-9_17

9. Kreft, S., Navarro, G.: On compressing and indexing repetitive sequences. Theor. Comput. Sci. **483**, 115–133 (2013). https://doi.org/10.1016/J.TCS.2012.02.006

10. Navarro, G.: Indexing highly repetitive string collections, part I: Repetitiveness measures. ACM Comput. Surv. **54**(2), 29:1–29:31 (2021). https://doi.org/10.1145/3434399

11. Rytter, W.: Application of Lempel-Ziv factorization to the approximation of grammar-based compression. Theor. Comput. Sci. **302**(1–3), 211–222 (2003). https://doi.org/10.1016/S0304-3975(02)00777-6

12. Storer, J.A., Szymanski, T.G.: Data compression via textual substitution. J. ACM **29**(4), 928–951 (1982). https://doi.org/10.1145/322344.322346

13. Ziv, J., Lempel, A.: A universal algorithm for sequential data compression. IEEE Trans. Inf. Theory **23**(3), 337–343 (1977). https://doi.org/10.1109/TIT.1977.1055714

14. Ziv, J., Lempel, A.: Compression of individual sequences via variable-rate coding. IEEE Trans. Inf. Theory **24**(5), 530–536 (1978). https://doi.org/10.1109/TIT.1978.1055934

On the Number of MUSs Crossing a Position

Hiroto Fujimaru[1]([✉]), Takuya Mieno[2][ID], and Shunsuke Inenaga[3][ID]

[1] Department of Information Science and Technology, Kyushu University,
Fukuoka, Japan
fujimaru.hiroto.134@s.kyushu-u.ac.jp
[2] Department of Computer and Network Engineering,
University of Electro-Communications, Chofu, Japan
tmieno@uec.ac.jp
[3] Department of Informatics, Kyushu University, Fukuoka, Japan
inenaga.shunsuke.380@m.kyushu-u.ac.jp

Abstract. A string w is said to be a *minimal unique substring* (*MUS*) of a string T if w occurs exactly once in T, and any proper substring of w occurs at least twice in T. It is known that the number of MUSs in a string T of length n is at most n, and that the set $\mathsf{MUS}(T)$ of all MUSs in T can be computed in $O(n)$ time [Ilie and Smyth, 2011]. Let $\mathsf{MUS}(T, i)$ denote the set of MUSs that contain a position i in a string T. In this short paper, we present matching $\Theta(\sqrt{n})$ upper and lower bounds for the number $|\mathsf{MUS}(T, i)|$ of MUSs containing a position i in a string T of length n.

1 Introduction

A string w is said to be a *minimal unique substring* (*MUS*) of a string T if w occurs exactly once in T, and any proper substring of w occurs at least twice in T. Let $\mathsf{MUS}(T)$ denote the set of MUSs in T. It is known [7] that the number $|\mathsf{MUS}(T)|$ of MUSs in any string T of length n is at most n, and that this bound is tight (e.g. consider a string consisting of n distinct characters). This number $|\mathsf{MUS}(T)|$ can also be bounded by other parameters that relate to compression: For any string T whose *run-length encoding* (*RLE*) size is m, $|\mathsf{MUS}(T)| \leq 2m - 1$ holds [9]. Also, if the number of edges of the *compact directed acyclic word graph* (*CDAWG*) [4] of a string T is e, then $|\mathsf{MUS}(T)| \leq e$ holds [8]. Computing $\mathsf{MUS}(T)$ is important preprocessing for the *shortest unique substring* (*SUS*) queries [1], which are motivated by bioinformatics applications [6,12].

Since each element $w \in \mathsf{MUS}(T)$ can be represented by a unique pair (i, j) such that $w = T[i..j]$, the set $\mathsf{MUS}(T)$ can be represented by $O(n \log n)$ bits of space, or in $O(\min\{m, e\} \log n)$ bits of space. Ilie and Smyth [7] showed how to compute $\mathsf{MUS}(T)$ in $O(n)$ time with $O(n \log n)$ bits of working space. There are also other space-efficient algorithms for computing $\mathsf{MUS}(T)$ [3,8–11].

In this paper, we are interested in the following: "How many MUSs in $\mathsf{MUS}(T)$ can contain the same position in T?" Namely, we evaluate the worst-case bounds

G. Badkobeh et al. (Eds.): SPIRE 2025, LNCS 16073, pp. 124–132, 2026.
https://doi.org/10.1007/978-3-032-05228-5_11

for the number $|\mathsf{MUS}(T, i)| = |\{T[k..j] \in \mathsf{MUS}(T) \mid k \leq i \leq j\}|$ of MUSs that contain position i in T. We prove that $|\mathsf{MUS}(T, i)| = O(\sqrt{n})$ holds for any string T of n and any position i ($1 \leq i \leq n$), and present a family of strings for which $|\mathsf{MUS}(T, i)| = \Omega(\sqrt{n})$ holds. Hence our bounds for $|\mathsf{MUS}(T, i)|$ are tight.

While the new combinatorial properties of MUSs presented in this paper are interesting in their own right, this study is also well motivated for applications including design of a dynamic data structure for storing the set $\mathsf{MUS}(T)$ of MUSs. Namely, our results can be seen as the first step toward a representation of MUSs allowing for sublinear-time updates. Another example of our motivation is to analyze the *sensitivity* [2] of MUSs, which is the worst-case increase in the number $|\mathsf{MUS}(T)|$ of MUSs after performing a single-character edit operation in T. Our $\Omega(\sqrt{n})$ lower bound instance for $|\mathsf{MUS}(T, i)|$ immediately leads to an $\Omega(\sqrt{n})$ lower bound for the sensitivity of $|\mathsf{MUS}(T)|$, and our $O(\sqrt{n})$ upper bound for $|\mathsf{MUS}(T, i)|$ suggests that the sensitivity of $|\mathsf{MUS}(T)|$ is also $O(\sqrt{n})$.

2 Preliminaries

Let Σ be an alphabet. An element of Σ is called a character. An element of Σ^* is called a string. The length of a string T is denoted by $|T|$. The empty string ε is the string of length 0. For a string $T = xyz$, x, y, and z are called a *prefix*, *substring*, and *suffix* of T, respectively. For a string T of length n, $T[i]$ denotes the ith character of T for $1 \leq i \leq n$, and $T[i..j]$ denotes the substring of T that begins at position i and ends at position j for $1 \leq i \leq j \leq n$. For convenience, let $T[i..j] = \varepsilon$ for $i > j$. We say that string w *occurs* in a string T iff w is a substring of T. An interval $[i..j]$ is said to be an *occurrence* of w in T if $w = T[i..j]$. We may identify an occurrence $[i..j]$ of w with the string $w = T[i..j]$ if there is no confusion. Let $\mathrm{occ}_T(w)$ denote the number of occurrences of w in T. For convenience, let $\mathrm{occ}_T(\varepsilon) = |T| + 1$. For substring $T[i..j]$ of T and position $1 \leq p \leq |T|$, we say that $T[i..j]$ contains p if $i \leq p \leq j$. An integer $p \geq 1$ is said to be a *period* of a string T if $T[i] = T[i + p]$ holds for all $1 \leq i \leq n - p$. We use the following fact, which can be shown with the *periodicity lemma* [5]:

Fact 1. *If a string S occurs at three positions i, j, and k in T with $i < j < k \leq i + |S| - 1$, then $\gcd(j - i, k - j)$ is a period of $T[i..k + |S| - 1]$, and hence of S. Thus, $T[j - 1] = T[k - 1]$ holds.*

A substring w of string T is called a *unique* substring in T if $\mathrm{occ}_T(w) = 1$, and it is called a *repeat* in T if $\mathrm{occ}_T(w) \geq 2$. A unique substring w in T is called a *minimal unique substring (MUS)* in T if $\mathrm{occ}_T(w[1..|w| - 1]) \geq 2$ and $\mathrm{occ}_T(w[2..|w|]) \geq 2$. Let $\mathsf{MUS}(T)$ denote the set of MUSs in string T. Let $\mathsf{MUS}(T, i) = \{T[k..j] \in \mathsf{MUS}(T) \mid k \leq i \leq j\}$ denote the set of MUSs that contain position i in T. By definition, MUSs do not nest, namely, for two distinct MUSs $T[i..j], T[i'..j'] \in \mathsf{MUS}(T)$, $[i..j]$ is not a subinterval of $[i'..j']$, and vice versa.

3   Bounds for Maximum of $|\mathsf{MUS}(T, i)|$

In this section, we show the following:

Theorem 1. *For any string T of length n and any position i in T, $|\mathsf{MUS}(T, i)| \in O(\sqrt{n})$ holds. Also, this upper bound is tight.*

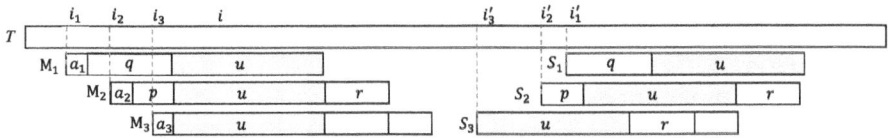

Fig. 1. Illustration for Lemma 1. This depicts one of several possible mutual locations of S_1, S_2, S_3.

3.1   Upper Bound

In this subsection, we fix a string T of length n arbitrarily. We first prove our key lemma to our $O(\sqrt{n})$ upper bound for $|\mathsf{MUS}(T, i)|$. Suppose that $\mathrm{M}_1 = [i_1..i_1 + |\mathrm{M}_1| - 1], \mathrm{M}_2 = [i_2..i_2 + |\mathrm{M}_2| - 1], \mathrm{M}_3 = [i_3..i_3 + |\mathrm{M}_3| - 1]$ are MUSs that contain the position i and occur in this order. Let $a_k = T[i_k]$ be the first character of M_k and $s_k = T[i_k+1..i_k+|\mathrm{M}_k|-1]$ be the rest for each $k \in \{1, 2, 3\}$. Since M_3 is the rightmost MUS among three MUSs $\mathrm{M}_1, \mathrm{M}_2, \mathrm{M}_3$, the overlap of them can be written as $a_3 u$ for some string $u \in \Sigma^*$. Let $q = T[i_1 + 1..i_3]$ and $p = T[i_2 + 1..i_3]$ be non-empty prefixes of s_1 and s_2 that immediately precede the occurrence $[i_3 + 1..i_3 + |u|]$ of u respectively (see also Fig. 1). Further let $r = T[i_2 + |pu| + 1..i_2 + |\mathrm{M}_2| - 1]$ be the suffix of s_2 that immediately follows the occurrence $[i_2 + |p| + 1..i_2 + |pu|]$ of u. By the definition of MUSs, s_k has an occurrence other than $[i_k + 1..i_k + |s_k|]$ for each $k \in \{1, 2, 3\}$. For s_1, s_2, and s_3, we have following lemma:

Lemma 1 (key lemma). *Let $S_1 = [i'_1..i'_1 + |s_1| - 1], S_2 = [i'_2..i'_2 + |s_2| - 1], S_3 = [i'_3..i'_3 + |s_3| - 1]$ be any occurrences of s_1, s_2, s_3 in T such that $i'_1 \neq i_1 + 1$, $i'_2 \neq i_2 + 1$, and $i'_3 \neq i_3 + 1$, respectively. Then, for the three occurrences $U_1 = [i'_1 + |q|..i'_1 + |q| + |u| - 1], U_2 = [i'_2 + |p|..i'_2 + |p| + |u| - 1], U_3 = [i'_3..i'_3 + |u| - 1]$ of u in S_1, S_2, S_3, at least two of them do not overlap (see Fig. 1).*

Proof. For the sake of contradiction, we assume that the three occurrences of u overlap. Let $a'_2 = T[i'_2 - 1]$ and $a'_3 = T[i'_3 - 1]$ be the characters that precede S_2 and S_3 respectively. By the definition of MUSs, $a_2 \neq a'_2$ and $a_3 \neq a'_3$ hold. While the character immediately before the occurrence U_3 of u is a'_3, the characters immediately before the occurrences U_1 and U_2 of u must be $T[i_3] = a_3$. For the overlapped three occurrences of u, due to Fact 1, the characters immediately before two occurrences except the leftmost one are the same. This implies that if U_3 is not the leftmost among U_1, U_2, U_3, then $a_3 = a_3'$, which is contradiction. Thus, U_3 is the leftmost occurrence among the three occurrences of u.

Let $S_1' = T[i_1' + |q| - |p|..i_1' + |s_1| - 1]$ be the occurrence of pu as a suffix of S_1, and $i_1'' = i_1' + |q| - |p|$ be the starting position of S_1'. Let $P_1 = [i_1''..i_1'' + |p| - 1]$ and $P_2 = [i_2''..i_2' + |p| - 1]$ be the occurrences of p as prefixes of S_1' and S_2 respectively. It holds that $i_1'' \neq i_2'$ since $T[i_1'' - 1] = a_2 \neq a_2' = T[i_2' - 1]$. In the following, we consider the four cases depending on i_1'', i_2', i_3', which are the starting positions of S_1', S_2, S_3 respectively; Case 1. $i_2' < i_3'$ or $i_1'' < i_3'$; Case 2. $i_3' < i_2'$ and $i_3' < i_1''$; Case 3. $i_3' = i_1'' < i_2'$; Case 4. $i_3' = i_2' < i_1''$.

Case 1: $i_2' < i_3'$ or $i_1'' < i_3'$. We further consider the sub-case where $i_2' < i_1''$. The other sub-case can be shown similarly. By assumptions, $i_2' < i_3'$ holds, and thus, $a_3' u$ that occurs at position $i_3' - 1$ is a substring of pu (see also Fig. 2). Since two occurrences $P_1 U_1, P_2 U_2$ of pu overlap, pu has period $i_1'' - i_2'$. Hence, its substring $a_3' u$ also has period $i_1'' - i_2'$. The periodicity implies that $a_3 = u[i_1'' - i_2'] = a_3'$, a contradiction.

Case 2: $i_3' < i_2'$ and $i_3' < i_1''$. We further consider the sub-case where $i_3' < i_2' < i_1''$. The other sub-case can be shown similarly. As in Case 1, pu has period $i_1'' - i_2'$. Also, its substring u has period $i_1'' - i_2'$. Since $i_3' < i_2' < i_1''$ holds and three occurrences U_1, U_2, U_3 overlap, $i_3' \leq i_2' - 1 < i_1'' - 1 < i_3' + |u| - 1$. This implies that both positions $i_2' - 1$ and $i_1'' - 1$ are within occurrence U_3 of u (see also Fig. 3). Since u has period $i_1'' - i_2'$, it holds that $a_2' = T[i_2' - 1] = T[i_2' - 1 + (i_1'' - i_2')] = T[i_1'' - 1] = a_2$, a contradiction.

Case 3: $i_1'' = i_3' < i_2'$. Let us consider periods of u in detail. Since three overlapping occurrences U_3, U_1, and U_2 of u appear in this order, u has period $\gcd((i_1'' + |p|) - i_3', i_2' + |p| - (i_1'' + |p|)) = \gcd(|p|, i_2' - i_1'')$ due to Fact 1. Let $\alpha = \gcd(|p|, i_2' - i_1'')$. Next, we consider periods of r. Let $R_2 = [i_2' + |pu|..i_2' + |s_2| - 1]$ and $R_3 = [i_3' + |u|..i_3' + |ur| - 1]$ be occurrences of r that immediately follow the occurrences U_2 and U_3 of u respectively. We consider two sub-cases 3-1 and 3-2 depending on the length of r: **Case 3-1.** If $|r| > i_2' - i_1'' + |p|$, then U_2 and R_3 overlap, and the length of their intersection is $i_2' - i_1'' + |p|$ (see also Fig. 4). Also, the length of the overlap of occurrences R_3 and R_2 of r is $|r| - (i_2' - i_1'' + |p|)$, and thus r has period $i_2' - i_1'' + |p|$. By the definition of $\alpha = \gcd(|p|, i_2' - i_1'')$, $i_2' - i_1'' + |p|$ is a multiple of α. Thus, the overlap of U_2 and R_3 forms an integer power of a string of length α, and r also has period α. To summarize, string ur also has period α, since both u and r have period α and their overlap is a multiple of α. **Case 3-2.** If $|r| \leq i_2' - i_1'' + |p|$, then r is a prefix of the length-$(i_2' - i_1'' + |p|)$ suffix of u, which is an integer power of some string of length α. Therefore, similar to the above, ur has period α.

In both sub-cases, ur has period α. Next, let us consider $x = T[i_1''..i_2' + |s_2| - 1]$. Since ur is a border of x and $|x| - |ur| = i_2' - i_1'' + |p|$ is a multiple of α, x has period α. Then, $s_2 = T[i_2'..i_2' + |s_2| - 1] = T[i_1''..i_1'' + |s_2| - 1]$ holds. Since $T[i_1'' - 1] = a_2$, $T[i_1'' - 1..i_1'' + |s_2| - 1] = a_2 s_2$,which is identical to the second MUS $T[i_2..i_2 + |M_2| - 1]$. Since a MUS must be unique in T, $i_1'' - 1 = i_2$ holds.

Then, considering the starting position i_1' of S_1, we have $i_1' = i_1'' + |p| - |q| = i_2 + 1 + |p| - |q| = i_1 + 1$, which contradicts the definition of i_1'.

Case 4: $i_2' = i_3' < i_1''$. Since two occurrences U_3 and U_2 of u overlap, u has period $|p|$ (see also Fig. 5). Also, $T[i_1'' + |p| - 1] = a_3$ holds since the character that precedes the suffix u of s_1 is a_3. Since U_3 and U_1 overlap and $i_3' < i_1''$, $T[i_1'' - 1..i_1'' + |p| - 1] = a_2 p$ is a substring of u, which has period $|p|$. Thus, $a_2 = T[i_1'' - 1] = T[i_1'' + |p| - 1] = a_3$ holds. Here, one of s_2 and s_3 is a prefix of the other since $i_2' = i_3'$ holds. Therefore, one of $a_2 s_2$ and $a_3 s_3 = a_2 s_3$ is a prefix of the other, which contradicts that MUSs $a_2 s_2$ and $a_3 s_3$ are unique in T. □

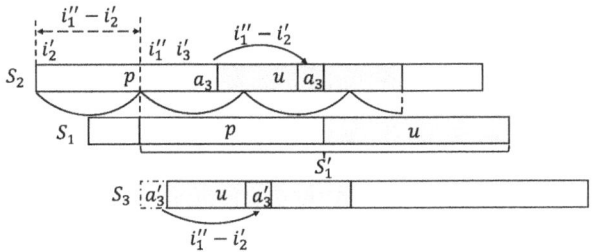

Fig. 2. Illustration for a contradiction in Case 1 of the proof of Lemma 1.

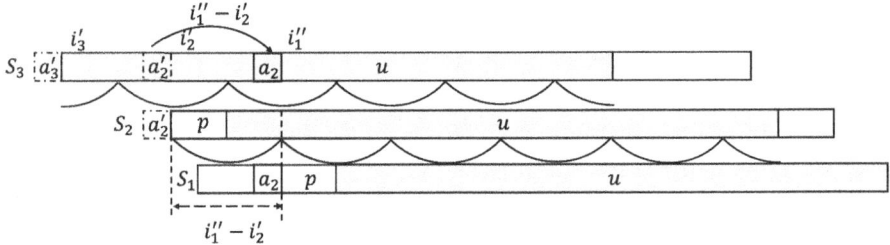

Fig. 3. Illustration for a contradiction in Case 2 of the proof of Lemma 1.

Based on the key lemma, we estimate an upper bound for $|\mathrm{MUS}(T, i)|$. Suppose that $h + 1$ MUSs contain the same position i in T. Let $M_1, M_2, \ldots, M_{h+1}$ be such MUSs from left to right. We focus only on the first h MUSs. For each $1 \leq k \leq h$, let $s_k = M_k[2..|M_k|]$ be the suffix of M_k of length $|M_k| - 1$. We *mark* the character at position i on s_k (see Fig. 6). Let S_k be an occurrence of s_k that is not the suffix of M_k for each $1 \leq k \leq h$. We denote by i_k the marked position on the occurrence S_k of s_k. Further let $I = \{i_k \mid 1 \leq k \leq h\}$ be the set of marked positions excluding i. We define a function $f : \{1, \ldots, h\} \to \{1, \ldots, h\}$ such that $f(x) = k$ iff i_k is the xth smallest value in I. In other words, $f(x) = k$ iff $x = |\{i_\ell \in I \mid i_\ell \leq i_k\}|$. For each $1 \leq x \leq h - 2$, we have the following:

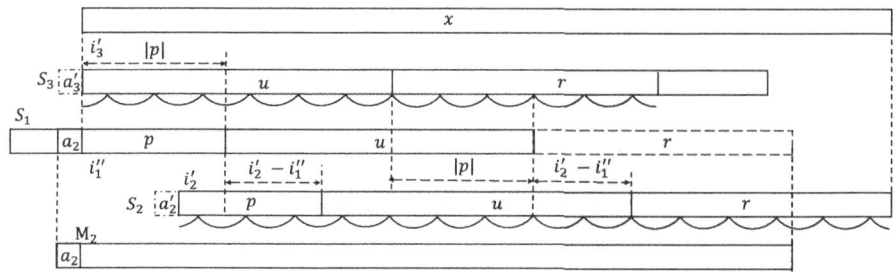

Fig. 4. Illustration for Case 3 of the proof of Lemma 1.

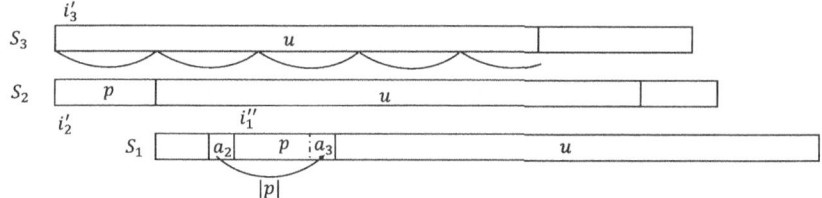

Fig. 5. Illustration for a contradiction in Case 4 of the proof of Lemma 1.

Lemma 2. $i_{f(x+2)} - i_{f(x)} > h - \omega$ *holds where* $\omega = \max\{f(x), f(x+1), f(x+2)\}$.

Proof. Let au be the overlap of $M_{f(x)}, M_{f(x+1)}$ and $M_{f(x+2)}$ where $a \in \Sigma$. By Lemma 1, at least two of the three occurrences of u in $S_{f(x)}, S_{f(x+1)}$, and $S_{f(x+2)}$ do not overlap. Thus, marked positions in $S_{f(x)}$ and $S_{f(x+2)}$ are separated by a distance of at least $|u|$ (see also Fig. 7). Namely, $i_{f(x+2)} - i_{f(x)} \geq |u|$ holds Next, we analyze the length of u. Let p be the starting position of M_ω. Since MUSs do not nest and there are $h + 1 - \omega$ MUSs that contain i whose starting positions are greater than p, $i - p \geq h - \omega + 1$ holds. Therefore, $|u| \geq i - p > h - \omega$. To summarize, we have $i_{f(x+2)} - i_{f(x)} \geq |u| > h - \omega$. □

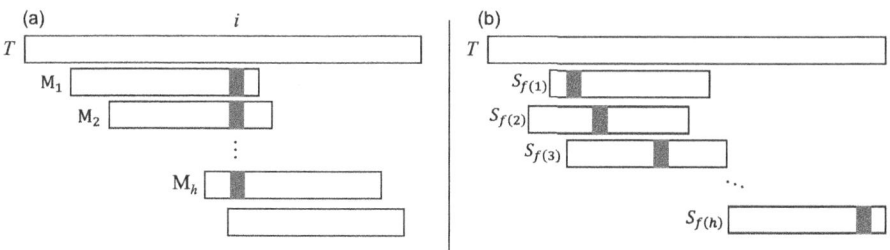

Fig. 6. (a) These are $h+1$ MUSs that contain position i. Markers are indicated in red. (b) Strings $S_{f(1)}, S_{f(2)}, \ldots, S_{f(h)}$ are sorted by their marked positions. (Color figure online)

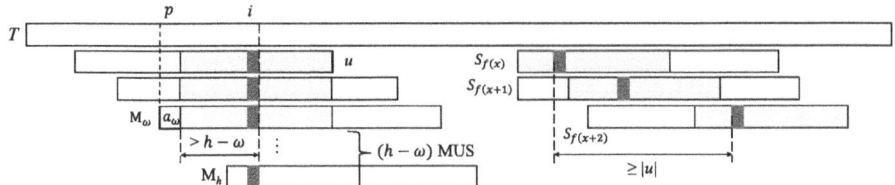

Fig. 7. Illustration for the proof of Lemma 2.

Let $\omega_x = \max\{f(x), f(x+1), f(x+2)\}$ for each $1 \le x \le h - 2$. Applying Lemma 2 for all $1 \le x \le h - 2$, we obtain $\sum_{1 \le x \le h-2}(h - \omega_x) < \sum_{1 \le x \le h-2}(i_{f(x+2)} - i_{f(x)}) = i_{f(h)} + i_{f(h-1)} - i_{f(2)} - i_{f(1)} < 2n$. To estimate a lower bound of $\sum_{1 \le x \le h-2}(h - \omega_x) = h(h-2) - \sum_{1 \le x \le h-2}\omega_x$, we consider an upper bound of $\sum_{1 \le x \le h-2}\omega_x$. Since ω_x is defined as the minimum among three function values, each value q with $1 \le q \le h$ can be selected as ω_x at most three times. Thus, $\sum_{1 \le x \le h-2}\omega_x \le 3\sum_{h-\lceil h/3 \rceil + 1 \le x \le h} x < 5h^2/6 + 3h + 3/2$ holds. Combining the inequalities, we have $2n > \sum_{1 \le x \le h-2}(h - \omega_x) = h(h-2) - \sum_{1 \le x \le h-2}\omega_x > h(h-2) - (5h^2/6 + 3h + 3/2) = h^2/6 + h - 3/2$, which implies $h + 1 \in O(\sqrt{n})$. This concludes the proof of $|\mathsf{MUS}(T, i)| \in O(\sqrt{n})$.

3.2 Lower Bound

Lemma 3. *There exists an infinite family of strings T_m that satisfies* $|\mathsf{MUS}(T_m, p)| \in \Omega(\sqrt{|T_m|})$ *for a position p in T_m.*

Proof. Consider the string

$$T_m = ab^{2\,m}ab^{2\,m+2}S_{m,1}S_{m,2}\cdots S_{m,k}\cdots S_{m,m-1},$$

where $S_{m,k} = ab^k ab^{2m-k}$ for each $1 \le k \le m-1$. Let $p = 2m+4$. See also Fig. 8 for a concrete example.

For each $2 \le i \le m - 1$, consider the substring

$$T_m[p - (2+i)..p + (2\,m - 1 - i)] = b^i ab^{2\,m-i+1}$$

of T_m. It is clear that each $T_m[p - (2+i)..p + (2m-1-i)]$ contains position p.

Now we show that each $b^i ab^{2m-i+1}$ is a MUS of T_m. First, $b^i ab^{2m-i+1}$ is unique in T_m. Second, $b^{i-1}ab^{2m-i+1}$ is a repeat in T_m since $S_{m,i-1}[2..2m+2] = b^{i-1}ab^{2m-i+1}$. Third, $b^i ab^{2m-i}$ is a repeat in T_m since $S_{m,i}[2..2m+2] = b^i ab^{2m-i}$. Thus, $\{b^i ab^{2m-i+1} \mid 2 \le i \le m-1\} \subset \mathsf{MUS}(T_m, p)$. Since $|T_m| = 2m^2 + 4m + 2 \in \Theta(m^2)$, $|\mathsf{MUS}(T_m, p)| = \Omega(\sqrt{|T_m|})$ holds. $\qed$

To summarize, we have shown Theorem 1.

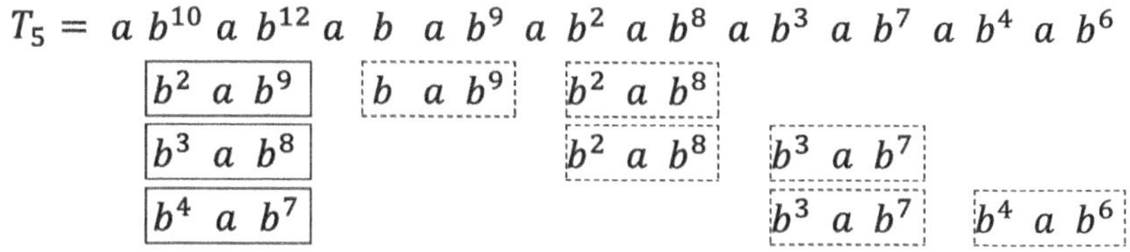

$$T_5 = a\ b^{10}\ a\ b^{12}\ a\ b\ a\ b^9\ a\ b^2\ a\ b^8\ a\ b^3\ a\ b^7\ a\ b^4\ a\ b^6$$

$b^2\ a\ b^9$	$b\ a\ b^9$	$b^2\ a\ b^8$		
$b^3\ a\ b^8$		$b^2\ a\ b^8$	$b^3\ a\ b^7$	
$b^4\ a\ b^7$			$b^3\ a\ b^7$	$b^4\ a\ b^6$

Fig. 8. An example of T_m for $m = 5$ in the proof of Lemma 3. There are $m - 2 = 3$ MUSs b^2ab^9, b^3ab^8, and b^4ab^7 that contain the position $2m + 4 = 14$.

4   Conclusions and Further Work

In this paper, we presented a tight $\Theta(\sqrt{n})$ bound on the number $|\mathsf{MUS}(T, i)|$ of the MUSs that contains a position i in any string T of length n.

One of our future work includes analysis of the sensitivity of MUSs, which is the worst-case increase in the number $|\mathsf{MUS}(T)|$ of MUSs after performing a single-character edit operation on T. Our $\Omega(\sqrt{n})$ lower bound instance for $|\mathsf{MUS}(T, i)|$ in Sect. 3.2 already gives us an $\Omega(\sqrt{n})$ lower bound for the sensitivity of MUSs in both additive and multiplicative senses. We conjecture that the worst-case additive/multiplicative sensitivity of $|\mathsf{MUS}(T)|$ is also $O(\sqrt{n})$. Our $O(\sqrt{n})$ upper bound for $|\mathsf{MUS}(T, i)|$ presented in this paper partly contributes to analyzing an upper bound of the sensitivity of $|\mathsf{MUS}(T)|$, since the number of new MUSs containing the edit position i is equal to $|\mathsf{MUS}(T, i)|$. What remains is to analyze the number $|\mathsf{MUS}(T) \setminus (\mathsf{MUS}(T, i) \cup \mathsf{MUS}(T'))|$ of new MUSs which do *not* contain the edited position i, where T' is the string before edit.

Acknowledgements. This work was supported by JST BOOST Grant Number JPMJBS2406 (HF), and by JSPS KAKENHI Grant Numbers JP24K20734 (TM), JP23K24808, and JP23K18466 (SI). We thank Hideo Bannai and Haruki Umezaki for discussions.

References

1. Abedin, P., Külekci, M.O., Thankachan, S.V.: A survey on shortest unique substring queries. Algorithms **13**(9), 224 (2020). https://doi.org/10.3390/A13090224
2. Akagi, T., Funakoshi, M., Inenaga, S.: Sensitivity of string compressors and repetitiveness measures. Inf. Comput. **291**, 104999 (2023). https://doi.org/10.1016/j.ic.2022.104999
3. Belazzougui, D., Cunial, F., Gagie, T., Prezza, N., Raffinot, M.: Composite repetition-aware data structures. In: Cicalese, F., Porat, E., Vaccaro, U. (eds.) CPM 2015. LNCS, vol. 9133, pp. 26–39. Springer, Cham (2015). https://doi.org/10.1007/978-3-319-19929-0_3
4. Blumer, A., Blumer, J., Haussler, D., McConnell, R.M., Ehrenfeucht, A.: Complete inverted files for efficient text retrieval and analysis. J. ACM **34**(3), 578–595 (1987). https://doi.org/10.1145/28869.28873

5. Fine, N.J., Wilf, H.S.: Uniqueness theorems for periodic functions. Proc. Am. Math. Soc. **16**(1), 109–114 (1965)
6. Haubold, B., Pierstorff, N., Möller, F., Wiehe, T.: Genome comparison without alignment using shortest unique substrings. BMC Bioinf. **6**, 123 (2005). https://doi.org/10.1186/1471-2105-6-123
7. Ilie, L., Smyth, W.F.: Minimum unique substrings and maximum repeats. Fund. Informaticae **110**(1–4), 183–195 (2011). https://doi.org/10.3233/FI-2011-536
8. Inenaga, S., Mieno, T., Arimura, H., Funakoshi, M., Fujishige, Y.: Computing minimal absent words and extended bispecial factors with CDAWG space. In: IWOCA 2024. Lecture Notes in Computer Science, vol. 14764, pp. 327–340. Springer, Heidelberg (2024). https://doi.org/10.1007/978-3-031-63021-7_25
9. Mieno, T., Inenaga, S., Bannai, H., Takeda, M.: Shortest unique substring queries on run-length encoded strings. In: 41st International Symposium on Mathematical Foundations of Computer Science, MFCS 2016, Kraków, Poland, 22–26 August 2016. LIPIcs, vol. 58, pp. 69:1–69:11. Schloss Dagstuhl - Leibniz-Zentrum für Informatik (2016). https://doi.org/10.4230/LIPICS.MFCS.2016.69
10. Mieno, T., Köppl, D., Nakashima, Y., Inenaga, S., Bannai, H., Takeda, M.: Space-efficient algorithms for computing minimal/shortest unique substrings. Theor. Comput. Sci. **845**, 230–242 (2020). https://doi.org/10.1016/J.TCS.2020.09.017
11. Nishimoto, T., Tabei, Y.: R-enum: Enumeration of characteristic substrings in bwt-runs bounded space. In: 32nd Annual Symposium on Combinatorial Pattern Matching, CPM 2021, Wrocław, Poland, 5–7 July 2021. LIPIcs, vol. 191, pp. 21:1–21:21. Schloss Dagstuhl - Leibniz-Zentrum für Informatik (2021). https://doi.org/10.4230/LIPICS.CPM.2021.21
12. Pei, J., Wu, W.C., Yeh, M.: On shortest unique substring queries. In: 29th IEEE International Conference on Data Engineering, ICDE 2013, Brisbane, Australia, 8–12 April 2013, pp. 937–948. IEEE Computer Society (2013). https://doi.org/10.1109/ICDE.2013.6544887

String Consensus Problems with Swaps and Substitutions

Estéban Gabory[1]([email]) [iD], Laurent Bulteau[2] [iD], Gabriele Fici[1] [iD],
and Hilde Verbeek[3] [iD]

[1] Dipartimento di Matematica e Informatica, Università di Palermo, Palermo, Italy
esteban.gabory@unipa.it
[2] LIGM, CNRS, Université Gustave Eiffel, F77454 Marne-la-vallée, France
[3] CWI, Amsterdam, The Netherlands

Abstract. String consensus problems aim at finding a string that minimizes some given distance with respect to an input set of strings. In particular, in the CLOSEST STRING problem, we are given a set of strings of equal length and a radius d. The goal is to find a new string that differs from each input string by at most d substitutions. We study a generalization of this problem where, in addition to substitutions, swaps of adjacent characters are also permitted, each operation incurring a unit cost. Amir et al. showed that this generalized problem is NP-hard, even when only swaps are allowed. In this paper, we show that it is FPT with respect to the parameter d. Moreover, we investigate a variant in which the goal is to minimize the *sum* of distances from the output string to all input strings. For this version, we present a polynomial-time algorithm.

Keywords: Closest String · Parameterized Algorithms · Swap Distances · String Consensus

1 Introduction

Consensus problems aim at finding, given a collection of objects and a distance function, an object that is as close as possible to each and every object in the collection. In the CLOSEST STRING problem, objects are strings and the distance constraint is given by an integer d: the solution must be at Hamming distance at most d (i.e., at most d substitutions away) from each input string. We call such a constraint a *radius constraint*, which is generally harder to satisfy than the corresponding *sum constraint*: minimize the sum of distances (indeed, CLOSEST STRING is NP-hard [12] and FPT for parameter d [14], whereas the sum of distances can easily be optimized by taking the most frequent character at each position).

Betzler et al. [5] give an algorithmic framework to show that the median problem (i.e., the sum-only variant) for various distances is FPT for the average distance between input objects.

CLOSEST STRING has many application domains. In bioinformatics, where strings such as genomes from the same species are often compared [8], the use of

G. Badkobeh et al. (Eds.): SPIRE 2025, LNCS 16073, pp. 133–147, 2026.
https://doi.org/10.1007/978-3-032-05228-5_12

the radius d as a parameter is justified by the similarity of the inputs. A possible extension of CLOSEST STRING consists of choosing, among all solution strings satisfying the radius constraint, the one minimizing the sum of distances to input strings [3]. Incidentally, most algorithms for variants of *radius-only* CLOSEST STRING can be adapted to this *radius and sum* problem (in particular, CLOSEST STRING is FPT for d in this setting as well [7]).

In this paper, we consider not only substitution operations, but also *swaps*, i.e., exchange of consecutive characters, and define the *Swap distance* (∂_S). Swaps are used as an extension to the Levenshtein distance (yielding the Damer-auLevenshtein distance), e.g., in spell checkers where they help improve predictions [15]. Importantly, their formalism enforces that swaps are pairwise disjoint, hence some pairs of strings may be incomparable even if they use the same multi-set of characters[1]. For example, strings abcd and badc are at distance 2, but abc and bca are incomparable, since swaps are pairwise disjoint. We further consider the *Swap+Hamming* distance (written ∂_{SH}), where permitted operations are both substitutions and swaps (on distinct characters).

The aim of this paper is to study the parameterized complexity of CLOSEST STRING for both Swap and Swap+Hamming measures, and for any of the radius-only, sum-only, or radius+sum objective functions. The main parameters we consider are the radius bound d and the alphabet size. For both distances, we show that the radius-only variant is FPT, and that the sum-only variant is in P. Additionally, for the Swap distance, we show that the radius+sum variant is FPT for d.

Problems. We consider the following three problems, where ∂ is any distance function.

$(\mathbf{r}, \partial)$-CONSENSUS
Input: A set of k strings $\mathcal{S} = \{s_1, .., s_k\}$ each of length n, integer d.
Question: Is there a length-n string s^* such that $\max_{s \in \mathcal{S}} \partial(s, s^*) \leq d$?

$(\mathbf{s}, \partial)$-CONSENSUS
Input: A set of k strings $\mathcal{S} = \{s_1, .., s_k\}$ each of length n, integer D.
Question: Is there a length-n string s^* such that $\sum_{s \in \mathcal{S}} \partial(s, s^*) \leq D$?

$(\mathbf{rs}, \partial)$-CONSENSUS
Input: A set of k strings $\mathcal{S} = \{s_1, .., s_k\}$ each of length n, integers d and D.
Question: Is there a length-n string s^* such that $\max_{s \in \mathcal{S}} \partial(s, s^*) \leq d$ and $\sum_{s \in \mathcal{S}} \partial(s, s^*) \leq D$?

---

[1] The variant where swaps may overlap corresponds to the Kendall-tau distance.

Note that previously-studied problems can be named within this formalism, e.g., CLOSEST STRING is $(\mathbf{r}, \partial_{\text{Ham}})$-CONSENSUS and $(\mathbf{r}, \text{Levenshtein})$-CONSENSUS is known as CENTER STRING.

Our Contributions. We summarize our main contributions in Table 1.

Additionally, we prove that neither $(\mathbf{r}, \partial_{\mathsf{S}}) - Consensus(d, n, |\Sigma|)$ $(\mathbf{rs}, \partial_{\mathsf{S}}) - Consensus(d, n, |\Sigma|)$ (Theorem 1), nor $(\mathbf{r}, \partial_{\mathsf{SH}}) - Consensus(d, n, |\Sigma|)$, nor $(\mathbf{rs}, \partial_{\mathsf{SH}}) - Consensus(d, n, |\Sigma|)$ (Theorem 3) admit a polynomial kernel unless $\mathsf{NP} \subseteq \mathsf{coNP/poly}$.

Table 1. Summary of our contributions.

	∂_{S}		Sect. 3	∂_{SH}	Sect. 4
Sum (s)	$\mathcal{O}(kn)$		Corollary 2	$\mathcal{O}(k^3 n^2)$	Theorem 4
Radius (r)	$\mathsf{FPT}(d)$		Theorem 1	$\mathsf{FPT}(d)$	Theorem 2
	Kernel of size $\mathcal{O}(k^2 \log d)$		Theorem 1		
Radius and	$\mathsf{FPT}(d)$		Theorem 1	Open	
Sum (rs)	Kernel of size $\mathcal{O}(k^2 \log d)$		Theorem 1		
	Kernel of size $\mathcal{O}(D^3 \log D)$		Theorem 1		

Related Work. Consensus problems under the Hamming distance have been extensively studied from both classical and parameterized complexity perspectives. It is known that $(\mathbf{r}, \partial_{\text{Ham}}) - Consensus$ is NP-complete [12]. However, when parameterized by the radius d, the problem becomes fixed-parameter tractable (FPT), and $(\mathbf{r}, \partial_{\text{Ham}}) - Consensus(d, k)$ admits a polynomial kernel of size $\mathcal{O}(k^2 d \log k)$ [14]. In the sum-of-distances variant $(\mathbf{s}, \partial_{\text{Ham}}) - Consensus$, the problem becomes much simpler: the optimal consensus string is the column-wise majority string, and can be computed in linear time [14]. The branching algorithm underlying the FPT result for $(\mathbf{r}, \partial_{\text{Ham}}) - Consensus$ can be extended to handle the combined radius-and-sum variant $(\mathbf{rs}, \partial_{\text{Ham}}) - Consensus$, which also lies in $\mathsf{FPT}(d)$ [7]. With respect to kernelization, while no polynomial kernel exists for $(\mathbf{r}, \partial_{\text{Ham}}) - Consensus$ and $(\mathbf{rs}, \partial_{\text{Ham}}) - Consensus$ when parameterized by (d, n) unless $\mathsf{NP} \subseteq \mathsf{coNP/poly}$, both $(\mathbf{r}, \partial_{\text{Ham}}) - Consensus(d, k)$ and $(\mathbf{rs}, \partial_{\text{Ham}}) - Consensus(d, k)$ have polynomial kernels of size $\mathcal{O}(k^2 d \log k)$, and $(\mathbf{rs}, \partial_{\text{Ham}}) - Consensus$ additionally admits a polynomial kernel of size $\mathcal{O}(D^3 \log D)$ [4,7]. The problem $(\mathbf{r}, \partial_{\mathsf{S}}) - Consensus$ was introduced in [2], which showed that $(\mathbf{r}, \partial_{\mathsf{S}})$-CONSENSUS is NP hard. Before that, pattern matching with swaps was introduced in [16], and solved in [1]. Note that other definitions of swap distance exist, e.g., in [9], where swap distance is considered between permutations and allows for swapping several times at a given position (namely, the Kendall-tau distance).

For space reasons, we reported some of the proofs to the full version of this paper [13].

2   Preliminaries

Parameterized Complexity. We use general notions of parameterized complexity and kernels [10,11]. In a nutshell (omitting some technical difficulties not irrelevant to our case), for a parameterized problem $P(k)$, an *FPT algorithm* is an exact algorithm solving P in $f(k)\text{poly}(n)$ time where n is the instance size and k is the parameter. A *kernel* is a polynomial-time algorithm reducing an instance of $P(k)$ to another instance with size $f(k)$. Such a kernel is *polynomial* if f is polynomial. It is known that there exists a kernel for $P(k)$ if and only if $P(k) \in$ FPT, but having a polynomial kernel is seemingly more restrictive. Finally, the *parameterized reductions* we use in this paper are polynomial-time reductions that also preserve the value of the parameter (up to some polynomial). In other words, a parameterized reduction from $P(k)$ to $Q(k')$ is a function f with: **(i)** (x, k) is a yes-instance for P if and only if $(x', k') = f(x, k)$ is a yes-instance for Q, and **(ii)** $k' \le \text{poly}(k)$. With such a parameterized reduction, any FPT (resp. polynomial kernel) for $Q(k')$ can be carried over as an FPT algorithm (resp. polynomial kernel) for $P(k)$ [6].

Strings. Let Σ^n be the set of *strings* $s = s[1] \cdots s[n]$ of *length* $|s| = n$ over an alphabet Σ, the elements of which are called *letters*. Given a letter $\mathsf{a} \in \Sigma$, we write $|s|_{\mathsf{a}} = |\{p \in [1 \mathinner{..} n] \mid s[p] = \mathsf{a}\}|$—. The *reversal* of a string s is the string $s^r = s[|s|] \cdots s[|1|]$. Given an arbitrary interval $I = [i \mathinner{..} j] \subseteq [1 \mathinner{..} n]$, we write $s[I] = s[i \mathinner{..} j]$ for the *substring* of s starting at position i and ending at position j, and we write $s[j \mathinner{..} i] = s[i \mathinner{..} j]^r$. A *prefix* of s is a substring of the form $s[1 \mathinner{..} p]$ and a *suffix* of s is a substring of the form $s[p \mathinner{..} |s|]$, for some p. By $s_1 \cdot s_2$ we denote the *concatenation* of two strings s_1 and s_2, i.e., $s_1 \cdot s_2 = s_1[1] \cdots s_1[|s_1|] \cdot s_2[1] \cdots s_2[|s_2|]$. We say that two strings s_1, s_2 are *matching* at a position p if $s_1[p] = s_2[p]$, and that they are *mismatching* otherwise. We say that s_2 is obtained from s_1 by a *substitution* at position p with letter a if $s_1[p] \ne s_2[p] = \mathsf{a}$ and $s_1[p'] = s_2[p']$ when $p' \ne p$. Given a set $\mathcal{S} = \{s_1, \mathinner{..}, s_k\}$ containing strings having the same length, we call *p-th column* the set $\mathcal{S}[p] = \{s[p] \mid s \in \mathcal{S}\}$. We say that $\mathcal{S}[p]$ is *dirty* if it contains at least two distinct elements; otherwise, we say that it is *clean*. We also write $\mathcal{S}[i \mathinner{..} j]$ for the *segment* of $\mathcal{S}$ between i and j, which is defined as the set $\{s[i \mathinner{..} j], s \in \mathcal{S}\}$. Given two binary strings h_1 and h_2, we write $h_1 \oplus h_2$ for the string obtained by taking the position-wise *XOR* of both input strings.

Swaps. Given an integer n, we denote by $\mathfrak{S}_n$ the set of all permutations of $[n] = [1 \mathinner{..} n]$. Given an integer p, the *swap at positions* $(p, p+1)$ (or simply at position p) is the operation transforming a string s into $s[1] \cdots s[p-1] \cdot s[p+1] \cdot s[p] \cdot s[p+2] \cdots s[|s|]$. Two swaps at positions p and q respectively are *disjoint* or *compatible* if $|p - q| \ge 2$ (in which case, the two operations commute). A *swap permutation* of size m is a set of m pairwise-compatible swaps (it can also be seen as a permutation of $\mathfrak{S}_n$ consisting of m support-disjoint adjacent transpositions). The *swap string* h_σ of a swap permutation σ is the length-$(n-1)$ binary string h such that $h[p] = 1$ if σ contains a swap at position $(p, p+1)$, and $h[p] = 0$ otherwise.

Two strings s_1 and s_2 are *matching* if there exists a swap permutation transforming s_1 into s_2. To ensure the uniqueness of the swap permutations between strings, we say that such a permutation is *valid* if it never swaps two identical letters: any swap permutation can be made valid by removing such superfluous swaps. In that case, we write $\mathfrak{S}_{s_1,s_2}$ for this permutation and h_{s_1,s_2} for the corresponding swap string. We state the following lemma:

Lemma 1. *Given two matching strings s_1 and s_2, $\mathfrak{S}_{s_1,s_2}$ and h_{s_1,s_2} are unique and can be computed in linear time.*

We say that matching strings s_1 and s_2 have a *swap* at position $(p, p+1)$ (or simply at position p) if $\mathfrak{S}_{s_1,s_2}$ contains the corresponding swap (i.e., $h_{s_1,s_2}[p] = 1$), and we write $(p, p+1) \in \mathfrak{S}_{s_1,s_2}$. Note that this is more precise than simply having $s_1[p] = s_2[p+1]$ and $s_2[p+1] = s_1[p]$, as can be seen in the following example.

Example 1. Let $s_1 =$ abab and $s_2 =$ baba. One has $h_{s_1,s_2} = 101$ (so their swap distance is 2), and s_1 and s_2 do not have a swap at position 2, even though $s_1[2,3] =$ ba and $s_2[2,3] =$ ab.

Remark 1. Given two strings s_1, s_2 of length n that are matching, one has $|s_1|_x = |s_2|_x$ for any letter $x \in \Sigma$. Furthermore, given a position p such that $h_{s_1,s_2}[p] = 0$, one has $|s_1[1 .. p]|_x = |s_2[1 .. p]|_x$, and $|s_1[p+1 .. n]|_x = |s_2[p+1 .. n]|_x$, for any letter $x \in \Sigma$. This can be verified by considering the swap permutation as two distinct ones, respectively on the prefixes and suffixes of s_1 and s_2.

The following technical Lemma is illustrated in Fig. 1.

Lemma 2. *Given three strings s_1, s_2, s_3 such that both s_1, s_2 and s_2, s_3 are matching, let $h = h_{s_1,s_2} \oplus h_{s_2,s_3}$. If h does not have two consecutive 1s, then s_1 and s_3 are matching and $h = h_{s_1,s_3}$. If h has two consecutive 1s at positions $p-1$ and p for some p (i.e., $h[p-1] = h[p] = 1$), then s_1 and s_3 do not match, and for any string t matching both s_1 and s_3 we have $t[p-1 .. p+1] = s_2[p-1 .. p+1]$.*

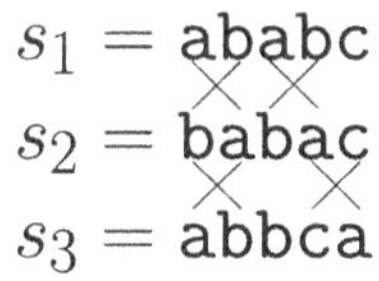

$$s_1 = \text{ababc}$$
$$s_2 = \text{babac}$$
$$s_3 = \text{abbca}$$

Fig. 1. Example illustrating Lemma 2. Black crosses represent swaps between pairs of adjacent letters. One has $h_{s_1,s_2} = 1010$, $h_{s_2,s_3} = 1001$ and $h_{s_1,s_2} \oplus h_{s_2,s_3} = 0011$. By Lemma 2, any string t that matches both s_1 and s_3 must have bac as a suffix—this is because it is the only suffix that can be swapped both into abc and bca. For example, $t =$ abbac matches both s_1 and s_3. In this case, one has $h_{t,s_1} = 0010$, $h_{t,s_3} = 0001$, and $h_{t,s_2} = h_{t,s_1} \oplus h_{s_1,s_2} = h_{t,s_3} \oplus h_{s_3,s_2} = 1000$.

Distances. Given two strings s_1 and s_2, the *Adjacent Swap Distance* $\partial_S(s_1, s_2)$ (or simply *Swap Distance* in this paper) is $+\infty$ if s_1 and s_2 are not matching, and the size of $\mathfrak{S}_{s_1, s_2}$ otherwise. The *Swap+Hamming distance* $\partial_{SH}(s_1, s_2)$ is the minimum over all strings t of $\partial_S(s_1, t) + \partial_{Ham}(t, s_2)$, where the Hamming distance ∂_{Ham} is the number of positions p, $1 \leq p \leq n$, such that $s'[p] \neq t[p]$. Note that ∂_{SH} is a lower bound on both the swap and the Hamming distances.

3   Algorithms for the Swap Distance only

1. Input strings

$s_1 = $ a b g a b c a h i d a b d e f e d a $\partial_S(s_1, s^*) \leq 4$

$s_2 = $ b a g c a a b i h d a b e f d d e a $\partial_S(s_2, s^*) \leq 4$

$s_3 = $ b a g c a b a i h d b a e f d e d a $\partial_S(s_3, s^*) \leq 4$

2. Tangled intervals (input strings are pairwise compatible outside these)

3. Any string in C_S is necessarily of the form:

$s^* = $? ? ? | a c b a | ? ? ? ? ? | e d f | ? ? ?

4. Disentanglement

$s_1' = $ a b g a c b a h i d a b e d f e d a $\partial_S(s_1', s^*) \leq 2$

$s_2' = $ b a g a c b a i h d a b e d f d e a $\partial_S(s_2', s^*) \leq 1$

$s_3' = $ b a g a c b a i h d b a e d f e d a $\partial_S(s_3', s^*) \leq 2$

5. With pairwise compatible strings, compute $h_i = h_{s_1', s_i'}$

$h_1 = $ 0 0 0 0 0 0 0 0 0 0 0 0 0 0 0 0 0 0 . $\partial_{Ham}(h_1, h^*) \leq 2$

$h_2 = $ 1 0 0 0 0 0 0 1 0 0 0 0 0 0 0 1 0 . $\partial_{Ham}(h_2, h^*) \leq 1$

$h_3 = $ 1 0 0 0 0 0 0 1 0 0 1 0 0 0 0 0 0 . $\partial_{Ham}(h_3, h^*) \leq 2$

6. Compute an (r, ∂_{Ham})-CONSENSUS of $\{h_1, h_2, h_3\}$

$h^* = $ 1 0 0 0 0 0 0 1 0 0 0 0 0 0 0 0 0 0 .

7. Retrieve s^* such that $h_* = h_{s_1', s^*}$

$s^* = $ b a g a c b a i h d a b e d f e d a

8. Check swap distance between s^* and input strings

$s_1 = $ | a b | g a b c | a h i | d a b | d e | f e d a $\partial_S(s_1, s^*) = 4$

$s_2 = $ b a g | c a | a b | i h d a b e | f d | d e | a $\partial_S(s_2, s^*) = 4$

$s_3 = $ b a g | c a | b a i h | d b | a e | f d | e d a $\partial_S(s_3, s^*) = 3$

Fig. 2. Illustration of the algorithm for $(r, \partial_S) - Consensus$ over length-18 input strings $\{s_1, s_2, s_3\}$, with $d = 4$.

We present our algorithm for distance ∂_S, i.e., for the swap distance only. See Fig. 2 for an illustration.

We first consider the case where all strings are pairwise matching (corresponding to steps 5–7 in the figure). In this case, we show that each of the studied consensus problems under the swap distance reduces to the same problem under the Hamming distance.

Given a set of strings $S = \{s_1, .., s_k\}$ all of length n, we call C_S the set of strings from Σ^n that are matching each and all strings in S. We use the following result.

Lemma 3. *Given a set of pairwise matching strings $S = \{s_1, .., s_k\}$, for any string $s^* \in C_S$, and any $s_i \in S$, we have $\partial_S(s^*, s_i) = \partial_{Ham}(h_{s_1, s^*}, h_{s_1, s_i})$.*

Corollary 1. *Given a set of pairwise matching strings $S = \{s_1, .., s_k\}$ all having length n, one can compute in linear time a set $S_h = \{h_{s_1, s_1} .. h_{s_1, s_k}\}$ of binary strings, all having length n, such that S admits a (r, ∂_S)-CONSENSUS (resp. $(s, \partial_S) - Consensus$, resp. $(rs, \partial_S) - Consensus$) for d (resp. for D, resp. for (d, D)) if and only if S_h admits a $(r, \partial_{Ham}) - Consensus$ (resp. $(s, \partial_{Ham}) - Consensus$, resp. $(rs, \partial_{Ham}) - Consensus$) for d (resp. for D, resp. for (d, D)).*

The latter result gives us a parameterized reduction to Hamming distance for pairwise matching strings.

We consider now the more general case where input strings are not pairwise matching. Note that this does not trivially lead to no-instances, since only the solution needs to match with all input strings (not necessarily other input strings). However, a necessary condition for a yes-instance is that $C_S \neq \emptyset$.

We show a two-step method for solving swap-distance consensus problems. First, in linear time we transform $S = \{s_1, \ldots, s_k\}$ into $S' = \{s'_1, \ldots, s'_k\}$ by a minimal set of swaps so that the s'_i are pairwise matching. Then, by Corollary 1, we reduce S' to a variant of the consensus problem under Hamming distance, and prove that this variant admits a polynomial reduction to the standard Hamming-distance consensus problem.

Let $S = \{s_1, .., s_k\}$. If for some $s \in S$ one has $(p, p+1) \in \mathfrak{S}_{s^*, s}$ for every $s^* \in C_S$, we say that the swap $(p, p+1)$ is *necessary* for s. We say that a set $S' = \{s'_1, .., s'_k\}$ is a *disentanglement* of S if **(i)** the strings from S' are pairwise matching, and **(ii)** for each i, the strings s_i and s'_i are matching and the permutation $\mathfrak{S}_{s_i, s'_i}$ contains only necessary swaps for s_i.

Lemma 4. *Let $S = \{s_1, .., s_k\}$ be a set of strings such that $C_S \neq \emptyset$. Then, S admits a disentanglement, which can be constructed in $\mathcal{O}(kn)$ time.*

Proof. Assume $C_S \neq \emptyset$, and let $s^* \in C_S$. Although we do not have access to s^*, its existence provides structural information about S.

An interval $I \subseteq [1 .. n]$ of length at least 2 is *tangled* if for every $i \in I$, there exists $s_j \in S$ such that $h_{s_j, s^*}[i] = 1$. If no tangled interval exists, then by Lemma 2 all strings in S are pairwise matching and there are no necessary swaps.

If I is tangled, Lemma 2 implies that all strings in C_S agree on I. Therefore, any difference between $s_j \in S$ and s^* within I corresponds to a necessary swap.

Importantly, this implies that the set of tangled intervals is independent of the choice of $s^* \in \mathcal{C}_\mathcal{S}$.

Assume a tangled interval exists, and let i_0 be the initial position of the first one. By definition, there exists $s_j \in \mathcal{S}$ with $h_{s_j, s^*}[i_0] = 1$, hence $s^*[i_0] \neq s^*[i_0+1]$. Similarly, $s^*[i_0 + 1] \neq s^*[i_0 + 2]$. Let $s^*[i_0, i_0 + 1] = \mathsf{ab}$. Furthermore, no $s_j \in \mathcal{S}$ admits a swap at position $i_0 - 1$ with s^*. We consider two cases:

- If $s^*[i_0 + 2] = \mathsf{c} \neq \mathsf{a}$, then $\mathcal{S}[i_0, i_0 + 1] = \{\mathsf{ba}, \mathsf{ac}\}$.
- If $s^*[i_0 + 2] = \mathsf{a}$, then $\mathcal{S}[i_0, i_0 + 1] = \{\mathsf{ba}, \mathsf{aa}\}$.

In either case, a necessary swap at position $i_0 + 1$ occurs for $s_j \in \mathcal{S}$ if and only if $s_j[i_0 + 1] \notin \mathcal{S}[i_0]$ or $s_j[i_0] = s_j[i_0 + 1]$. In particular, for such s_j, one has $s'[i_0 + 1, i_0 + 2] = s^*[i_0 + 1, i_0 + 2] = s_j[i_0 + 2, i_0 + 1]$ for every $s' \in \mathcal{C}_\mathcal{S}$.

We now describe how to compute $\mathcal{S}'$ (see Fig. 3 for an example). First, we can detect whether a tangled interval exists by checking if the set of all dirty columns can be grouped in pairs $i, i+1$ such that $\mathcal{S}[i, i+1] = \{\mathsf{ab}, \mathsf{ba}\}$ for distinct letters a, b. If this holds for all dirty columns, then there is no tangled interval and we set $\mathcal{S}' = \mathcal{S}$.

Otherwise, let i_0 be the first column violating this condition. Namely, the column i_0 is dirty and there is at least one $s_j \in \mathcal{S}$ with $s_j[i_0 + 1] \notin \mathcal{S}[i_0]$ or $s_j[i_0 + 1] = s_j[i_0]$. Then, i_0 is the initial position of a tangled interval. Based on the earlier characterization, we can deduce $s^*[i_0 + 1, i_0 + 2]$ for any $s^* \in \mathcal{C}$, hence all necessary swaps at i_0: namely, we swap s_j at i_0 if and only if $s_j[i_0, i_0 + 1] = s^*[i_0 + 1, i_0 + 2]$. After applying these swaps, we identify the strings requiring swaps at i_0–namely, those not matching at $i_0 + 1$ with the updated strings. At this step, we computed $s^*[i_0, i_0 + 1, i_0 + 2]$ and every necessary swap at position i_0 and $i_0 + 1$. If, after those swaps, there are still strings that do not match with s^* at position $i_0 + 2$, we know that these have a necessary swap at position $i_0 + 2$, and hence we also know $s^*[i_0 + 3]$. By continuing this procedure as long as unmatched strings remain–progressively resolving each next position–we eventually reach the end of the tangled interval, or the end of $\mathcal{S}$. Let $\tilde{\mathcal{S}}$ be the set obtained after performing each of those necessary swaps. Then $\tilde{\mathcal{S}}$ has no tangled interval before position $i_0 + 2$.

This process takes $\mathcal{O}(k)$ time, and by repeating it at most $\mathcal{O}(n)$ time, we eventually construct $\mathcal{S}'$ with no tangled intervals.   □

The following result is deduced from Lemma 4:

Corollary 2. (s, ∂_S)-CONSENSUS *can be solved in* $\mathcal{O}(kn)$ *time.*

For $(\mathbf{r}, \partial_S) - Consensus$ and $(\mathbf{rs}, \partial_S) - Consensus$, one needs an additional result in order to use Corollary 1 and reduce the problems to their Hamming versions. Indeed, the strings in the disentanglement have been swapped a different number of times, and we need to consider this when computing the consensus. We give a formal reduction in the extended version of this paper.

We can now state the main theorem of the section:

Input strings

$s_1 =$ | g | a b c a h i

$s_2 =$ | g | c a a b i h

$s_3 =$ | g | c a b a i h

Column $i_0 = 1$ is clean, so not part of a tangled interval

$s^* = $?

$s_1 =$ g | a | b | c a h i

$s_2 =$ g | c | a a b i h

$s_3 =$ g | c | a b a i h

Column $i_0 = 2$ is dirty with $s_1[3] \notin S[2] = \{a, c\}$, $\rightarrow s^*[3, 4] = s_1[4, 3] = cb$

$s^* = $? ? | c | b

$s'_1 =$ g a | c | b a h i

$s'_2 =$ g c | a | a b i h

$s'_3 =$ g c | a | b a i h

Column $i_0 = 3$: $s'_2[3] = s'_3[3] = a \neq s^*[3] \rightarrow s^*[2] = a$.

$s^* = $? a c | b

$s''_1 =$ g a c | b | a h i

$s''_2 =$ g a c | a | b i h

$s''_3 =$ g a c | b | a i h

Column $i_0 = 4$: $s''_2[4] = a \neq s^*[4] \rightarrow s^*[5] = a$.

$s^* = $? a c b | a

$s'''_1 =$ g a c b | a | h i

$s'''_2 =$ g a c b | a | i h

$s'''_3 =$ g a c b | a | i h

Column $i_0 = 5$: end of tangled interval

Columns $i_0 = 6$, $i_0 = 7$: no new tangled interval

$s^* = $? a c b a ? ?

Fig. 3. Step-by-step disentanglement algorithm. We start with $S = \{s_1, s_2, s_3\}$. Column 2 is the first dirty column and we have $s_1[3] \notin S[2] = \{a, c\}$. This means that there is a necessary swap at position 3 for s_1, and in particular $s^*[3, 4] = s_1[4, 3]$ for *every* $s^* \in C_S$. After swapping s_1 at position 3, we obtain a new set s'_1, s'_2, s'_3, and we deduce the necessary swaps at position 2, as the ones that do not match $s^*[2]$: namely, s_2 and s_3, and we deduce $s^*[2] = a$. After swapping s'_2 and s'_3 we obtain $S'' = s''_1, s''_2, s''_3$. We now resolve necessary swaps at position 4, since $s^*[4] = b$ is known and $s''_2[4] = a \neq b$. We finally obtain the last set $S''' = \{s'''_1, s'''_2, s'''_3\}$, and we see that every string matches s^* at the new computed position $s^*[5] = a$. This means that we reached the end of the first tangled interval. Finally, we notice that the last set does not contain any other tangled interval, so it is a disentanglement for S.

Theorem 1. *The problems $(r, \partial_S) - Consensus$ (resp. $(rs, \partial_S) - Consensus$ and $(r, \partial_{Ham}) - Consensus$ (resp. $(rs, \partial_{Ham}) - Consensus$) both admit a parameterized reduction to each other. We have the following:*

- $(r, \partial_S) - Consensus(d) \in$ FPT *and* $(rs, \partial_S) - Consensus(d) \in$ FPT .
- $(r, \partial_S) - Consensus(k, d)$ *and* $(rs, \partial_S) - Consensus(k, d)$ *both have poly-nomial kernels of size* $\mathcal{O}(k^2 d \log k)$.
- $(rs, \partial_S) - Consensus(D)$ *has a polynomial kernel of size* $\mathcal{O}(D^3 \log D)$.
- *Neither* $(r, \partial_S) - Consensus(d, n, |\Sigma|)$ *nor* $(rs, \partial_S) - Consensus$ $(d, n, |\Sigma|)$ *have polynomial kernels unless* NP $\subseteq$ coNP/poly.

4 Algorithms for the Swap+Hamming Distance

4.1 Fixed Parameter Complexity of (r, ∂_{SH})-CONSENSUS

In this section, we study the parameterized complexity of $(r, \partial_{SH}) - Consensus$. We start by stating two basic results.

Lemma 5. *Given two strings* s_1, s_2 *both of length* n, *the distance* ∂_{SH} *can be computed in linear time as follows: If* $s_1[i, i+1] = s_2[i+1, i]$, *we count a swap at position* i *if and only if we did not count a swap at position* $i - 1$. *The distance* ∂_{SH} *is then the number of mismatches not resolved by a swap.*

Lemma 6. *Given two strings* s_1, s_2 *both of length* n, *the following inequality holds:*

$$\partial_{SH}(s_1, s_2) \leq \partial_{Ham}(s_1, s_2) \leq 2\partial_{SH}(s_1, s_2).$$

As a consequence, given 3 *strings* s_1, s_2, s_3 *all of length* n, *a weakened triangle inequality holds:*

$$\partial_{SH}(s_1, s_3) \leq \min\{2\partial_{SH}(s_1, s_2) + \partial_{SH}(s_2, s_3), \partial_{SH}(s_1, s_2) + 2\partial_{SH}(s_2, s_3)\}.$$

Our result is mainly based on the following Lemma, which is already central in the proof of the corresponding theorem for the Hamming distance.

Lemma 7 ([14]). *Let* $\mathcal{S} = \{s_1, \ldots, s_k\}$ *be a set of strings having length* n, s^* *a string satisfying* $\max_{s \in \mathcal{S}} \partial_{Ham}(s^*, s) \leq d$, *and* $\tilde{s}$ *a string such that for at least one* $s \in \mathcal{S}$ *we have* $\partial_{Ham}(\tilde{s}, s) \geq d + 1$. *Let us fix a set* P *of* $d + 1$ *positions* p *such that* $\tilde{s}[p] \neq s[p]$ *for each* $p \in P$. *Then, for at least one* $p \in P$, $s[p] = s^*[p]$.

We now prove our main result on $(r, \partial_{SH}) - Consensus$:

Theorem 2. $(r, \partial_{SH}) - Consensus(d) \in$ FPT.

Proof. We adapt the algorithm from [14], with additional branchings that take swapped pairs of letters from one of the strings in $\mathcal{S}$. Namely, let us assume that we are given a candidate string $\tilde{s}$ such that, for some fixed $s \in \mathcal{S}$, we have $\partial_{SH}(\tilde{s}, s) > d$, and that there exists a consensus string s^* with $\max_{s \in \mathcal{S}} \partial_{SH}(s, s^*) \leq d$. We first show that, similarly to the Hamming distance, we can always find a set of $\mathcal{O}(d)$ strings where at least one is strictly closer (in Hamming distance) from s^* than $\tilde{s}$ is.

Let us first assume that $\partial_{\mathtt{Ham}}(\tilde{s}, s) \geq 2d + 1$. Then, from Lemma 6, we know that $\max_{s \in \mathcal{S}} \partial_{\mathtt{Ham}}(s, s^*) \leq 2d$, hence from Lemma 7, for any fixed set M of $2d+1$ mismatch positions between $\tilde{s}$ and s, the strings s and s^* must match on at least one position from M. We can then obtain a suitable set by defining for each $p \in M$ a string obtained from $\tilde{s}$ by replacing $\tilde{s}[p]$ by $s[p]$. Let us now assume that $\partial_{\mathtt{Ham}}(\tilde{s}, s) \leq 2d$. By hypothesis and by Lemma 6, we also have $\partial_{\mathtt{Ham}}(\tilde{s}, s) \geq d + 1$. Let M be the set of mismatching positions between $\tilde{s}$ and s, with $|M| = \mathcal{O}(d)$. If for some $p \in M$, one has $s[p] = s^*[p]$, we obtain the set of $\mathcal{O}(d)$ strings as before. On the other hand, if for every $p \in M$, one has $s[p] \neq s^*[p]$, then the strings s and s^* have $d+1$ mismatches, and since $\partial_{\mathtt{SH}}(s, s^*) \leq d$, there is at least one swap between s and s^* involving a position from M. So, for some pair $(p, p+1)$ with p or $p+1$ in M, one has $s[p, p+1] = s^*[p+1, p]$. What remains to show is that for at least one such pair, one has $\tilde{s}[p, p+1] \neq s^*[p, p+1]$: if this is not the case, then each mismatch (resp. swap) from $\tilde{s}$ to s is also a mismatch (resp. swap) from s and s^*, hence $\partial_{\mathtt{SH}}(s, s^*) \geq \partial_{\mathtt{SH}}(s, s') \geq d + 1$, a contradiction. In that case, the $\mathcal{O}(d)$ strings are constructed by replacing $\tilde{s}[p, p+1]$ with $s[p+1, p]$, or $\tilde{s}[p-1, p]$ with $s[p, p-1]$ for each $p \in M$.

We can now describe an FPT algorithm for $(\mathtt{r}, \partial_{\mathtt{SH}})$-Consensus: we start with a candidate $\tilde{s} \in \mathcal{S}$ and explore each branch as described (if $\partial_{\mathtt{Ham}}(\tilde{s}, s) \geq 2d + 1$, we consider only $2d + 1$ substitutions, otherwise we also consider the swaps as detailed above). There are $\mathcal{O}(d)$ of them since we impose $|M| \leq 2d + 1$. The depth of the branching tree is also in $\mathcal{O}(d)$: since $\tilde{s} \in \mathcal{S}$, one has $\partial_{\mathtt{Ham}}(\tilde{s}, s^*) \leq 2\partial_{\mathtt{SH}}(\tilde{s}, s^*) \leq 2d$ (Lemma 6), and we can conclude because this distance decreases for at least one branch at each step. Therefore, after i recursive calls, the best candidate string is at Hamming distance at most $2d - i$ from a solution, hence at Hamming distance at most $4d - i$ from any string in $\mathcal{S}$ (by Lemma 6 and the standard triangle inequality for the Hamming distance). Therefore, we can ignore the search branches that cannot lead to a solution by trimming any instance such that $\partial_{\mathtt{Ham}}(\tilde{s}, s) \geq 4d - i + 1$, where i is the depth of the current recursion in the search tree. $\qquad\square$

Finally, we study the kernelization of the consensus problems under $\partial_{\mathtt{SH}}$. Intuitively, problems for the $\partial_{\mathtt{SH}}$ distance are at least as hard as the corresponding problems for the Hamming distance; hence, we can adapt the results from [4,7]. In the full version of this paper, we show the following:

Theorem 3. *Neither $(\mathtt{r}, \partial_{\mathtt{SH}})$ - Consensus$(d, n, |\Sigma|)$, nor $(\mathtt{rs}, \partial_{\mathtt{SH}})$ - Consensus$(d, n, |\Sigma|)$ admits a polynomial kernel, unless $\mathsf{NP} \subseteq coNP/poly$.*

4.2 Polynomial Algorithm for $(\mathtt{s}, \partial_{\mathtt{SH}})$-Consensus

We conclude by showing that $(\mathtt{s}, \partial_{\mathtt{S}})$-Consensus can be solved in polynomial time, using a dynamic programming algorithm.

Let $\mathcal{S} = \{s_1, .., s_k\}$ be a set of strings all having length n, and let s^* be an arbitrary string of length n. We write $SW(\mathcal{S}, s^*, i)$ for the set of $j \in [1..k]$ such that there is a swap at position i between s^* and s_j (swaps are taken greedily, from left to right, as in Lemma 5). We say that $W = SW(\mathcal{S}, s^*, i)$ is *reached* at position i (or that (i, W) is *reached*) by s^*. Given a set $W \subseteq [1 .. k]$, and a position $i \in [1 .. n]$, we say that (i, W) is *reachable* from $\mathcal{S}$, or that W is *reachable* at position i from $\mathcal{S}$ if it is reached by some s^*.

We start with the following Lemma. Informally, it implies that a solution for $(\mathbf{s}, \partial_{\text{SH}})$-CONSENSUS can be obtained from a solution for $(\mathbf{s}, \partial_{\text{Ham}})$-CONSENSUS by the repeated operations of replacing its letters by swapped letters taken from strings in $\mathcal{S}$.

Lemma 8. *Let $\mathcal{S} = \{s_1, .., s_k\}$ be a set of strings all having length n and admitting a solution to $(\mathbf{s}, \partial_{\text{SH}}) - Consensus$ with sum of distances D. Let s_H^* be the lex-minimal solution to $(\mathbf{s}, \partial_{\text{Ham}}) - Consensus$. Then, if s_{SH}^* is the lex-minimal solution to $(\mathbf{s}, \partial_{\text{SH}}) - Consensus$ for $\mathcal{S}$, the following holds: for each $1 \leq i \leq n$, if one has $s_{SH}^*[i] \neq s_H^*[i]$, then one has $SW(\mathcal{S}, s^*, i-1) \neq \emptyset$ or $SW(\mathcal{S}, s^*, i) \neq \emptyset$.*

Given $\mathcal{S}$, we consider the set $\mathcal{S}_i$ of length $i + 1$ prefixes of strings in $\mathcal{S}$. We define a dynamic programming table $\mathbf{T}$ as follows: Let $i \in [n]$ and $W \subseteq [k]$ such that (i, W) is reachable. Then, $\mathbf{T}[i, W]$ is, among all strings $t \in \Sigma^{i+1}$ reaching (i, W), the one minimizing $\sum_{s \in \mathcal{S}_i} \partial_{\text{SH}}(s, t)$. In case of ties, we pick the lex-minimal such string. Additionally, we set $\mathbf{T}[0, \emptyset] = s_H^*[1]$, where s_H^* is a lex-minimal solution for $(\mathbf{s}, \partial_{\text{Ham}}) - Consensus(\mathcal{S})$.

Remark 2. Consider a set of n-length strings $\mathcal{S} = \{s_1, .., s_k\}$, and a pair (i, W) reached by some $t \in \Sigma^{i+1}$. Furthermore, we assume $W \neq \emptyset$. By definition, one must have $t[i, i+1] = s_j[i+1, i]$ for every $j \in W$. In particular, when $W \neq \emptyset$, the pair (i, W) uniquely determines the length-2 suffix of $\mathbf{T}[i, W]$. However, the converse does not hold, as illustrated in the following example:

Example 2. Let $\mathcal{S} = \{s_1, s_2\}$, with $s_1 = bab$ and $s_2 = aab$ and consider $t_1 = aba$, $t_2 = bba$. One can observe that in this case, $SW(\mathcal{S}, t_1, 2) = \{2\}$, and $SW(\mathcal{S}, t_2, 2) = \{1, 2\}$, even if $t_1[2, 3] = t_2[2, 3]$.

Although there are 2^k distinct subsets $W \subseteq [k]$, the following lemma shows that at most $\mathcal{O}(kn)$ of them are reachable at a given position i; hence, the size of the table $\mathbf{T}$ remains polynomial.

Lemma 9. *Let $\mathcal{S} = \{s_1, .., s_k\}$ be a set of strings all having length n, and $1 \leq i \leq n$. At each position i, there are at most ki reachable sets.*

The following lemma describes the dynamic programming relations for $\mathbf{T}$. We first define, for every reachable (i, W), $i \geq 2$, the set $\text{pre}(i, W)$ of strings that are used to compute $\mathbf{T}[i, W]$.

If $W \neq \emptyset$, from Remark 2, the length-2 suffix of $\mathbf{T}[i, W]$ is entirely determined. On the other hand, if $W = \emptyset$, we can assume, from Lemma 8, that the last letter of $\mathbf{T}[i, W]$ is $s_H^*[i + 1]$, where s_H^* is a lex-minimal solution for $(\mathbf{s}, \partial_{\text{Ham}}) - Consensus(\mathcal{S})$. In both cases, let us write $\mathbf{b}$ for the last letter of $\mathbf{T}[i, W]$. We then set $t = \mathbf{T}[i - 1, W'] \cdot \mathbf{b} \in \text{pre}(i, W)$ if t reaches W at position i. Furthermore, if $W \neq \emptyset$, writing $\mathbf{a}$ for the penultimate letter of $\mathbf{T}[i, W]$

(which is entirely determined from W), we also consider strings of the form $t = \mathbf{T}[i-2, W'] \cdot \mathsf{ab}$ where $(i-2, W')$ is reachable, and such that t reaches W at i and $\emptyset$ at $i-1$, if such strings exist.

Lemma 10. *Let $\mathcal{S} = \{s_1, \ldots s_k\}$ be a set of strings all having length n. For $i \geq 2$, the table $\mathbf{T}$ satisfies the following recurrence relation:*

$$\mathbf{T}[i, W] = \underset{t \in \mathrm{pre}(i, W)}{\arg\min} \sum_{s \in \mathcal{S}_i} \partial_{\mathit{SH}}(s, t).$$

Furthermore, the states $\mathbf{T}[1, W]$ can be computed for every W reachable at position 1 by setting $\mathbf{T}[0, \emptyset] = s_H^*[1]$, $\mathbf{T}[1, W] = \mathsf{ba}$ where $W = \{j \in [k] \mid s_j[1, 2] = \mathsf{ab}\}$ for every $\mathsf{ab} \in \mathcal{S}[1, 2]$, and $\mathbf{T}[1, \emptyset] = s_H^*[1, 2]$, if it reaches $\emptyset$, where s_H^* is chosen to be the lexicographically minimal solution for $(\mathbf{s}, \partial_{\mathit{Ham}}) - \mathit{Consensus}$.

Theorem 4. *One can solve $(\mathbf{s}, \partial_{\mathit{SH}})$-CONSENSUS in $\mathcal{O}(k^3 n^2)$ or $\mathcal{O}(|\Sigma| k^2 n^2)$ time.*

Proof. We use dynamic programming with the table $\mathbf{T}[i, W]$. We first compute s_H^*, a lexicographically minimal solution for $(\mathbf{s}, \partial_{\mathit{Ham}})$-CONSENSUS, and we initialize the lines 0 and 1 of $\mathbf{T}$ as described in Lemma 10. We then compute $\mathbf{T}$ inductively using the relation from Lemma 10. More precisely, for each reachable (i, W) we construct all the strings that are extensions of $t = \mathbf{T}[i, W]$ and are contained in $\mathrm{pre}(i', W')$ for some W' that is reachable at position $i' = i + 1$ or $i' = i + 2$. By definition, those strings are exactly the strings of the form $t \cdot \mathsf{a}$ for $\mathsf{a} \in \mathcal{S}[i+1]$ that reach $W \neq \emptyset$ at position $i+1$, the string $t \cdot s_H^*[i+2]$ if it reaches $\emptyset$ at position $i+1$, and the strings of the form $t \cdot \mathsf{ab}$ reaching $\emptyset$ at position $i+1$, with $\mathsf{ba} \in \mathcal{S}[i+2, i+3]$. Note that for the special case $\mathbf{T}[0, \emptyset] = s_H^*[1]$, we compute only this last type of extensions, as the other ones would correspond to elements of the first line of the table, which is already computed during initialization.

For each such t', we compute its partial score $d_{t'}$ (the sum of distances up to $i + 1$), and store the tuple $(i', W', t', d_{t'})$, where W' is the set reached by t' on position $i' = |t'| - 1$. We maintain only the best-scoring string for each (i', W'), and for every position i' we sort tuples lexicographically by the indicator vector of $W' \subseteq [k]$. By construction, and by Lemma 10, after processing every $i < i'$, the only tuple corresponding to (i', W') contains $\mathbf{T}[i', W']$.

From Lemma 9, we have at most $\mathcal{O}(kn)$ partial solutions per step. Each of them can be extended in at most $\mathcal{O}(k)$ or $\mathcal{O}(|\Sigma|)$ ways (extensions are always taken from $\mathcal{S}[i, i+1]$, or $s_H[i]$). Updating distances and computing swap sets both take $\mathcal{O}(k)$ time. Thus, the total time complexity is $\mathcal{O}(k^3 n^2)$, or $\mathcal{O}(|\Sigma| k^2 n^2)$. □

Example 3. Let $s_1 = \mathsf{baba}$, $s_2 = \mathsf{cabc}$, $s_3 = \mathsf{abca}$ and $\mathcal{S} = \{s_1, s_2, s_3\}$, with $s_H^* = \mathsf{aaba}$. The initial states are $\mathbf{T}[0, \emptyset] = \mathsf{a}$, $\mathbf{T}[1, \emptyset] = \mathsf{aa}$, $\mathbf{T}[1, \{1\}] = \mathsf{ab}$, $\mathbf{T}[1, \{2\}] = \mathsf{ac}$, $\mathbf{T}[1, \{3\}] = \mathsf{ba}$. When processing, for example, $\mathbf{T}[1, \{3\}]$, we first compute $t = \mathsf{ba} \cdot s_H^*[3] = \mathsf{bab}$ and add the tuple $(2, \emptyset, t, d_t) = (2, \emptyset, \mathsf{bab}, 3)$ only if t reaches $\emptyset$ at position 2. This is the case here (in particular, t does not reach

$\{1\}$ because t and s_1 already have a swap at position 1). Then, we compute the strings $\mathsf{ba} \cdot x$ for $x \in \mathcal{S}[2]$, namely baa and bab, but both reach $W = \emptyset$ at position 2 so we do nothing. Finally, we compute $\mathsf{ba} \cdot y \cdot x$ where $x \cdot y \in \mathcal{S}[3,4]$, obtaining baab, bacb, baac. Among those, only bacb reaches $\emptyset$ at position 2, and it reaches $W = \{2\}$ at position 3, hence we create the tuple $(3, \{2\}, \mathsf{bacb}, 6)$ if we do not already have a tuple $(3, \{2\}, t', d_t)$ with $d_t \leq 6$.

In this example, the final solution $\mathsf{baba} = \mathbf{T}[3, \emptyset]$ is found by extending $\mathsf{bab} = \mathbf{T}[2, \emptyset]$, and $\sum_{s \in \mathcal{S}} \partial_{\mathsf{SH}}(s, \mathsf{baba}) = 4$. See Table 2 for the full DP-table explored by the algorithm. Note that $(2, \{1,2\})$ is also reachable, for example by caba, but this string is also reaching $\emptyset$ at position 1, and since $\mathsf{c} \neq s_H^*[1]$ it can be ignored by Lemma 3, and is indeed never computed in the algorithm.

Table 2. DP-states $T[i, W]$ for $S = \{\mathsf{baba}, \mathsf{cabc}, \mathsf{abca}\}$, $s_H^* = \mathsf{aaba}$.

$W \backslash i$	0	1	2	3
$\emptyset$	a	aa	$\mathsf{bab} = T[1, \{3\}] \cdot s_H^*[3]$	$\mathsf{baba} = T[2, \emptyset] \cdot s_H^*[4]$
$\{1\}$		ab		$\mathsf{abab} = T[2, \{2\}] \cdot \mathsf{b}$
$\{2\}$		ac	$\mathsf{aba} = T[1, \{1\}] \cdot \mathsf{a}$	$\mathsf{bacb} = T[1, \{3\}] \cdot \mathsf{cb}$
$\{3\}$		ba	$\mathsf{abc} = T[1, \{3\}] \cdot \mathsf{c}$	$\mathsf{abac} = T[2, \{2\}] \cdot \mathsf{c}$

5   Open Questions

We leave the following questions open:

- Is $(\mathsf{rs}, \partial_{\mathsf{SH}}) - Consensus(d) \in \mathsf{FPT}$?
- Can one find conditional lower bounds, or a more efficient algorithm, for $(\mathsf{s}, \partial_{\mathsf{SH}}) - Consensus$
- Can our algorithms be generalized to allow deleting a certain number of outlier strings or columns in $\mathcal{S}$ such that the resulting substrings admit a solution (cf. [7])?
- As the FPT algorithms for those problems (even the classical algorithm from [14] for the Hamming distance) start with some given candidate, could they be studied from the perspective of learning-augmented algorithms?

References

1. Amir, A., Aumann, Y., Landau, G.M., Lewenstein, M., Lewenstein, N.: Pattern matching with swaps. J. Algorithms **37**(2), 247–266 (2000). https://doi.org/10.1006/jagm.2000.1120. https://www.sciencedirect.com/science/article/pii/S0196677400911209
2. Amir, A., Paryenty, H., Roditty, L.: On the hardness of the consensus string problem. Inf. Process. Lett. **113**, 371–374 (2013). https://doi.org/10.1016/j.ipl.2013.02.016

3. Amir, A., Landau, G.M., Na, J.C., Park, H., Park, K., Sim, J.S.: Consensus optimizing both distance sum and radius. In: Karlgren, J., Tarhio, J., Hyyrö, H. (eds.) String Processing and Information Retrieval, pp. 234–242. Springer, Heidelberg (2009)
4. Basavaraju, M., Panolan, F., Rai, A., Ramanujan, M.S., Saurabh, S.: On the kernelization complexity of string problems. Theor. Comput. Sci. **730**, 21–31 (2018)
5. Betzler, N., Guo, J., Komusiewicz, C., Niedermeier, R.: Average parameterization and partial kernelization for computing medians. J. Comput. Syst. Sci. **77**(4), 774–789 (2011)
6. Bodlaender, H.L., Thomassé, S., Yeo, A.: Kernel bounds for disjoint cycles and disjoint paths. Theor. Comput. Sci. **412**(35), 4570–4578 (2011)
7. Bulteau, L., Schmid, M.L.: Consensus strings with small maximum distance and small distance sum. Algorithmica **82**(5), 1378–1409 (2020)
8. Chen, Z.Z., Ma, B., Wang, L.: A three-string approach to the closest string problem. J. Comput. Syst. Sci. **78**(1), 164–178 (2012). jCSS Knowledge Representation and Reasoning. https://doi.org/10.1016/j.jcss.2011.01.003
9. Cunha, L., Lopes, T., Mary, A.: Complexity and algorithms for swap median and relation to other consensus problems (2025). https://arxiv.org/abs/2409.09734
10. Downey, R.G., Fellows, M.R.: Parameterized Complexity. Springer (2012)
11. Fomin, F.V., Lokshtanov, D., Saurabh, S., Zehavi, M.: Kernelization: Theory of Parameterized Preprocessing. Cambridge University Press (2019)
12. Frances, M., Litman, A.: On covering problems of codes. Theor. Comput. Syst. **30**, 113–119 (1997)
13. Gabory, E., Bulteau, L., Fici, G., Verbeek, H.: String consensus problems with swaps and substitutions (2025). https://arxiv.org/abs/2507.19139
14. Gramm, J., Niedermeier, R., Rossmanith, P.: Fixed-parameter algorithms for closest string and related problems. Algorithmica **37**, 25–42 (2003). https://doi.org/10.1007/s00453-003-1028-3
15. Viny Christanti, M., Rudy, R., Naga, D.S.: Fast and accurate spelling correction using Trie and Damerau-levenshtein distance bigram. TELKOMNIKA **16**(2), 827–833 (2018). https://doi.org/10.12928/TELKOMNIKA.v16i2.6890
16. Muthukrishnan, S.: New results and open problems related to non-standard stringology. In: Galil, Z., Ukkonen, E. (eds.) CPM 1995. LNCS, vol. 937, pp. 298–317. Springer, Heidelberg (1995). https://doi.org/10.1007/3-540-60044-2_50

Two-Player Communication Complexity
of Pattern Matching

Paweł Gawrychowski$^{(\boxtimes)}$ ⓘ and Wojciech Janczewski ⓘ

Institute of Computer Science, University of Wrocław, Wrocław, Poland
gawry@cs.uni.wroc.pl

Abstract. Porat and Porat [FOCS 2009] described the first efficient
randomised algorithm for the pattern matching problem in the stream-
ing model, using $\mathcal{O}(\log m \log n)$ bits for a text of length n and a pattern
of length m, against the lower bound of $\Omega(\log n)$ bits. Since then, multiple papers considered many variants of this problem, but with virtually
no progress on the lower bounds side, leaving a logarithmic gap in the
space complexity for the very basic variant.

We discuss a modification of the lower bound of Ergun, Jowhari, and
Saglam [RANDOM 2010] against a restricted class of algorithms for the
streaming pattern matching problem. Then, we show that the standard
approach via communication complexity does not suffice to obtain a better lower bound, by presenting an efficient communication protocol.

Keywords: Pattern matching · Streaming · Communication
complexity

1 Introduction

The problem of pattern matching, that is, finding occurrences of a given pattern
in a text, is fundamental in the field of string algorithms. There are many efficient solutions, such as the Knuth-Morris-Pratt [14] (KMP) or Boyer-Moore [2]
algorithms. We consider the streaming model of computation, in which the input
arrives one symbol at a time, and the goal is to minimise the space complexity of the algorithm. In the context of the pattern matching problem, we may
allow some initial preprocessing of the pattern, after which the text is given as a
stream. We denote the length of the pattern and text as m and n, respectively,
and assume that n is known in advance. As deterministic exact pattern matching
in this model requires space linear in m, see for example [6], we focus on randomised algorithms. Among the classic solutions, the Karp-Rabin algorithm [13]
and fingerprint functions are the most interesting to us.

Porat and Porat [16] described a randomised algorithm using $\mathcal{O}(\log m \log n)$
bits. In fact, their solution can report occurrences of all prefixes of the pattern

Partially supported by the Polish National Science Centre grant number
2023/51/B/ST6/01505.

G. Badkobeh et al. (Eds.): SPIRE 2025, LNCS 16073, pp. 148–155, 2026.
https://doi.org/10.1007/978-3-032-05228-5_13

with lengths 2^i (roughly speaking), for all relevant values of i, as the algorithm uses a hierarchy of fingerprints. Breslauer and Galil [3] improved this result to work in real-time (using worst-case constant time per input symbol). Their algorithm also reports occurrences of a logarithmic number of selected prefixes of the pattern. We call this the extended pattern matching.

For the standard streaming pattern matching, the trivial lower bound is $\Omega(\log m)$ bits for algorithms working with constant probability of error, by a straightforward reduction from EQUALITY [15]. Here we assume that random bits are also given as a stream, and the algorithm cannot read the previously generated random bits, unless they have been stored in memory, so we can work with the private-coin communication complexity. We later observe that a stronger $\Omega(\log n)$ lower bound also holds.

Ergun, Jowhari, and Saglam [8] showed that the extended pattern matching needs $\Omega(\log m \log n)$ bits, but we have no knowledge of any non-trivial lower bound for the standard pattern matching. Meanwhile, multiple variants of the problem were considered, for example pattern matching with mismatches or wildcards [7,11], dictionary matching [5,9,12], or the model with multiple input streams [6,10]. For all of them, we have only straightforward lower bounds by a reduction from the communication complexity of equality or indexing [15].

Therefore, we are still left with a quadratic gap between the $\Omega(\log n)$ lower bound and $\mathcal{O}(\log m \log n)$ upper bounds for the standard pattern matching. In particular, the lower bound does not really make a full use the definition of the matching problem, where we must be able to report an occurrence starting at any position in the text. The most popular approach to lower bounds in streaming algorithms is via reductions to communication complexity, often one-way two-player communication problems [1,4,18].

In this paper, we show that the two-player communication problem corresponding to pattern matching can be solved with a one-way message of logarithmic size. Therefore, the straightforward approach cannot work, and we need either more players or a different framework. Three-player communication complexity of approximate pattern matching was considered by Starikovskaya [17], but in the context of deterministic upper bounds. Additionally, we slightly modify the lower bound from [8] to remove an unnecessary assumption.

2 Lower Bounds

For the standard streaming pattern matching, the only known lower bound is $\Omega(\log n)$ bits, by reducing from EQUALITY. Ergun, Jowhari, and Saglam [8] considered algorithms working as the one in [16], which outputs not only all occurrences of the pattern in the text, but also occurrences of all prefixes of the pattern of length 2^i, for every $i = 0, 1, \ldots$. They showed that $\Omega(\log m \log n)$ bits are necessary for any streaming algorithm outputting the positions of all such prefixes, if $m^{1+\epsilon} < n$, for some constant $\epsilon > 0$.

To achieve that, Ergun et al. use a variant of the indexing problem. In the *augmented indexing*, there are two players, Alice and Bob. Alice is given a string

S of length n over the alphabet Σ with $|\Sigma| = k$. Bob is given an index $i \in [n]$ and also a prefix $S[1, i-1]$ (thus *augmented* indexing). The goal of a protocol is for Bob to output a single letter $S[i]$. If we denote this problem by IND_k^n, the following was proven:

Lemma 1 ([8]). *The one-way randomised communication complexity of IND_k^n is $\Omega((1-\delta)n\log k)$, for algorithms correct with probability at least $1-\delta$.*

We denote by $T[i, j]$ the substring of T from position i to j, and the concatenation of strings A, B by $A \circ B$. Let us consider one-way two-player communication problem for *extended pattern matching*, in which Alice is given a pattern P of length m, while Bob has a text T of length n, with both strings over the alphabet Σ. After Alice sends a message to Bob, he should output $\log m + 1$ binary vectors, each of length n, with vector v_i indicating occurrences of $P[1, 2^i]$, that is, $v_i[j] = 1$ whenever $T[j, j + 2^i - 1] = P[1, 2^i]$. Using Lemma 1, Ergun et al. showed that the communication complexity of this problem is $\Omega(\log m \log n)$ if $n > m^{1+\epsilon}$, for some constant $\epsilon > 0$. Breslauer and Galil [3] stated *We suspect that the same lower bound should hold even when $n = cm$, for a constant c.* Indeed, we can modify the methods of Ergun, Jowhari, and Saglam to obtain the following:

Lemma 2. *The one-way randomised communication complexity of the extended pattern matching, where Bob with probability $1-\delta$ correctly reports occurrences of all prefixes of the pattern with lengths being powers of 2, is $\Omega((1-\delta)\log m \log n)$ even for $n = m$.*

Proof. We reduce from $\mathrm{IND}_{\sqrt{n}}^{(\log n)/2}$, so Alice receives a string S of length $\frac{1}{2}\log n$ over alphabet Σ of size $\sqrt{n}$. She then constructs the pattern P as

$$P = S[1] \circ S[2] \circ S[3]^2 \circ S[4]^4 \circ \ldots \circ S[\tfrac{1}{2}\log n]^{\sqrt{n}/4} \circ \mathsf{a}^{n - \sqrt{n}/2},$$

where a is any letter from Σ. This pattern has the first $\frac{1}{2}\log n$ prefixes of exponential lengths depending on S, and is further padded to have length n. Alice runs the preprocessing phase of the extended pattern matching algorithm on P and sends to Bob the full state of the algorithm.

Recall that Bob receives index i and $S[1, i-1]$. Let $P_i = S[1] \circ S[2] \circ S[3]^2 \circ \ldots \circ S[i-1]^{2^{i-3}}$, and $P_{i,\ell} = P_i \circ \ell^{2^{i-2}}$, for any $\ell \in \Sigma$. Bob constructs the text as a concatenation of $P_{i,\ell}$ for all letters ℓ from the alphabet. This is at most $|\Sigma|2|P_i| = \sqrt{n}(2^{i-1}) < n$ letters, so then Bob pads it to have length n.

With probability at least $1-\delta$ after resuming the computation of the extended pattern matching algorithm on this text with state of the algorithm obtained in the message from Alice, occurrences of $P[1, 2^{i-1}]$ are reported correctly. As letters ℓ used to define all $P_{i,\ell}$ are different, and they are at positions being multiples of 2^{i-1}, there will be exactly one occurrence of $P[1, 2^{i-1}]$ reported ending at position being multiple of 2^{i-1} (before the padding), and letter at this position is $S[i]$. By Lemma 1, we obtain the desired lower bound. $\square$

With slight modifications, the lower bound works also for the extended pattern matching over the binary alphabet.

Moreover, we observe that using $\text{IND}^1_{n/m}$, we can prove $\Omega(\log{(n/m)})$ lower bound for the standard pattern matching. Indeed, when Alice is given a single letter $\ell \in [n/m]$, she constructs the pattern $\mathbf{a}^{m-1}\ell$, then it is enough for Bob to concatenate n/m strings of length m to find out ℓ. Combining both lower bounds $\Omega(\log{(n/m)})$ and $\Omega(\log m)$, we obtain a lower bound of $\Omega(\log n)$.

3   Preliminaries

A string $S = s_1 s_2 \ldots s_l$ is a sequence of symbols over the alphabet Σ. We will consider strings of length $\mathcal{O}(n)$ over $\Sigma = [n^t]$ for constant t, that is, the alphabet consists of integers and its size is polynomial in n^1.

We will use the Karp and Rabin [13] fingerprint function, which is standard in this area. When working with a fixed alphabet of size k, the fingerprint function f is parametrised by two integer values: a fixed prime number $p > k$ and a base $r \in \mathbb{Z}_p$ chosen uniformly at random. Then for any string S over the alphabet we define $f_{p,r}(S) = \Sigma_{i=1}^{|S|} S[i] r^{i-1} \pmod{p}$. When the parameters are known, we will write just f without a subscript. Usually, we store a pair of values as a fingerprint of S, the length of S and the value of $f_{p,r}(S)$.

We have the following properties, proven in previous works [3, 16]:

- If $p = \mathcal{O}(n^c)$ for some constant c, then for any string S with $|S| = \mathcal{O}(n)$ we can store its fingerprint on $\mathcal{O}(\log n)$ bits.
- For two different strings S, T of length at most d, fixed p, and uniformly chosen r, we have $f_{p,r}(S) = f_{p,r}(T)$ (a collision) with probability at most $d/(p-1)$. This means that for $p = \Theta(n^c)$ and strings of length $\mathcal{O}(n)$, the collision probability is $\mathcal{O}(n^{-c+1})$.
- When parameters p, r are fixed, given the fingerprints of any two strings from $\{S, T, S \circ T\}$, we can always obtain the fingerprint of the third string.

A period of string S is a positive number p such that $S[i] = S[i + p]$, for all $i \leq |S| - p$. The period of S is the smallest such number. We will use the well known fact that whenever S has periods p, q with $p + q \leq |S|$, then $\gcd(p, q)$ is also a period of S (the so-called *Periodicity Lemma/Theorem*).

For a string S of length n, its prefix $S[1, k]$ is called a border if $S[1, k] = S[n - k + 1, n]$. It is known that if S has a border of length k, then $n - k$ is a period of S. We will use the word *period* to mean either the substring of S or its length, depending on the context.

4   Two-Player Communication Protocol

We consider one-way two-player communication complexity.

---

[1] A larger alphabet can be always hashed down to polynomial at the expense of adding $\mathcal{O}(\log |\Sigma|)$ bits to describe a 2-universal hash function.

Definition 1. *In the two-player pattern matching problem there is a pattern P of length $n+1$, text T of length $2n$, and two players named Alice and Bob. Alice is given P and $T' = T[1, n]$, Bob is given $T'' = T[n + 1, 2n]$. Alice can send a message $m(P, T')$ to Bob, then Bob should output all starting positions of the occurrences of P in T.*

Note that there are no occurrences entirely contained in the first half of T, nor the second. The value of n is clearly known to both players.

Theorem 1. *There is a communication protocol for the two-player pattern matching problem using $\mathcal{O}(\log n)$ bits, correct with high probability.*

Proof. Let us first describe the message $m(P, T')$ which Alice sends to Bob, given her input. It would be enough to send the fingerprints of P and all prefixes of P being suffixes of T', but there can be too many such prefixes to enumerate them one by one. Instead, Alice can use the fingerprints. Let $\Sigma = [n^t]$. Alice picks a fingerprint function $f_{p,r}$ with the parameters being any prime number $p \in [n^{\max(3,t)}, 2n^{\max(3,t)}]$ and $r \in \mathbb{Z}_p$ chosen uniformly at random; this will give us a failure probability of $\mathcal{O}(n^{-1})$. We can also pick a larger p to obtain a failure probability of $\mathcal{O}(n^{-c})$, for any constant c.

Then Alice finds the largest index k such that $P[1, k]$ is a suffix of T'. We assume $k > 0$, as otherwise there are no occurrences of P in T. Notice that it would be sufficient for Alice to send information about all borders of $P[1, k]$, as these describe all possible matches of P in T to be considered by Bob.

Let d be the (smallest) period of $P[1, k]$, and ℓ be the length of the longest border of $P[1, k]$ with $2\ell \leq k$. Alice sends as her message $m(P, T')$ the following:

1. Parameters p, r of the fingerprint function f.
2. The fingerprint $f(P)$ of the pattern.
3. The fingerprint $f(P[1, k])$ of the prefix.
4. The fingerprint $f(P[1, d])$ of the period.
5. The fingerprint $f(P[1, \ell])$ of the border.

Recall that we include the length of a string in its fingerprint. All five components consist of fixed numbers of bits, $\mathcal{O}(\log n)$ in total. Bob, after receiving the message, proceeds in two phases. First, Bob can combine $f(P[1, k])$ with $f(T''[1, n - k + 1])$ and check if this is equal to $f(P)$; if so, an occurrence is reported. Then, using the fingerprint of the period, for all relevant j he iteratively computes the fingerprints $f(P[1, k - jd])$ and $f(T''[1, n - k + 1 + jd])$, combines them and compare with the fingerprint $f(P)$ of the pattern, reporting occurrences whenever the obtained values are equal.

In the second phase, Bob computes the fingerprints of all substrings of T'' of length ℓ and compares them with the fingerprint of the border sent by Alice. This is done one by one, and if at any point there is equality between two values, Bob assumes the underlying strings are also equal. That is, if a substring $T''[i, j]$ was considered, he assumes that $T''[i, j] = P[1, \ell]$; this will hold with probability $1 - \mathcal{O}(n^{-1})$, as $\mathcal{O}(n)$ substrings are checked. If such a substring is

$$T'\qquad\qquad\qquad\qquad\qquad\qquad T''$$

abacaabaabacaabaabacaabaaba

abacaabaabacaabaaba

abacaabaaba

aba

abacaabaabacaabaabacaabaaba

abacaabaaba

abacaabaaba

Fig. 1. All borders for $P[1,k]$ = abacaabaabacaabaabacaabaaba are checked by Bob. The period $d = 8$ is marked in blue. In the first phase, Bob will be able to check for the four possible occurrences defined by the shifts by multiples of d, these are shown accordingly aligned in the top part of the figure. ℓ, the length of the longest border of $P[1,k]$ of length at most $k/2$, equals 11, with the corresponding border marked in red. If there is an occurrence of P anywhere in $T'[n-\ell+1,2n]$, then $P[1,\ell]$ must have an occurrence in T'', that is T'' contains a full occurrence of the red border. In such a case, in the second phase Bob will be able to leverage the fingerprints to identify this border and then check for the additional possible occurrences, shown in the lower part of the figure. This is done by iterating through all the borders of $P[1,\ell]$, in this case $P[1,11]$, $P[1,3]$ and $P[1,1]$. The first two have already been checked as shifts by multiples of d, but there is also a new one corresponding to extending $P[1,1] = $ a.

found (and assuming there are no collisions), it is easy to iterate through all the borders of $T''[i,j]$. Say one of these borders is $T''[i,i+q]$, then using $f(P)$ and $T''[i,j] = P[1,\ell]$, Bob can compute $f(P[q+2,n+1])$ and compare it with $f(T''[1,n-q])$, declaring a new occurrence whenever the values are equal. See also Fig. 1.

Let us check the correctness of the protocol described above. Bob makes $\mathcal{O}(n)$ comparisons of fingerprints, thus with high probability there are no collisions, and in the following we assume zero collisions. We need to argue that for all q such that $P[1,q]$ is a suffix of T', Bob computes and compares the fingerprints of $P[q+1,n+1]$ and $T''[1,n-q+1]$. Clearly, there are no false positives, that is, Bob does not consider values of q for which $P[1,q]$ is not a suffix of T'.

Consider $q \leq \ell$. If in the second phase Bob found no substring $T''[i,j]$ with a fingerprint equal to the one from the last part of the message sent by Alice, there can be no occurrences of P in T starting at positions $[n-l+1,n]$, in particular $n-q+1$. This is since we have $P[1,\ell] = P[k-\ell+1,k]$ and $k \geq 2\ell$, so for $T[n-q+1,2n-q+1] = P$ to hold there must be an occurrence of $P[1,\ell]$ in T''. If in the second phase Bob found a matching substring, then he iterated through all the correct lengths of the borders, in particular q, and checked for an occurrence in T.

For $q > \ell$, it must be that $k < 2q$, since otherwise a larger ℓ would have been chosen. As $P[1,q]$ is a border of $P[1,k]$, $P[1,k]$ has a period of length $k-q$. It holds that $k - q + d \leq 2(k - q) < k$ and so it must be that $k - q = cd$, for some integer c. Thus, in the first phase of Bob's algorithm, q must be checked. See also Fig. 2.   □

$$T'\qquad\qquad\qquad\qquad\qquad\qquad T''$$

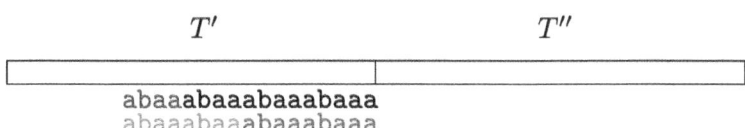

abaaabaaabaaabaaa
abaaabaaabaaabaaa

Fig. 2. For $P[1, k = 17]$ = abaaabaaabaaabaaa, a red border of P has length $q = 9$, with $k < 2q$. This means that P has a green period of length $k - q = 8$, a multiple of the blue period $d = 4$, so the occurrence corresponding to q is checked by Bob in the first phase when considering all shifts by multiples of d.

5   Conclusion

After the initial paper of Porat and Porat [16], there have been many works published about variants of the streaming pattern matching problem, but none of the results is tight as for all variants only simple lower bounds exist. For example, in the dictionary pattern matching problem, we have a collection of d patterns and need to check for occurrences of all of them. Breslauer and Galil [3] note that *In private communications [Alon 2010] pointed out an $\Omega(d \log n)$ space streaming lower bound argument for dictionary matching with d patterns.* Similarly, for multiple stream pattern matching with s streams of independent texts (and a single pattern), we can design $\Omega(s \log n)$ lower bound via basic communication complexity methods. This leaves a gap of $\Theta(\log n)$ between the lower bounds and the algorithms. We have shown that the basic two-player approach is not sufficient for obtaining a better lower bound for the standard streaming pattern matching problem.

References

1. Alon, N., Matias, Y., Szegedy, M.: The space complexity of approximating the frequency moments. J. Comput. Syst. Sci. **58**(1), 137–147 (1999)
2. Boyer, R.S., Moore, J.S.: A fast string searching algorithm. Commun. ACM **20**(10), 762–772 (1977)
3. Breslauer, D., Galil, Z.: Real-time streaming string-matching. ACM Trans. Algorithms **10**(4), 22:1–22:12 (2014)
4. Chakrabarti, A., Regev, O.: An optimal lower bound on the communication complexity of gap-hamming-distance. SIAM J. Comput. **41**(5), 1299–1317 (2012)
5. Clifford, R., Fontaine, A., Porat, E., Sach, B., Starikovskaya, T.: Dictionary matching in a stream. In: Bansal, N., Finocchi, I. (eds.) ESA 2015. LNCS, vol. 9294, pp. 361–372. Springer, Heidelberg (2015). https://doi.org/10.1007/978-3-662-48350-3_31

6. Clifford, R., Jalsenius, M., Porat, E., Sach, B.: Pattern matching in multiple streams. In: Kärkkäinen, J., Stoye, J. (eds.) CPM 2012. LNCS, vol. 7354, pp. 97–109. Springer, Heidelberg (2012). https://doi.org/10.1007/978-3-642-31265-6_8
7. Clifford, R., Kociumaka, T., Porat, E.: The streaming k-mismatch problem. In: SODA. pp. 1106–1125. SIAM (2019)
8. Ergun, F., Jowhari, H., Sağlam, M.: Periodicity in Streams. In: Serna, M., Shaltiel, R., Jansen, K., Rolim, J. (eds.) APPROX/RANDOM -2010. LNCS, vol. 6302, pp. 545–559. Springer, Heidelberg (2010). https://doi.org/10.1007/978-3-642-15369-3_41
9. Gawrychowski, P., Starikovskaya, T.: Streaming dictionary matching with mismatches. Algorithmica 84(4), 896–916 (2022)
10. Golan, S., Kopelowitz, T., Porat, E.: Towards optimal approximate streaming pattern matching by matching multiple patterns in multiple streams. In: ICALP. LIPIcs, vol. 107, pp. 65:1–65:16. Schloss Dagstuhl - Leibniz-Zentrum für Informatik (2018)
11. Golan, S., Kopelowitz, T., Porat, E.: Streaming pattern matching with d wildcards. Algorithmica 81(5), 1988–2015 (2019)
12. Golan, S., Porat, E.: Real-time streaming multi-pattern search for constant alphabet. In: ESA. LIPIcs, vol. 87, pp. 41:1–41:15. Schloss Dagstuhl - Leibniz-Zentrum für Informatik (2017)
13. Karp, R.M., Rabin, M.O.: Efficient randomized pattern-matching algorithms. IBM J. Res. Dev. 31(2), 249–260 (1987)
14. Knuth, D.E., Jr., J.H.M., Pratt, V.R.: Fast pattern matching in strings. SIAM J. Comput. 6(2), 323–350 (1977)
15. Kushilevitz, E., Nisan, N.: Communication complexity. Cambridge University Press (1997)
16. Porat, B., Porat, E.: Exact and approximate pattern matching in the streaming model. In: FOCS, pp. 315–323. IEEE Computer Society (2009)
17. Starikovskaya, T.: Communication and streaming complexity of approximate pattern matching. In: CPM. LIPIcs, vol. 78, pp. 13:1–13:11. Schloss Dagstuhl - Leibniz-Zentrum für Informatik (2017)
18. Woodruff, D.P.: Optimal space lower bounds for all frequency moments. In: SODA. pp. 167–175. SIAM (2004)

REINDEER2: Practical Abundance Index at Scale

Yohan Hernandez–Courbevoie[1]🆔, Mikaël Salson[1]🆔, Chloé Bessière[3,4]🆔,
Haoliang Xue[5]🆔, Daniel Gautheret[2]🆔, Camille Marchet[1(✉)]🆔,
and Antoine Limasset[1(✉)]🆔

[1] Univ. Lille, CNRS, Centrale Lille, UMR 9189 CRIStAL, 59000 Lille, France
`{camille.marchet,antoine.limasset}@univ-lille.fr`
[2] I2BC, Université Paris-Saclay, CNRS, CEA, Gif sur Yvette, France
[3] IRMB, INSERM U1183, Hopital Saint-Eloi, Universite de Montpellier,
Montpellier, France
[4] CRCT, Inserm, CNRS, Université Toulouse III-Paul Sabatier, Centre de
Recherches en Cancérologie de Toulouse, Toulouse, France
[5] Gif sur Yvette, France

Abstract. Recent advances in biological sequence indexing have
enabled the efficient querying of sequence presence across massive genomic data repositories. While presence queries have become
tractable at petabyte scale, retrieving quantitative information such as
sequence abundances remains a significant algorithmic challenge. Existing abundance-aware indexes are mostly static, difficult to scale, and
often trade off completeness, precision, or updatability.

We describe a novel discrete abundance index designed for scalability,
dynamic updates, and tunable precision. We combine an inverted index
with probabilistic and exact structures to support fast, memory-efficient
construction and precise high-throughput queries across thousands of
RNA datasets.

Our experiments demonstrate that our method REINDEER2 achieves
one to two orders of magnitude speedup in construction compared
to existing methods, while maintaining comparable or better memory
use. Despite using approximate structures for scalability, REINDEER2
achieves sub-1% error on abundance recovery and correlates strongly
with reference quantifiers like Kallisto. It also supports sequence-level
queries in seconds over thousands of datasets.

Code and experiments github.com/Yohan-HernandezCour
bevoie/REINDEER2

Keywords: Data retrieval · Bioinformatics · Lookup index ·
Abundance index

H. Xue—Independent researcher.
C. Marchet and A. Limasset—Co-last authors.

G. Badkobeh et al. (Eds.): SPIRE 2025, LNCS 16073, pp. 156–171, 2026.
https://doi.org/10.1007/978-3-032-05228-5_14

1   Introduction

Significant progress in algorithms and data structures has benefitted the fields of bioinformatics and molecular biology in parallel to progress in DNA and RNA sequencing. The exponential growth of high-throughput sequencing technologies, combined with a dramatic reduction in cost, has enabled data generation overnight in many platforms and hospitals; but poses scalability challenges.

The reuse of sequencing data from previous studies beeing encouraged, sequencing datasets are stored in large numbers in public repositories, which now hold petabytes of data. In order to enable reanalysis, a fundamental operation is to retrieve in which datasets a query sequence matches. Tools like BLAST [2], based on pairwise sequence alignment, have been instrumental for such queries. However, their reliance on quadratic-time operations makes them computationally prohibitive at current scales. More recently, the problem has been reframed as data retrieval, and methods have shifted toward the use of fixed-length substrings, known as k-mers [17,18]. The core idea behind current strategies is to represent each dataset as a set of k-mers and estimate query-dataset similarity through k-mer intersections [29], using inverted indexes based on different types of data-structures and compression strategy (e.g. [1,8,21]). These k-mer-based approaches have proven scalable and versatile but must scale to billions of k-mers and thousands of datasets at least.

Fewer abundance-aware indexing methods extend this principle. In the context of RNA data, querying k-mer abundance is as important as detecting presence, since abundance comparison is fundamental to understanding gene regulation [4,5]. Therefore, methods specifically adapted to indexing abundance are required. Tools such as REINDEER [19], Needle [12], and Metagraph [14] represent the state of the art, enabling exact or approximate k-mer abundance queries across massive sequence collections. However, REINDEER remains computationally expensive to build at scale. Metagraph achieves scalability through aggressive k-mer filtering, which may miss important signals, especially in cases involving small sequence variants. Entire petabytes-scale databases also exist [9,14]; but while these methods provide a comprehensive overview of our current data, they are resource-intensive to build, making them impractical for new data or clinical settings that need to be built on private servers. Furthermore, their queries remain slow due to the sheer volume of data being scanned. For all these tools, new datasets require complete rebuilding of the index, which is impractical in settings with continuous data inflow. We demonstrate how, using appropriate modeling and data structures, our method can scale to thousands of datasets, support abundance tracking, and allow efficient insertion of new data.

Needle introduced a probabilistic query system indexing a subset of k-mers with coarse abundance estimates. While it excels as a filter, its approximation strategy may be unsuitable for applications requiring base-level accuracy, such as rare variant detection or complex isoform analysis. In our work, we aim to provide a more precise and tunable approximate scheme that balances scalability with accurate abundance estimation.

2   Methods

We introduce the REINDEER2 index, a scalable inverted index for retrieving RNA abundance information across large-scale sequencing datasets. In Subsect. 2.1, we outline the data model and approximations underlying our design. Subsection 2.2 describes strategies for improving scalability, enabling efficient memory usage, and supporting dynamic dataset insertion.

2.1   Underlying Model

Input Representation. The primary input to RNA abundance studies consists of sequencing datasets in textual formats composed of four characters or bases, each line representing a fragment of the molecule, called a *read*. To handle the redundancy inherent in sequencing data, where RNA molecules, and *reads*, appear with varying abundance, we first decompose each *read* at all possible positions into substrings of length k, called k-mers.

It is frequent (and a requirement in many recent indexing methods) to then group k-mers into longer superstrings called *unitigs*, which capture contiguous, biologically relevant sequences [7]. Notably, millions of such *unitig* datasets are readily accessible to the community via the Logan project [10], which removes the burden of *unitig* construction by providing pre-processed, reusable representations of publicly available data.

This transformation reduces storage and computational burden at the indexing level, especially because it usually filters out low-abundance, likely-noisy k-mers. Previous studies observed that all k-mers in a *unitig* have similar abundances [19], therefore standard formats approximate *unitig*-level abundance by averaging their counts, smoothing minor fluctuations at the individual k-mer level. These *unitigs* and their counts are the actual input to REINDEER2. Finally, to reduce index size further, we introduce a novel discretization scheme for abundance values, using a logarithmic encoding with tunable precision (see Subsect. 2.2).

K-mer Set and Abundance Representation. We use Bloom filters to represent each k-mer sets efficiently. The use of discretized abundances enables k-mer abundances to be encoded in a matrix-like superposition of Bloom filters, where each row - Bloom filter - is associated with a dataset and an abundance level, as illustrated in Fig. 1. When a pair (k-mer, abundance) is parsed in a given dataset, the k-mer sequence is hashed to point to its column in the matrix and its minimizer indicates the partition to use (minimizers are the smallest m-mer with $m < k$ in a given k-mer, according to a defined order [25]), then it is added in the row corresponding to its dataset and abundance level. We detail in Subsect. 2.2 how abundance levels are computed.

As described below; the rationale for using Bloom filters is their support for fast streaming k-mer insertion and real-time updates and representing each dataset independently allows easier updates to the index. Following practices from prior works [6,29], we use a single hash function to accelerate queries while controlling the false positive rate through appropriate filter sizing. To minimize space, we encode filters with Roaring bitmaps [16], avoiding explicit storage of all bits.

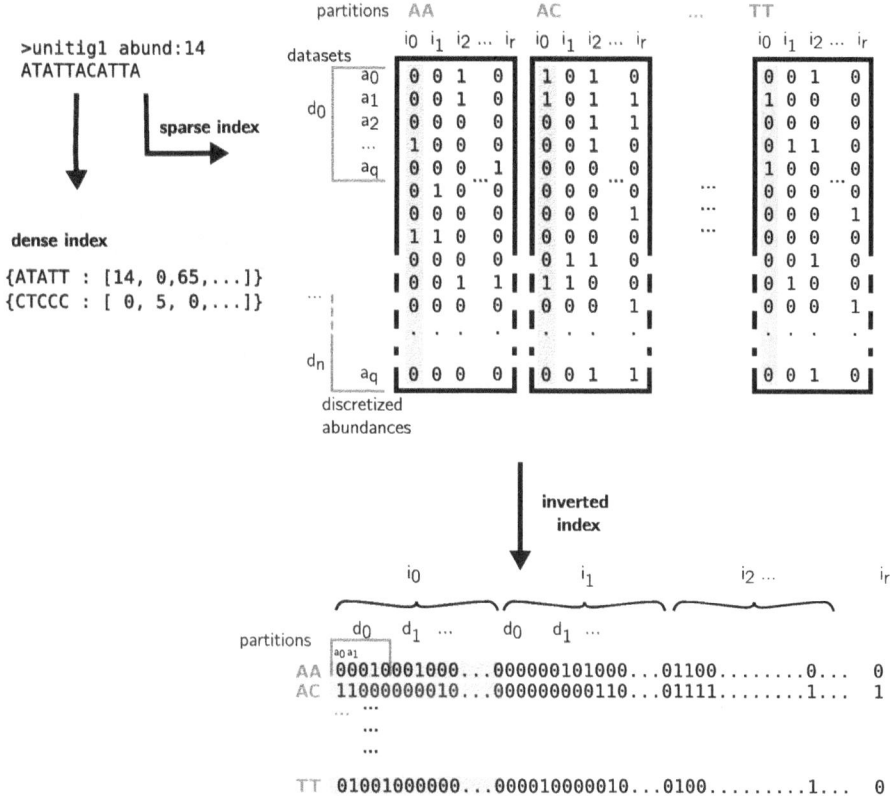

Fig. 1. Simplified view of the REINDEER2 index. Left: dense index, where k-mer occurring in all or most datasets are stored with explicit and exact abundances. Top-right: abstract idea of the sparse structure, where we have one Bloom filter per dataset per abundance (row-wise Bloom filters). Each position i in the Bloom filters range from 0 to r, the size of the partitionned Bloom filters. Bottom: actual inverted index of our sparse structure. Each line is a single Bloom filter containing the information of several datasets. Partitioning is shown with lexicographic minimizers of size 2. The first slice of each partition is shown in gray. (Color figure online)

Normalization. Sequencing depth in RNA-seq indicates how thoroughly the RNA molecules have been sampled, and can vary between experiments. Therefore, *read* abundances or k-mer abundances must be normalized to account for differences in depth across experiments [27]. TPM (Transcripts per million) is a normalization commonly used in RNA studies to account for molecules and sequencing depth abundances when genes are well-known [31].

We work agnostically from known genes, directly on raw data, and with k-mers, therefore, we apply a normalization by dividing abundance values by the total number of (non distinct) k-mers and using per million as a scaling factor to resemble TPM.

Inverted Index and Shared Data Optimization. REINDEER2 stores abundance data in an inverted index, mapping each k-mer to the datasets in which it appears, along with its abundance. This structure allows direct access to abundance values across datasets with a single random access, rather than querying each dataset separately.

A naive per-dataset k-mer index leads to redundant storage of shared sequences, especially common in RNA-seq from the same species (e.g., house-keeping genes). In contrast to prior literature, REINDEER2 instead separates k-mers into two categories: *dense k-mers* found in most datasets, stored in an exact index, and *sparse k-mers* found in fewer datasets, stored in Bloom filters. This hybrid design reduces Bloom filter load, improving accuracy and reducing false positives by decreasing the Bloom filters load factor. *Dense k-mers* are identified on-the-fly during construction (details in Subsect. 2.2).

Query Model and Output. A typical query consists of a *read* or RNA molecule (often >100 bases). Queries are decomposed into overlapping k-mers, which are intersected with the index, and results are reported if the intersection size is larger than a threshold (default 50% of the query set). This threshold is important to account for k-mer content that may differ due to errors or small mutations. For each dataset, the median abundance across all matching k-mers of the query is returned (see Fig. 2) in a csv file where rows are (query, abundance, dataset) triplets.

Two types of errors are possible. The first type occurs when two or more k-mers in the same dataset but with different abundance values have their hashes collide. In this case, several abundance values are available for a single k-mer hash in a dataset. At query time, we keep the minimum, therefore the abundance values may be underestimated in some cases. The second type occurs when the hash of a queried k-mer collides with another k-mer hash that has been indexed. In this case, we report an abundance despite the k-mer absence (see Fig. 2 right). Importantly, false positives from Bloom filters are mitigated by query length: the likelihood that all k-mers in a query are false positives decreases geometrically with the number of k-mers [6, 29]. We analyze query performance and accuracy in Subsects. 3.4 and 3.6.

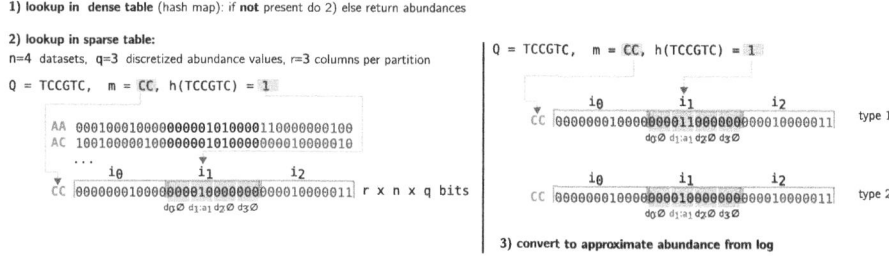

Fig. 2. Examples of query at the k-mer level. Left: query mechanism in REINDEER2. If it is not found in the dense index, queried k-mers are hashed to find their indices in the filter and their partitions are determined by minimizers. Then all datasets are screened to find bits indicating abundances (here abundance level 1 in dataset 1). Right: two types of possible errors. The first type can lead to an under estimation of the abundance level when different abundance levels are reported for a given dataset because only the lowest possible value is retained. The second type is a false positive, reporting an abundance in a dataset where the k-mer is absent.

2.2 Strategies for Improved Scalability

Bloom Filter Factorization. To mitigate the significant memory overhead from using up to billions of Roaring bitmaps, we encode multiple data slices (columns in the matrix-like structure) into a single, larger bitmap. This is achieved in a collision-free manner by assigning a unique offset to the identifiers within each slice. This strategy effectively amortizes the storage overhead across the combined slices while still permitting independent queries on each original slice within the aggregated bitmap, as illustrated in Fig. 3.

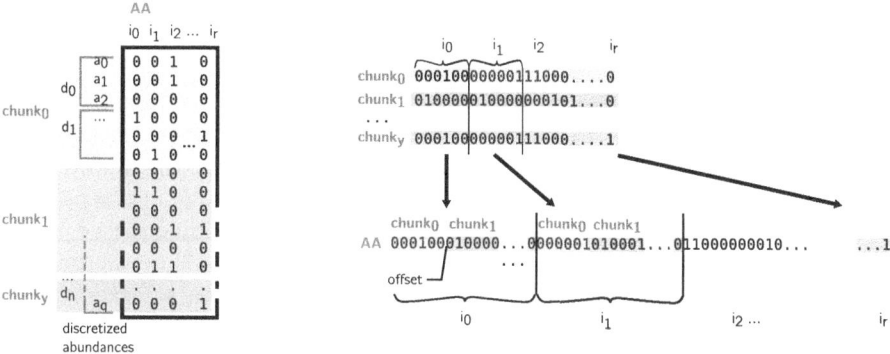

Fig. 3. Schematic view of the sparse index division and merge. The example is shown on a single partition for simplicity. On the left, an abstract view of the structure divided in chunks. On the right, bits of each chunks are added to a single Bloom filter per partition, by doing a union of chunks with offsets.

Partitioning. Minimizer-based partitioning has demonstrated its efficiency for k-mer indexing [13, 20, 21]; therefore, we assign k-mers to partitions based on their minimizer. Partitions correspond to a vertical splitting of the matrix. Thus, each partition represents a subset of the columns in the original index. Figure 1 shows a simplified example with lexicographic minimizers.

This strategy leverages the observation that consecutive k-mers often share the same minimizer, which enables buffered operations that improve both construction and query times by optimizing cache coherence and allow coarse-grained parallelism [24]. Indeed, using hundreds or thousands of partitions reduces the size of the loaded substructure by orders of magnitude, both during construction and at query time. This allows the structure to utilize lower-level caches, thereby accelerating performance. Although the distribution of minimizers in genomic data is known to be skewed [22], we uniformly shuffle a large set of minimizers within each partition to mitigate this bias, resulting in relatively uniform partition sizes in practice.

Merge and Update of REINDEER2 Indexes. Indexing billions of k-mers from very large collections can cause a peak in RAM usage that exceeds available memory, even when using compressed structures. To manage this, REINDEER2 relies on its partition system to build multiple sub-indexes sequentially, effectively lowering this memory peak. Once constructed, these sub-indexes are merged into a final, global index. This merging process is also memory-efficient as it operates on a partition-by-partition basis. At the partition level, merging involves a simple bitmap remapping. For each sub-index, local column positions are converted to global ones by adding an offset. All these remapped bits are then combined into the final, unified bitmap for that partition.

Furthermore, this merging mechanism makes the index inherently updatable. A new dataset can be indexed separately and then efficiently merged with the main index, enabling seamless, incremental updates.

Logarithmic Quantization Scheme for Abundances. One distinctive feature of Needle is its ability to compute optimal quantization levels based on the observed distribution within the indexed collection. However, this approach is computationally expensive and does not scale well when a large number of levels, necessary for high precision, is required. Moreover, maintaining these optimal levels becomes non-trivial when incorporating novel datasets. Here, we propose a more general approach that leverages logarithmic scales to achieve tunable precision over an arbitrarily large input range. Once the maximum abundance is specified, our method guarantees that the quantization system remains fixed even as new datasets are added, thereby enabling efficient updates.

REINDEER introduced the idea of approximating the abundances with the base-2 logarithm. However, this has the drawback of being a rough approximation. We adopt a strategy based on this idea, but adapting the logarithmic base to achieve a tunable precision.

To determine the base B that provides Q levels for handling abundances up to $Max\_Abundance$, we start by considering the formula: $Q \leq \log_B(Max\_Abundance) < Q + 1$. Note that this quantization is often suboptimal because certain abundance values are not observed in practice, $i.e.$ when $\lfloor \log_B(q) \rfloor + 1 < \lfloor \log_B(q + 1) \rfloor$, with $q \in \mathbb{N}^*$.

To address this issue where some values would not be used, we employ an identity function for lower abundances while applying logarithmic scaling to higher values. The threshold t, at which we switch from the identity function to logarithmic scaling, is defined as follows: $\log_B(t+1) - \log_B(t) \leq 1 \iff t \geq \frac{1}{B-1}$, for $B > 1$. Converting an abundance x to its log-transformed value is achieved with:

$$quantization(x) = \begin{cases} x & \text{if } x < t \\ \left\lfloor \log_B(x) + \frac{1}{B-1} - \log_B\left(\frac{1}{B-1}\right) \right\rfloor & \text{otherwise} \end{cases}$$

Using this function, one can determine the optimal base B, given the maximal abundance ($Max\_Abundance$) and the number of abundance levels Q. In practice, as shown in Fig. 4 (left), we observe that using 256 abundance levels allows us to achieve high precision (near 1%), and that the overall abundance range to be handled has a significant impact on the final precision.

Handling of Dense k-mers. Inserting ubiquitous k-mers into thousands of filters is suboptimal with respect to indexing time, index size, and query efficiency. As illustrated in Fig. 4 (right) using a human cancer sample collection (CCLE, detailed in Subsect. 3.1), the prevalence of commonly shared k-mers justifies a specialized handling strategy. Notably, this approach is optional and is particularly beneficial when a significant portion of the dataset content is highly shared.

To address this issue, we propose to handle *dense k-mers*, those appearing in most datasets, separately from *sparse k-mers*, which occur in fewer datasets. In an online procedure, sparse k-mers are inserted into the primary index, while dense k-mers are stored in a dedicated hash table that maps each dense k-mer to its abundance vector. At regular intervals, we re-evaluate the dense k-mers to ensure they continue to satisfy the density criterion by iterating on the abundance vectors. Since all k-mers are initially stored in the dense table, dense k-mers are reevaluated very frequently at the beginning (after each dataset) in order to filter out sparse k-mers as quickly as possible. If a k-mer no longer qualifies as dense, it is removed from the dense table and reinserted into the sparse index according to its occurrences and abundances. After a few dozen datasets, dense k-mers are no longer reevaluated. This method guarantees the absence of dense k-mers in the sparse index while accepting a certain proportion of sparse k-mers in the dense table. Overall, this strategy avoids redundant insertions of k-mers common to thousands of datasets and enables more efficient storage of their abundance vectors.

At query time, each k-mer is first searched for in the dense table; if it is not found there, the search proceeds to the sparse index. In practice, this strategy incurs negligible overhead in terms of query time and index size, while enhancing precision via two mechanisms. First, dense k-mers are stored exactly, eliminating the possibility of false positives caused by the use of Bloom filters. Second, by excluding dense k-mers from the Bloom filters in the sparse index, the overall false positive rate for sparse k-mers is reduced.

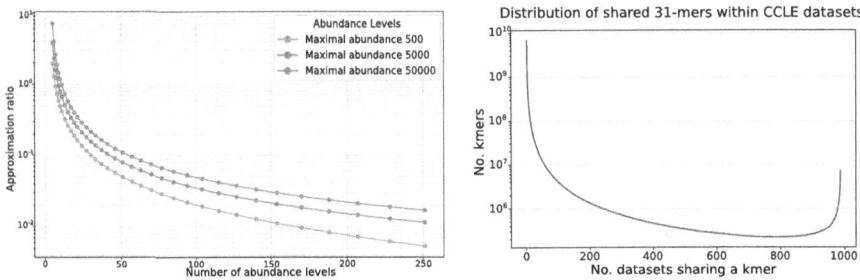

Fig. 4. Left: abundance precision respective to the amount of abundance levels used and the maximal abundance handled. For the maximal number of abundance levels (256), the base B is 1.010 for $Max\_Abundance = 500$, 1.023 for $Max\_Abundance = 5,000$ and 1.033 for $Max\_Abundance = 50,000$. Right: distribution of 31-mers according to the number of CCLE datasets in which they appear.

3   Results

3.1   Datasets and Sequences Description

We used pre-computed *unitigs* from Logan Project [10] constructed from the Cancer Cell Line Encyclopedia (CCLE) collection [3] that includes curated sequencing data from more than 1,000 different human cancer lab-grown cells. We used *unitigs* from the Logan initiative, that provides cleaned *unitigs* of 987 of the CCLE datasets, built for $k = 31$ for any k-mer seen more than twice. Therefore all our indexes are built for the same k value, and very rare k-mers may be absent. The whole collection represents a rough 11.6 billion *unitigs*, with 164 billion 31-mers (statistics on sequences were computed with seqkit [28]).

In addition, we selected a second dataset to demonstrate our scalability capabilities. We filtered and downloaded 10,000 human RNA-seq accessions from the European Nucleotide Archive so that the datasets contain at least 10 million *reads*, and are associated with melanoma/melanocyte meta-data. The corresponding *unitigs* represent a total of 67 billion sequences and 3.02 trillion bases or 992 billion 31-mers.

	REINDEER2				REINDEER				Needle (w=41)				Needle (w=51)				Metagraph			
Files	CPU	Sys	RAM	Disk	CPU	Sys	RAM	Disk	CPU	Sys	RAM	Disk	CPU	Sys	RAM	Disk	CPU	Sys	RAM	Disk
1	6	1	0.1	0.05	674	1118	4.3	1.0	603	16	10.6	0.09	444	7	5.3	0.05	6292	365	25.4	2.7
4	49	6	0.7	0.24	1912	2189	7.8	2.3	2523	68	35.4	0.09	1950	42	17.7	0.05	21178	957	25.5	8.6
16	551	38	3.5	1.17	5446	5068	16.4	6.4	11499	452	80.1	0.09	9209	245	45.3	0.05	×	×	×	×
32	2433	111	9.1	2.9	11955	8834	29.3	14.8	×	×	×	×	22487	786	94.1	0.05	×	×	×	×
64	6628	275	13.3	14	–	–	–	–	×	×	×	×	49182	1986	106.4	–	×	×	×	×
128	16237	612	18.5	29	–	–	–	–	×	×	×	×	96636	3591	103.3	0.10	×	×	×	×
256	38310	1332	21.7	61	–	–	–	–	×	×	×	×	185629	6736	97.9	0.19	×	×	×	×
987	218776	5917	30.5	258	–	–	–	–	×	×	×	×	×	×	×	×	×	×	×	×

CPU time (s) — System time (s) — Max RAM (GB) — Disk usage (GB)

Fig. 5. Resource usage during sparse index construction on CCLE collection. "×" stands for tools that failed to run and "–" for tools that were not run by lack of time.

3.2 REINDEER2 Outperforms State-of-the Art in Resource Usage for Construction

First, we wanted to describe the inherent advantages and drawbacks of using state-of-the-art abundance indexes on a server with 128 GB of RAM, namely REINDEER, Needle, and Metagraph, in comparison to REINDEER2. We excluded recent structures dedicated to indexing abundances of a single dataset, such as [23].

We constructed indexes representing subsets of the CCLE collection with growing numbers of datasets. In order to align with Needle capabilities, we used a low number of quantization levels (16) with REINDEER2 and a discretization in logarithm base 2 with REINDEER. The Bloom filters size parameter used for REINDEER2 was 28 and the minimizer size was 11. Needle indexes was built with different values for the parameter w which is the window size in which the smallest k-mer is sampled. We reported the results in Fig. 5. We observed that REINDEER2 is one or several orders of magnitude faster to construct than all its competitors. It is also memory-efficient, thanks to its merging system, which prevents high memory usage peaks. While index size optimization is not the primary focus, the resulting index remains practical and can even be competitive with state-of-the-art approaches, except for Needle achieving smaller index sizes because of sampling. Indeed, Needle only indexing k-mers with the smallest hash in windows of 10 or 20 positions (corresponding respectively to the parameter values w = 41 and w = 51), the amount of information to store is lower.

It is worth noting that this benchmark is not entirely fair for several reasons. Needle indexes only a subset of k-mers with false positives, Metagraph performs stringent graph cleaning before indexing, while REINDEER indexes all k-mers of *unitigs* exactly. More importantly, REINDEER2, like its predecessor, relies on pre-constructed *unitigs*, which can be orders of magnitude smaller than the original FASTQ files used by Needle or Metagraph. However, the pre-processing steps of Metagraph and Needle cannot be isolated from their own usage, making a direct comparison challenging.

We still argue that this benchmark effectively highlights the practical usage cost of these tools, as *unitigs* are emerging as efficient alternatives to FASTQ files and extremely efficient techniques now exist to compute them rapidly and in a memory-efficient manner [11,15] (typically around 1 h for a whole genome human dataset, usually larger than RNA samples) when they are not already available [9].

3.3 Increased Precision for a Marginal Overhead

In the previous section, we used a low number of quantization levels (16) to adapt to the other tools. In REINDEER2, however, we advocate for high-precision abundance quantization, which involves using a substantially larger number of levels. A natural concern is that this increased precision might significantly increase the cost of both index construction and query operations. Similarly, employing larger Bloom filters to achieve lower false positive rates could result in higher resource consumption, potentially limiting the method's practicality in some applications. To evaluate these concerns, we experimented with various settings for the number of abundance levels and a wide range of Bloom filter sizes (from $2^{26} \approx 67M$ bits to $2^{34} \approx 17G$ bits). Figure 6 reports the corresponding construction times and index sizes. Despite exponential increases in both Bloom filter size and abundance levels, resulting in the number of implicit bits increasing by more than four orders of magnitude, the overall index size increases by less than an order of magnitude. This highlights the efficiency of the sparse data structure relying on Roaring and Bloom filters factorization (as described in Subsect. 2.2), and the fact that high precision indexing is affordable in practice in terms of resources required.

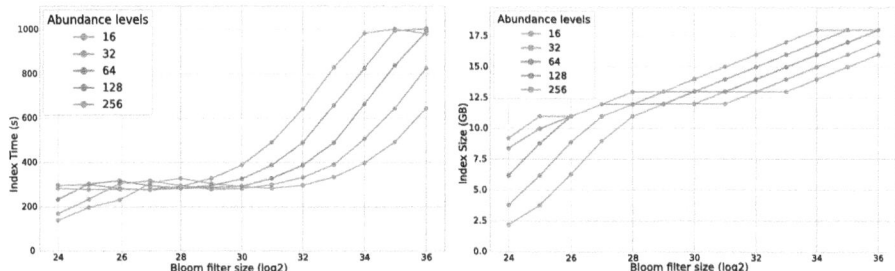

Fig. 6. Construction time (left) and index size (right) according to the number of abundance levels on various Bloom filter size of 2^b on 128 files from CCLE collection (sparse index).

3.4 High Throughput Queries

Query performance is known to degrade as indexes grow larger. Here we wanted to assess the impact of index size, driven both by Bloom filter size and number of abundance levels, on query performances (Fig. 7). We queried human full-length

RNAs (10,851 Refseq human transcripts) as positive queries, and sequences unrelated to the index content (randomly generated), as negative queries to have a better idea of real performance.

In a first experiment we increased Bloom filter sizes and abundances levels on 128 files of the CCLE collection. We queried 10,851 human transcripts to observe that the Bloom filter size has little impact on query throughput, in contrast to the number of abundance levels.

In a second experiment, for the same query, we selected a large Bloom filter size (2^{32}) and a high number of abundance levels (256). The index showed high efficiency, processing millions of checks per second. As expected, the running time increases linearly with the number of documents in the index, with negative queries being noticeably faster to execute. Since each query must be evaluated across all indexed datasets, the effective query throughput is inversely proportional to the number of documents in the index.

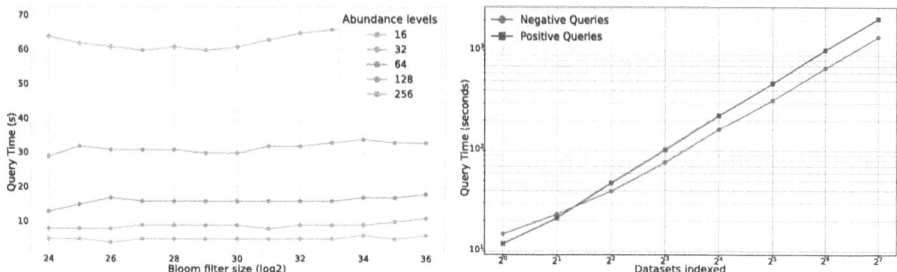

Fig. 7. Query time in the sparse index, according to the number of abundance levels on various Bloom filter size of 2^b on 128 files from CCLE collection (left). Time to perform a batch of queries on a growing index size (right). The positives queries are 10,851 human transcripts totalling 40 millions bases. The negatives queries are a file with matching number of sequences and length distribution composed of random sequences.

3.5 Scalability on RNA-Seq Datasets

To assess the scalability of REINDEER2, we indexed a large collection of melanoma datasets. This collection was composed of ten thousand files covering a total of 67 billion sequences and 3.02 trillion bases with each file containing 86.5 million 31-mers on average. This indexing step was completed in 2 days and 10 h using 64 threads.

3.6 Cross-Tool Quantification Accuracy of REINDEER2

Here we wanted to validate that REINDEER2's abundance estimations could be trusted for research experiments. We compared REINDEER2's results to REINDEER to assess the impact of false positives in comparison to an exact structure.

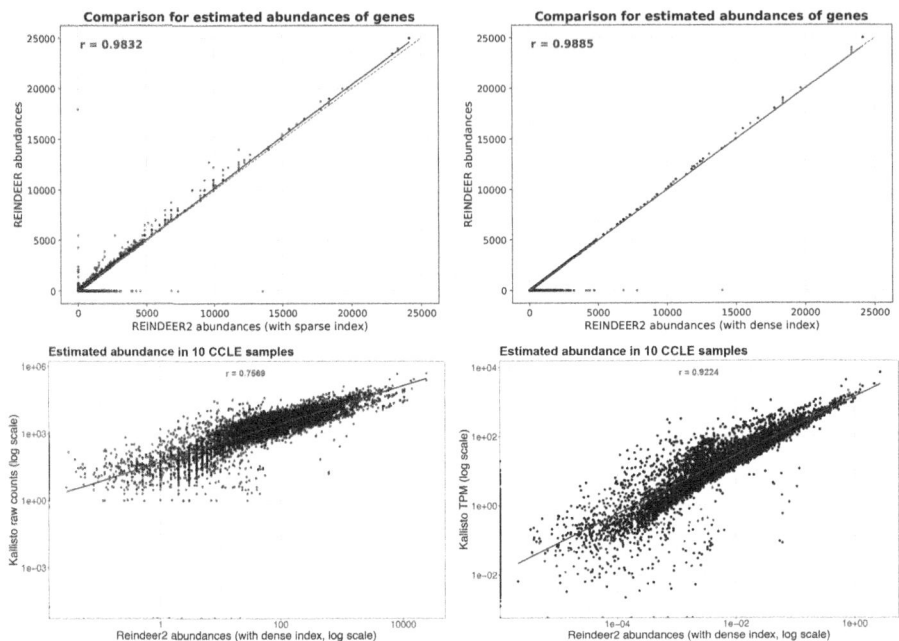

Fig. 8. Pearson correlations for quantifying 1000 genes in a CCLE subset. In the top row, REINDEER and REINDEER2 are compared for queries at the gene level with top right figure showing REINDEER2's regular strategy where dense k-mers are handled outside of Bloom filters and top left figure shows the impact on the query accuracy when these dense k-mers are inserted in filters. Bottom row shows comparisons to Kallisto's TPM and raw counts with REINDEER2's regular index. Bottom figures are shown in log-log scale to better visualize small and medium values.

A complete review of REINDEER's quantification accuracy was also performed in [4]), showing REINDEER yields results well correlated even to *in vitro* precise quantification techniques, therefore it can serve as a direct mean of comparison for our approach. We also added Kallisto [30] to the comparison, that is one of the most established RNA quantification tool, designed to work at the single dataset level and using well known RNA molecule models. Kallisto's index is built on a fundamentally different input, making direct performance comparisons with our method inappropriate. We selected a subset of 10 CCLE datasets where 1,000 human genes were quantified using both Kallisto and REINDEER. REINDEER was built on the Logan unitigs of the 10 CCLE samples and queried using default parameters for indexation and retrieval ($k = 31$). For raw counts and transcripts per million (TPM) that were extracted we kept similar settings to those in [4]. REINDEER2 was used to query directly the CCLE indexes built in the previous sections, with 255 abundance levels and a Bloom filter size of 2^{32}.

We expected differences between approaches, since Kallisto assigns sequences to the different molecules of the database based on an expectation-maximization algorithm to resolve multiple assignments. Kallisto can yield raw counts and normalized counts using TPM (see Subsect. 2.1). REINDEER and REIN-DEER2's result must be closer since they index the samples and quantify a query sequence by seeking its k-mers in the index, and their normalizations are the same.

REINDEER2 showed strong correlation with the exact index of REINDEER. It is worth noting that handling dense k-mers separately enhances accuracy since their abundances are stored exactly and this reduces the overall false positives of the sparse index. This strategy improves results compared to a sparse-only insertion approach, as illustrated in Fig. 8, top row, right. Query sequences longer than 1,000 bases were less impacted by false positives, leading to improved correlations for the gene queries compared to k-mer queries (middle row, right with dense k-mers handling). The correlation with Kallisto confirms that REIN-DEER2 is in the same ballpark as REINDEER (based on [4]'s results) when comparing Kallisto's results, especially for normalized counts.

In all experiments, querying the sequences corresponding to 1,000 genes took approximately 6 to 8 s.

4   Discussion

We presented REINDEER2, a scalable index enabling fast, accurate k-mer abundance queries across large datasets via an inverted index structure with tunable precision and dense k-mer handling. Compared to prior methods, REINDEER2 offers superior speed, dynamicity, and maintains strong agreement with established quantifiers like Kallisto. Our design benefits from Bloom filter factorization and partitioning. While current probabilistic structures don't fully exploit k-mer consecutiveness, future improvements could reduce redundancy and enhance compression. Abundance estimates are robust overall, though short-sequence queries and high-multiplicity k-mers remain challenging. Alternative aggregation and collision-handling strategies (e.g. [26]) may improve accuracy further. Finally, our discretization scheme opens paths for richer metadata indexing and dynamic updates, supporting future extensions to transactional and real-time operations on large, evolving datasets.

Acknowledgments. The authors have no competing interests to declare that are relevant to the content of this article.

References

1. Alanko, J.N., Vuohtoniemi, J., Mäklin, T., Puglisi, S.J.: Themisto: a scalable colored k-mer index for sensitive pseudoalignment against hundreds of thousands of bacterial genomes. Bioinformatics **39**(Supplement_1), i260–i269 (2023)
2. Altschul, S.F., Gish, W., Miller, W., Myers, E.W., Lipman, D.J.: Basic local alignment search tool. J. Mol. Biol. **215**(3), 403–410 (1990). https://doi.org/10.1016/S0022-2836(05)80360-2
3. Barretina, J., et al.: The Cancer Cell Line Encyclopedia enables predictive modelling of anticancer drug sensitivity. Nature **483**(7391), 603–607 (2012)
4. Bessière, Cet al.: Transipedia. org: k-mer-based exploration of large RNA sequencing datasets and application to cancer data. Genome Biol. **25**(1), 266 (2024)
5. Bessière, C., et al.: A strong internal promoter drives massive expression of yeats-domain devoid mllt3 transcripts in hsc and most lethal aml. Cancer Commun. (London, England) (2025)
6. Bradley, P., Den Bakker, H.C., Rocha, E.P., McVean, G., Iqbal, Z.: Ultrafast search of all deposited bacterial and viral genomic data. Nat. Biotechnol. **37**(2), 152–159 (2019)
7. Cairo, M., Medvedev, P., Obscura Acosta, N., Rizzi, R., Tomescu, A.I.: Optimal omnitig listing for safe and complete contig assembly. In: 28th Annual Symposium on Combinatorial Pattern Matching (CPM 2017), pp. 29–1. Schloss Dagstuhl–Leibniz-Zentrum für Informatik (2017)
8. Campanelli, A., Pibiri, G.E., Fan, J., Patro, R.: Where the patterns are: repetition-aware compression for colored de bruijn graphs. J. Comput. Biol. **31**(10), 1022–1044 (2024)
9. Chikhi, R., Raffestin, B., Korobeynikov, A., Edgar, R., Babaian, A.: Logan: Planetary-Scale Genome Assembly Surveys Life's Diversity. bioRxiv (2024). https://doi.org/10.1101/2024.07.30.605881. https://www.biorxiv.org/content/early/2024/07/31/2024.07.30.605881
10. Chikhi, R., Raffestin, B., Korobeynikov, A., Edgar, R.C., Babaian, A.: Logan: planetary-scale genome assembly surveys life's diversity. In: bioRxiv, pp. 2024–07 (2024)
11. Cracco, A., Tomescu, A.I.: Extremely fast construction and querying of compacted and colored de bruijn graphs with ggcat. Genome Res. **33**(7), 1198–1207 (2023)
12. Darvish, M., Seiler, E., Mehringer, S., Rahn, R., Reinert, K.: Needle: a fast and space-efficient prefilter for estimating the quantification of very large collections of expression experiments. Bioinformatics **38**(17), 4100–4108 (2022)
13. Fan, J., Khan, J., Singh, N.P., Pibiri, G.E., Patro, R.: Fulgor: a fast and compact k-mer index for large-scale matching and color queries. Algor. Molec. Biol. **19**(1), 3 (2024)
14. Karasikov, M., et al.: Indexing all life's known biological sequences. In: bioRxiv (2024). https://doi.org/10.1101/2020.10.01.322164. https://www.biorxiv.org/content/early/2024/06/07/2020.10.01.322164
15. Khan, J., Kokot, M., Deorowicz, S., Patro, R.: Scalable, ultra-fast, and low-memory construction of compacted de bruijn graphs with cuttlefish 2. Genome Biol. **23**(1), 190 (2022)
16. Lemire, D., et al.: Roaring bitmaps: Implementation of an optimized software library. Softw. Pract. Exp. **48**(4), 867–895 (2018)
17. Marchet, C.: Advances in colored k-mer sets: essentials for the curious. arXiv preprint arXiv:2409.05214 (2024)

18. Marchet, C., Boucher, C., Puglisi, S.J., Medvedev, P., Salson, M., Chikhi, R.: Data structures based on k-mers for querying large collections of sequencing data sets. Genome Res. **31**(1), 1–12 (2021)
19. Marchet, C., Iqbal, Z., Gautheret, D., Salson, M., Chikhi, R.: REINDEER: efficient indexing of k-mer presence and abundance in sequencing datasets. Bioinformatics **36**(Supplement_1), i177–i185 (2020)
20. Marchet, C., Kerbiriou, M., Limasset, A.: Blight: efficient exact associative structure for k-mers. Bioinformatics **37**(18), 2858–2865 (2021)
21. Marchet, C., Limasset, A.: Scalable sequence database search using partitioned aggregated bloom comb trees. Bioinformatics **39**(Supplement_1), i252–i259 (2023)
22. Pibiri, G.E.: Sparse and skew hashing of k-mers. Bioinformatics **38**(Supplement_1), i185–i194 (2022)
23. Pibiri, G.E.: On weighted k-mer dictionaries. Algor. Molec. Biol. **18**(1), 3 (2023)
24. Pibiri, G.E., Shibuya, Y., Limasset, A.: Locality-preserving minimal perfect hashing of k-mers. Bioinformatics **39**(Supplement_1), i534–i543 (2023)
25. Roberts, M., Hayes, W., Hunt, B.R., Mount, S.M., Yorke, J.A.: Reducing storage requirements for biological sequence comparison. Bioinformatics **20**(18), 3363–3369 (2004)
26. Robidou, L., Peterlongo, P.: `findere`: fast and precise approximate membership query. In: Lecroq, T., Touzet, H. (eds.) SPIRE 2021. LNCS, vol. 12944, pp. 151–163. Springer, Cham (2021). https://doi.org/10.1007/978-3-030-86692-1_13
27. Robinson, M.D., Oshlack, A.: A scaling normalization method for differential expression analysis of rna-seq data. Genome Biol. **11**, 1–9 (2010)
28. Shen, W., Le, S., Li, Y., Hu, F.: Seqkit: a cross-platform and ultrafast toolkit for fasta/q file manipulation. PLoS ONE **11**(10), e0163962 (2016)
29. Solomon, B., Kingsford, C.: Fast search of thousands of short-read sequencing experiments. Nat. Biotechnol. **34**(3), 300–302 (2016)
30. Sullivan, D.K., et al.: kallisto, bustools and kb-python for quantifying bulk, single-cell and single-nucleus RNA-seq. Nat. Protocols. 1–21 (2024)
31. Wagner, G.P., Kin, K., Lynch, V.J.: Measurement of mrna abundance using rna-seq data: Rpkm measure is inconsistent among samples. Theory Biosci. **131**, 281–285 (2012)

Efficient Computation of Closed Substrings

Samkith K. Jain and Neerja Mhaskar(✉)

Department of Computing and Software, McMaster University, Hamilton, Canada
{kishors,pophlin}@mcmaster.ca

Abstract. A *closed string* u is either of length one or contains a border that occurs only as a prefix and as a suffix in u and nowhere else within u. In this paper, we present a fast $\mathcal{O}(n \log n)$ time algorithm to compute all $\mathcal{O}(n^2)$ closed substrings by introducing a compact representation for all closed substrings of a string $w[1..n]$, using only $\mathcal{O}(n \log n)$ space. We also present a simple and space-efficient solution to compute all maximal closed substrings (MCSs) using the suffix array (SA) and the longest common prefix (LCP) array of $w[1..n]$. Finally, we show that the exact number of MCSs $(M(f_n))$ in a Fibonacci word f_n, for $n \geq 5$, is $\approx \left(1 + \frac{1}{\phi^2}\right) F_n \approx 1.382 F_n$, where ϕ is the golden ratio.

Keywords: Closed Strings · Maximal Closed Substrings · Fibonacci Words

1 Introduction

A *closed string* u is either of length one or contains a border that occurs only as a prefix and as a suffix in u and nowhere else within u; otherwise, it is called *open*. In other words, the longest (possibly overlapping) border of a closed substring u does not have internal occurrences in u. For example, in the string $abcab$, ab occurs only as a prefix and a suffix. An occurrence of a substring u of a string w is said to be *maximal right (left)-closed* if it is a closed string and cannot be extended to the right (left) by a character to form another closed string. A *maximal closed substring (MCS)* of a string w is a substring that is both maximal left and maximal right-closed. An MCS of length one is called a *singleton* MCS; otherwise, it is called a *non-singleton* MCS. For example, in string $w[1..8] = abaccaba$, the non-singleton MCSs are $w[1..8] = abaccaba$, $w[3..6] = acca$, $w[1..3] = w[6..8] = aba$, $w[4..5] = cc$, and the singleton MCSs are all occurrences of a and b. Observe that $w[2..7] = baccab$, $w[2..8] = baccaba$, $w[1..7] = abaccab$, and both occurrences of c are closed but not maximal.

N. Mhaskar—Research funded by the Natural Sciences & Engineering Research Council of Canada [Grant Number RGPIN-2024-06915].

G. Badkobeh et al. (Eds.): SPIRE 2025, LNCS 16073, pp. 172–187, 2026.
https://doi.org/10.1007/978-3-032-05228-5_15

The concept of open and closed strings was introduced by Fici in [12] to study the combinatorial properties of strings, with connections to Sturmian and Trapezoidal words. Then concepts of *Longest Closed Factorization* and *Longest Closed Factor Array* were introduced in [2], with improved solutions later proposed in [7]. Later, [3] introduced efficient algorithms for minimal and shortest closed factors. Related problems such as the k-closed string problem [1] and combinatorial problems on closed factors in Arnoux-Rauzy and m-bonacci words [14,24] have also been explored.

It has been shown that a string contains at most $\mathcal{O}(n^2)$ *distinct* closed factors [5]. More recently (in 2024), [23] showed that the maximal number of distinct closed factors in a word of length n is $\frac{n^2}{6}$. In early 2025, [20] presented a linear algorithm to count the number of distinct closed factors of a string in $\mathcal{O}(n \log \sigma)$ time and another in $\mathcal{O}(n)$ time. They also described methods to enumerate all closed and distinct closed factors, achieving a complexity of $\mathcal{O}(n^2)$, which is further optimized to $\mathcal{O}(n\sqrt{\log n} + \text{output} \cdot \log n)$ for distinct closed substrings, using weighted ancestor queries.

In [4] Badkobeh et al. introduced the notion of MCSs and proposed a $\mathcal{O}(n\sqrt{n})$ loose upper bound on the number of MCSs in a string of length n. They also proposed an algorithm using suffix trees to compute the MCSs for strings on a binary alphabet in $\mathcal{O}(n \log n + n\sqrt{n})$ time. Badkobeh et al. in [6] extended their algorithm to a general alphabet with a time complexity of $\mathcal{O}(n \log n)$, implicitly improving the bound on the total number of MCSs, this was later directly proved by Kosolobov in [18].

In this paper, we propose solutions to the following problems:

Compact Representation of All $\mathcal{O}(n^2)$ Closed Substrings: Given a string $w[1..n]$ we define a compact representation of all $\mathcal{O}(n^2)$ closed substrings of w using $\mathcal{O}(n \log n)$ space.

Compute All $\mathcal{O}(n^2)$ Closed Substrings: Given a string $w[1..n]$, we compute all $\mathcal{O}(n^2)$ closed substrings of w in a compact representation using the suffix array (SA) and the longest common prefix array (LCP), in $\mathcal{O}(n \log n)$ time.

Compute All MCSs: Given a string $w[1..n]$ we find all MCSs of w using suffix arrays (SA) and longest common prefix arrays (LCP) in $\mathcal{O}(n \log n)$ time.

Exact Number of MCSs in a Fibonacci Word: Given a Fibonacci word f_n, we give an exact equation to find the number of MCSs in f_n is denoted by $M(f_n)$ and is defined in Equation (8). We also show that $M(f_n) \approx 1.382 F_n$.

In Sect. 2, we introduce the terminology and data structures used in the paper. In Sect. 3 we define a compact representation of all closed substrings of a given string $w[1..n]$. In Sect. 4, we present algorithms to efficiently compute all closed substrings in a compact representation, as well as all MCSs, in $\mathcal{O}(n \log n)$ time. In Sect. 5 we provide a formula to compute the exact number of MCSs in a Fibonacci word. Finally, we present experimental analysis in Sect. 6.

2   Preliminaries

An alphabet Σ, of size σ, is a set of symbols that is totally ordered. A ***string (word)***, written as $w[1..n]$, is an element of Σ^*, whose length is represented by

n. The **empty string** of length zero is denoted by ε. A **substring (factor)** $w[i..j]$ of w is a string, where $1 \leq i, j \leq n$, if $j > i$ then the substring is ε. The substring $w[i..j]$ is called a **proper substring**, if $j - i + 1 < n$. A substring $w[i..j]$ where $i = 1$ ($j = n$) is called a **prefix** (**suffix**) of w. If a substring u of w is both a proper prefix and a proper suffix of w, then it is called a **border**. A **cover** of a string w is a substring u of w such that w can be constructed by overlapping and/or adjacent instances of u. If $|u| < |w|$, then it is said to be a **proper cover** (for more details on covers, see the survey [19]).

A string $w[1..n]$ is **periodic** with period t' if $w[i] = w[i + t']$ holds for all $1 \leq i \leq n-t'$. The **shortest period** of w is the smallest positive integer $t \leq t'$ for which this condition holds. A periodic substring $w[i..j]$ of string $w[1..n]$ with shortest period t is called a **run** if its length $j - i + 1 \geq 2t$ and its periodicity cannot be extended to the left or right, that is, $i = 1$ or $w[i-1] \neq w[i+t-1]$, and $j = n$ or $w[j + 1] \neq w[j - t + 1]$.

A **gapped repeat** is a substring of $w[1..n]$ of the form uvu, where u and v are non-empty strings. The substring u is called the **arm** of the gapped repeat. The period[1] of the gapped repeat is $|u| + |v|$. An occurrence of a gapped repeat is called a **maximal gapped repeat** if it cannot be extended to the left or to the right by at least one symbol while preserving its period. A maximal gapped repeat, which is also closed, is referred to as a **gapped-MCS**.

The **suffix array** $\mathsf{SA}[1..n]$ of $w[1..n]$ is an integer array of length n, where each entry $\mathsf{SA}[i]$ points to the starting position of the i-th lexicographically least suffix of w. The **longest common prefix array** $\mathsf{LCP}[1..n]$ of $w[1..n]$ is an array of integers that represent the length of the longest common prefix of suffixes $\mathsf{SA}[i-1]$ and $\mathsf{SA}[i]$ for all $2 \leq i \leq n$, we assume $\mathsf{LCP}[1] = 0$.

A closed substring $u = w[i..j]$ of a string $w[1..n]$ is said to be **right-extendible** to a maximal right-closed substring $r = w[i..j']$ if $j < j' \leq n$ and every prefix $w[i..k]$ of r with $j < k < j'$ is also closed. In this case, r is the unique maximal right-closed extension of u.

3   Compact Representation of All Closed Substrings

A string $w[1..n]$ contains $\mathcal{O}(n^2)$ closed substrings, which can naively be computed in $\mathcal{O}(n^2)$ time by computing the border of all substrings of w. In this section, we define a compact representation for all $\mathcal{O}(n^2)$ closed substrings in Theorem 1 that requires only $\mathcal{O}(n \log n)$ space. Algorithm 2 in Sect. 4 presents an $\mathcal{O}(n \log n)$ time solution to compute all closed substrings of w in a compact representation of size $\mathcal{O}(n \log n)$.

Let u and v be the proper closed prefixes of the maximal right-closed substring r. Then, by definition of right-extendibility we define an equivalence relation R as follows:

$$R = \{(u, v) \mid u \text{ and } v \text{ are both right-extendible to } r\}.$$

---
[1] Note that the term period, as used in the definition of a gapped repeat, differs from its earlier usage in the context of periodic strings. Throughout this paper, the term period is consistently used in reference to periodic strings or runs.

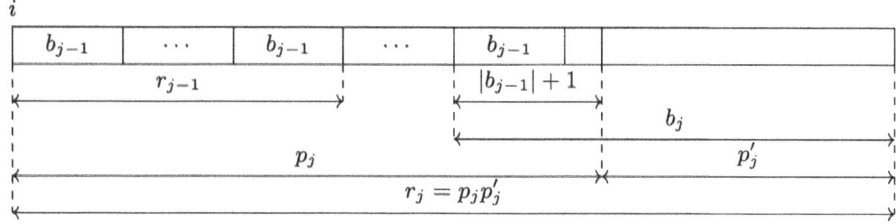

Fig. 1. Illustration of the second case in the proof of Lemma 2, showing that $|p_j|$ can be computed from $|b_j|$, $|b_{j-1}|$ and $|r_j|$, such that $p_j = r_j$ or $p_j \in E_{r_j}$. Note that when $p_j = r_j$, $p'_j = \varepsilon$.

Next, we define the equivalence class E_r of r under the relation R as follows:

$$E_r = \{u \mid u \text{ is right-extendible to } r\}. \tag{1}$$

Lemma 1. *Given a string $w[1..n]$, if $w[1..m]$ is the shortest maximal right-closed prefix of w, then $w[1..m] = \lambda^m$, where $\lambda \in \Sigma$ and $1 \leq m \leq n$. Moreover, each prefix of λ^m is also closed.*

Proof. We prove the lemma in two parts. First part by contradiction. Suppose that $w[1..m]$ is the shortest maximal right-closed prefix of w, and that $w[1..m] \neq \lambda^m$. W.l.o.g let $w[1] = \lambda$. Let $1 < j \leq m$ be the smallest index such that $w[j] = \lambda_1$ and $w[1..j-1] = \lambda^{j-1}$, where $\lambda_1 \neq \lambda \in \Sigma$. Then, $w[1..j] = \lambda^j \lambda_1$. However, by definition of maximal right-closed string, $w[1..j-1] = \lambda^{j-1}$ is a shorter maximal right-closed string of w—a contradiction.

Second, for any string of the form $u = \lambda^m$, consider its prefixes $p = u[1..j]$, where $1 \leq j \leq m$. If $j = 1$, then p is closed trivially. If $j \geq 2$, then the border length of each prefix p is equal to $j - 1$, which is maximum and unique. Thus, all the prefixes of $u = \lambda^m$ are closed.   □

Let r denote a maximal right-closed substring, b its border, and p its shortest closed prefix such that $p = r$ or $p \in E_r$. Let $r_1, r_2, \ldots, r_{K_i}$ be all the maximal right-closed substrings starting at index i of a string $w[1..n]$, where $1 \leq i \leq n$, and K_i is the number of maximal right-closed substrings starting at index i, such that $|r_1| < |r_2| < \cdots < |r_{K_i}|$.

In Eq. (2) we present a formula to compute the length of the shortest closed prefix p_j of r_j such that $p_j = r_j$ or $p_j \in E_{r_j}$, where $1 \leq j \leq K_i$. Lemma 2 proves Eq. (2) and is illustrated in Fig. 1.

Lemma 2. *Let $1 \leq j \leq K_i$. The length of the shortest closed prefix p_j of r_j, such that $p_j = r_j$ or $p_j \in E_{r_j}$, is given by:*

$$|p_j| = \begin{cases} 1, & \text{if } j = 1, \\ |r_j| - |b_j| + |b_{j-1}| + 1, & \text{if } 1 < j \leq K_i. \end{cases} \tag{2}$$

where b_j and b_{j-1} are the longest borders of r_j and r_{j-1}, respectively.

Proof. We prove the Lemma by considering two cases. For $j = 1$, by Equation (1) and Lemma 1, all prefixes of r_1 are closed and $|p_1| = 1$.

For $1 < j \le K_i$, by definition of a closed substring, the shortest closed prefix p_j starting at index i and longer than r_{j-1} (whose longest border has length $|b_{j-1}|$) must have the longest border of length exactly $|b_{j-1}| + 1$. Either $p_j = r_j$ or by definition of right-extendibility p_j extends to $r_j = p_j p'_j$, with the longest border b_j, and each prefix of r_j with length at least $|p_j|$ is closed. Since $|p'_j| = |b_j| - (|b_{j-1}| + 1)$, we have $|r_j| = |p_j| + |p'_j| = |p_j| + |b_j| - (|b_{j-1}| + 1)$. Solving for $|p_j|$, we obtain the desired formula: $|p_j| = |r_j| - |b_j| + |b_{j-1}| + 1$. □

Lemma 3 (Theorem 1 in [18]). *A string $w[1..n]$ contains at most $\mathcal{O}(n \log n)$ maximal right (left)-closed substrings.*

Theorem 1 follows directly from Eq. (1), Lemma 2 and 3.

Theorem 1. *The compact representation of all closed substrings of $w[1..n]$ requires at most $\mathcal{O}(n \log n)$ space and is given by:*

$$\mathcal{C}(w) = \{(i, |p_j|, |r_j|) \mid 1 \le i \le n, \ 1 \le j \le K_i\}, \tag{3}$$

where K_i is the number of maximal right-closed substrings starting at position i, and for each j-th such substring r_j, starting at index i, p_j is the shortest closed prefix of r_j such that $p_j = r_j$ or $p_j \in E_{r_j}$.

The set of all closed substrings of $w[1..n]$, denoted by Closed(w), can be enumerated using the following equation:

$$\text{Closed}(w) = \bigcup_{(i,|p_j|,|r_j|) \in \mathcal{C}(w)} \{w[i \ldots i + \ell - 1] \mid \ell \in [|p_j|, |r_j|]\}. \tag{4}$$

An example of $\mathcal{C}(w)$ and the enumeration Closed(w) for the string $w[1..11] = mississippi$ is given in Table 2 in the Appendix.

4   Algorithms for Closed Substrings and MCSs

In this section, we introduce the primary data structure –the $\mathcal{MRC}$ array– which is used to compute all closed substrings in a compact representation and MCSs in $\mathcal{O}(n \log n)$ time. See Table 1 for an example of the $\mathcal{MRC}$ array for the string $w[1..11] = mississippi$ in the Appendix.

Definition 1 ($\mathcal{MRC}$ Array[2]). *Given a string $w[1..n]$, the **maximal right-closed array** ($\mathcal{MRC}$) is an array of size n, where $\mathcal{MRC}[i]$, $1 \le i \le n$, stores the list of ordered pairs $(|r_j|, |b_j|)$, where $j = K_i$ to 1.*

Algorithm 1 computes the $\mathcal{MRC}$ array by first identifying all the maximal right-closed substrings of length greater than one. It uses a stack to store AVL trees created during its execution. The algorithm scans the LCP array from left to

---

[2] This definition uses the notation introduced in Sect. 3.

right. Beginning with LCP[1] = 0, and for increasing LCP[i] values, the algorithm creates a single node (root) AVL tree with $key = $ SA[i] and $value = $ LCP[i] and pushes it onto the stack, until either the LCP value decreases or the end of the array is reached.

We denote by LCP_{max} the value stored in the root node of the AVL tree on the top of the stack. When a decrease in the LCP value (or the end of the array) is encountered, the algorithm pops a collection of AVL trees from the stack, denoted as $\mathcal{L}_{suffixSets}$, such that all suffixes represented by an AVL tree key share the same longest common prefix equal to LCP_{max}. Since LCP_{max} is the maximum LCP value of all the AVL trees on the stack, popping this collection, $\mathcal{L}_{suffixSets}$, preserves the strictly decreasing order of ordered pairs in the $\mathcal{MRC}$ array. This collection is then merged into a single AVL tree using the $Union$ operation presented in [16][3], resulting in $\mathcal{T}_{suffixes} = \bigcup \mathcal{L}_{suffixSets}$. The $Union$ operation also computes a list of ordered pairs representing maximal right-closed substrings via the $ChangeList(\mathcal{L}_{suffixSets}, \mathcal{T}_{suffixes})$ operation defined as follows:

$$ChangeList(\mathcal{P}, \mathcal{P}') = \{(x, next_{\mathcal{P}'}(x)) : next_{\mathcal{P}}(x) \neq next_{\mathcal{P}'}(x)\}, \qquad (5)$$

where $\mathcal{P}$ is a partition of the set $\mathcal{P}'$, and $next_X(x) = \min\{y \in X \mid y > x\}$, $x \in X$ and $X \subseteq \mathbb{Z}^+$. In particular, if $x \in P_i$ for some $P_i \in \mathcal{P}$, then $next_{\mathcal{P}}(x) = next_{P_i}(x)$. Each pair $(x, y) \in ChangeList(\mathcal{L}_{suffixSets}, \mathcal{T}_{suffixes})$ identifies a maximal right-closed substring $r_j = w[x..y + LCP_{max} - 1]$, where LCP_{max} is the length of its longest border. Hence, the pair $(y + LCP_{max} - x, LCP_{max})$ is added to $\mathcal{MRC}[x]$. All such pairs from the $ChangeList$ are added to the $\mathcal{MRC}$ array. The algorithm then assigns to the root node of the AVL tree, $\mathcal{T}_{suffixes}$, the LCP value of the AVL tree at the top of the stack and pushes it onto the stack, only if the stack is not empty–thus maintaining the lexicographic ordering of the set of suffixes–and continues scanning the LCP array.

To compute the maximal right-closed substrings of length one, it performs a simple linear scan of the string w and adds $w[i]$ to $\mathcal{MRC}[i]$, if $i = n \lor (1 \leq i < n \land w[i] \neq w[i + 1])$.

It is known that LCP array identifies all the occurrences of the repeating substrings that cannot be extended to the right. Moreover, the suffix array groups all these occurrences within specific ranges. Algorithm 1 identifies such ranges by tracking the rise and fall of values in the LCP array, and stores the corresponding indices in $\mathcal{L}_{suffixSets}$ as keys in AVL trees, starting with the substring of length LCP_{max}. The $Union$ operation then merges these single node AVL trees to form one AVL tree ($\mathcal{T}_{suffixes}$) that contains all the SA values; that is, all the starting positions of the substrings of length LCP_{max} in the string w.

The $ChangeList(\mathcal{L}_{suffixSets}, \mathcal{T}_{suffixes})$ operation then returns every pair of consecutive occurrences of substrings in the $\mathcal{L}_{suffixSets}$. These consecutive occurrences form the border of the maximal right-closed substring and are therefore added to the $\mathcal{MRC}$ array. Finally, the simple linear scan adds maximal

---

[3] The $Union$ operation, defined in [16], builds upon foundational methods from [8,9].

Algorithm 1. Compute $\mathcal{MRC}$ Array

Input: A string $w[1..n]$, its $\mathsf{SA}[1..n]$ and $\mathsf{LCP}[1..n]$.
Output: The $\mathcal{MRC}$ array of string $w[1..n]$.

1: $i \leftarrow 0$, $\mathcal{MRC} \leftarrow \{\emptyset\}^n$, $stack \leftarrow \emptyset$
2: **while** $stack \neq \emptyset$ **or** $i < n$ **do**
3:    **while** $i < n$ **and** $(stack = \emptyset$ **or** $\mathsf{LCP}[i] \geq LCP(top(stack)))$ **do**
4:       $push(stack, AVLTree(\{\mathsf{SA}[i]\}, \mathsf{LCP}[i]))$
5:       $i \leftarrow i + 1$
6:    $\mathcal{L}_{suffixSets} \leftarrow \emptyset$
7:    $LCP_{max} \leftarrow LCP(top(stack))$
8:    **while** $stack \neq \emptyset$ **and** $LCP(top(stack)) = LCP_{max}$ **do**
9:       $insert(\mathcal{L}_{suffixSets}, pop(stack))$
10:   **if** $LCP_{max} \neq 0$ **then**
11:      $previousLCP \leftarrow LCP(top(stack))$
12:      $insert(\mathcal{L}_{suffixSets}, pop(stack))$
13:      $\mathcal{T}_{suffixes} = \bigcup \mathcal{L}_{suffixSets}$ $\triangleright \bigcup \mathcal{L}_{suffixSets}$ returns an AVL tree
14:      **foreach** $(x, y) \in ChangeList(\mathcal{L}_{suffixSets}, \mathcal{T}_{suffixes})$ **do** $\triangleright$ Equation (5)
15:         $insert(\mathcal{MRC}[x], (y + LCP_{max} - x, LCP_{max}))$
16:      $LCP(\mathcal{T}_{suffixes}) \leftarrow previousLCP$
17:      $push(stack, \mathcal{T}_{suffixes})$
18: **for** $i \leftarrow 1$ **to** n **do**
19:    **if** $i = n$ **or** $(i < n$ **and** $w[i] \neq w[i+1])$ **then**
20:       $insert(\mathcal{MRC}[i], (1, 0))$
21: **return** $\mathcal{MRC}$

---

right-closed substrings of length one. The algorithm clearly computes the maximal right-closed substrings in strictly decreasing order of length as required in the $\mathcal{MRC}$ array. Thus, the correctness of Algorithm 1 follows.

We note that the algorithm in the worst case performs $4n - 2$ stack operations for a string $w = \lambda^n$ and in the best case performs $2n$ stack operations for a string with all distinct characters.

Lemma 4 ([16]). *The Union operation correctly computes the ChangeList, and any sequence of Union operations takes $\mathcal{O}(n \log n)$ time in total.*

Theorem 2. *Algorithm 1 correctly computes the $\mathcal{MRC}$ array of a string $w[1..n]$ using SA and LCP in $\mathcal{O}(n \log n)$ time.*

Proof. We prove the total time complexity of Algorithm 1 in three parts. First, for stack operations, we introduce a potential function $\Phi(S) = |S| \geq 0$, where $|S|$ is the number of elements in the stack. The actual cost of each push and pop operation is 1. Since we push all n single node AVL trees corresponding to each suffix onto the stack, the amortized cost for each push operation is $c'_{push} = 1 + \Delta\Phi(S) = 2$, resulting in a total amortized cost of $2n$. Additionally, in the algorithm, we pop at least two AVL trees and push their *Union* back onto the stack, and because the amortized cost of this sequence of pops and push in

---

Algorithm 2. Compute $\mathcal{C}(w)$

Input: $\mathcal{MRC}$ array of strings of $w[1..n]$
Output: Compact representation $\mathcal{C}(w)$ as in Eq. (3)

1: $\mathcal{C} \leftarrow \emptyset$
2: **for** $i \leftarrow 1$ **to** n **do**
3:     **for** $j \leftarrow 1$ **to** length of $\mathcal{MRC}[i]$ **do**
4:         $(\ell r_j, \ell b_j) \leftarrow \mathcal{MRC}[i][j]$ $\triangleright \ell r_j = |r_j|, \ell b_j = |b_j|, \ell p_j = |p_j|$
5:         **if** $j = 1$ **then**
6:             $\ell p_j \leftarrow 1$ $\triangleright$ Eq. (2)
7:         **else**
8:             $(\ell r_{j-1}, \ell b_{j-1}) \leftarrow \mathcal{MRC}[i][j-1]$
9:             $\ell p_j \leftarrow \ell r_j - \ell b_j + \ell b_{j-1} + 1$ $\triangleright$ Eq. (2)
10:         insert$(\mathcal{C}, (i, \ell p_j, \ell r_j))$
11: **return** $\mathcal{C}$

---

the worse case is $c'_{pops-push} = 3 + \Delta\Phi(S) = 2$. The maximum number of such operations is $n - 1$, leading to a total amortized cost of $2(n - 1)$. Therefore, the total cost of all stack operations in the worst case is $4n - 2$, which is $\mathcal{O}(n)$. Second, by Lemma 4, the time complexity of the sequence of *Union* operations is $\mathcal{O}(n \log n)$. Finally, the single-characters that are maximal right-closed are computed by a linear scan on w, taking $\mathcal{O}(n)$ time. Therefore, the total time complexity is $\mathcal{O}(n \log n)$.   □

Algorithm to Compute $\mathcal{C}(w)$: Algorithm 2 computes the compact representation $\mathcal{C}(w)$, as in Equation (2), of all closed substrings of a string $w[1..n]$, by simply traversing all n lists in the $\mathcal{MRC}$ array (see example in Table 2 in the Appendix). Since the $\mathcal{MRC}$ array is of size $\mathcal{O}(n \log n)$ we get Theorem 3.

Theorem 3. *Algorithm 2 correctly computes all $\mathcal{O}(n^2)$ closed substrings of a string $w[1..n]$ in a compact representation in $\mathcal{O}(n \log n)$ time.*

Algorithm to Compute All MCSs: Algorithm 3 computes all MCSs of a string $w[1..n]$ by iterating through the $\mathcal{MRC}$ array (see example in Table 2 in the Appendix). For each index i in w (where $1 \leq i \leq n$), the algorithm examines each maximal right-closed substring represented in $\mathcal{MRC}[i]$ and verifies whether the substring is also maximal left-closed using a constant time check $i = 1$ or $(i > 1$ and $w[i - 1] \neq w[i + |r| - |b| - 1])$. If the condition is satisfied, the substring is added to the MCS list. Since scanning through the $\mathcal{MRC}$ array takes $\mathcal{O}(n \log n)$ time, and each check is performed in constant time, the algorithm computes all MCSs in $\mathcal{O}(n \log n)$ time, establishing Theorem 4.

Theorem 4. *Algorithm 3 correctly computes all MCSs of a string $w[1..n]$ in $\mathcal{O}(n \log n)$ time.*

5   MCSs in a Fibonacci Word

In this section, we study the occurrences of MCSs in a Fibonacci word (f_n), and provide a formula to compute the exact number of MCSs in f_n.

Algorithm 3. Compute all MCSs

Input: A string $w[1..n]$ and $\mathcal{MRC}$ array
Output: All MCSs

1: $mcsList \leftarrow \emptyset$
2: **for** $i \leftarrow 1$ **to** n **do**
3:     **for** $j \leftarrow 1$ **to** length of $\mathcal{MRC}[i]$ **do**
4:         $(\ell r, \ell b) \leftarrow \mathcal{MRC}[i][j]$ $\triangleright \ell r = |r|, \ell b = |b|$
5:         **if** $i = 1$ **or** $(i > 1$ **and** $w[i-1] \neq w[i + \ell r - \ell b - 1])$ **then**
6:             $insert(mcsList, w[i..i + \ell r - 1])$
7: **return** $mcsList$

Let us recall that a **Fibonacci** word f_n is defined recursively by $f_0 = 0$, $f_1 = 1$, and $f_n = f_{n-1}f_{n-2}$ for $n \geq 2$. We denote $F_n = |f_n|$, where F_n satisfies the Fibonacci recurrence relation $F_n = F_{n-1} + F_{n-2}$, with $F_0 = 1$ and $F_1 = 1$. We refer to the logical separation between f_{n-1} and f_{n-2}, in the string f_n, as the **boundary**.

Bucci et al. [10] study the open and closed prefixes of Fibonacci words and show that the sequence of open and closed prefixes of a Fibonacci word follows the Fibonacci sequence. De Luca et al. [11] explore the sequence of open and closed prefixes of a Sturmian word. In [14], the authors define and find the closed Ziv–Lempel factorization and classify closed prefixes of infinite m-bonacci words. In this section, we extend previous work on closed substrings in Fibonacci words by deriving a formula to compute the number of MCSs they contain.

For any integer $n \geq 2$, let $p_n = f_n[1..F_n - 2]$ denote the prefix of f_n excluding its last two symbols. Then, $f_n = p_n \delta_n$, where $\delta_n = 10$ if n is even and $\delta_n = 01$, otherwise. Let $P_n = F_n - 2$ denote the length of p_n. The golden ratio is defined as $\phi = \frac{1+\sqrt{5}}{2}$, and it is known that the ratio of consecutive Fibonacci numbers converges to ϕ, i.e., $\lim_{n \to \infty} \frac{F_n}{F_{n-1}} = \phi$.

The non-singleton MCSs are either runs or maximal gapped repeats in which the arm occurs exactly twice [4]. For a Fibonacci word f_n, let $SM(f_n)$, $R(f_n)$, and $GM(f_n)$ denote the number of singleton MCSs, runs, and gapped-MCSs, respectively. Therefore, the total number of MCSs in f_n is given by $M(f_n) = SM(f_n) + R(f_n) + GM(f_n)$.

Lemma 5. *The no. of singleton MCSs in a Fibonacci word f_n, for $n \geq 4$, is:*

$$SM(f_n) = \begin{cases} F_{n-2} + F_{n-4} + 2, & \text{if } n \text{ is odd,} \\ F_{n-2} + F_{n-4}, & \text{if } n \text{ is even.} \end{cases} \tag{6}$$

Proof. We prove the Lemma by induction on n. The base cases, $SM(f_4) = 3$ and $SM(f_5) = 6$, match the direct counts.

For the inductive step, assume that Eq. (6) holds for $k \geq 6$. By definition, $f_{k+1} = f_k f_{k-1}$. Every Fibonacci word f_k begins with a 10; if k is odd, it ends with 01 and if k is even, it ends with 10. By definition, the first and last character in f_k are singleton MCSs. Therefore, we examine whether the singleton MCSs at the boundary are retained in f_{k+1}.

Suppose $k+1$ is even. By inductive hypothesis, $SM(f_k) = F_{k-2} + F_{k-4} + 2$ and $SM(f_{k-1}) = F_{k-3} + F_{k-5}$. Since k is odd, f_k ends with 1 and f_{k-1} begins with 1, their concatenation results in the string 11 at the boundary, which forms a new non-singleton MCS, and reduces the count of the singleton MCSs in f_{k+1} by two. Thus, we get $SM(f_{k+1}) = SM(f_k) + SM(f_{k-1}) - 2 = F_{k-1} + F_{k-3}$.

Suppose $k+1$ is odd. By inductive hypothesis, $SM(f_k) = F_{k-2} + F_{k-4}$ and $SM(f_{k-1}) = F_{k-3} + F_{k-5} + 2$. Since k is even, f_k ends with 0 and f_{k-1} begins with 1, their concatenation results in the string 01 at the boundary, which does not reduce the count of singleton MCSs in f_{k+1}. Thus, we get $SM(f_{k+1}) = SM(f_k) + SM(f_{k-1}) = F_{k-1} + F_{k-3} + 2$. $\qquad\square$

Lemma 6 (Theorem 2 in [17]). *For $n \geq 4$, the number of runs in the Fibonacci word f_n, is given by $R(f_n) = 2F_{n-2} - 3$.*

To compute $GM(f_n)$, we count the number of maximal gapped repeats whose arms occur exactly twice, as these are, by definition gapped-MCSs. In Theorem 5 we first show that the only gapped-MCSs in a Fibonacci word are occurrences of the substring 101 that are also maximal gapped repeats. Then, we count all such occurrences in f_n.

Lemma 7 (Theorem 1 in [27]). *Suppose uvu is a maximal gapped repeat in f_n that is also a suffix of f_n. Then, the arm u is a Fibonacci word.*

Lemma 8. *For $n \geq 5$, suppose uvu is a maximal gapped repeat in f_n that is also a suffix of f_n. Then uvu is not a gapped-MCS.*

Proof. By Lemma 7, the arm of the maximal gapped repeat is a Fibonacci word $u = f_k$. Clearly, $k \neq n$.

If n is even, then the proper suffixes of f_n that are Fibonacci words are in the set $\mathcal{S} = \{f_k \mid k \mod 2 = 0 \wedge 0 \leq k < n\}$. For any proper suffix $f_k \in \mathcal{S} \setminus \{f_0, f_2\}$, f_{k+2} is a suffix of f_n. By definition, $f_{k+2} = f_{k+1}f_k = f_k f_{k-1} f_k = f_k f_k f_{k-3} f_{k-2}$, the suffix $f_{k-1}f_k = f_k f_{k-3} f_{k-2}$ has length $F_k + F_{k-1} < 2F_k$ implying that the last two occurrences of f_k overlap. Therefore, any uvu with f_k as its arm, will have at least three occurrences of f_k making it an open string. For suffix $f_k \in \{f_0, f_2\}$ of f_n, for $n \geq 6$, $f_6 = 1011010110110$ is a suffix of f_n, and the last two occurrences of f_k are left-extendible, and so, it is not maximal. Moreover, any longer uvu with the arm f_k will have at least three occurrences of f_k making it an open string.

If n is odd, then $\mathcal{S} = \{f_k \mid k \mod 2 = 1 \wedge 1 \leq k < n\}$. The argument is analogous for $f_k \in \mathcal{S} \setminus \{f_1\}$ and $f_k \in \{f_1\}$, respectively. $\qquad\square$

Lemma 9 (Theorem 2 in [27]). *For $n \geq 3$, suppose uvu is a maximal gapped repeat in f_n that is not a suffix of f_n. Then the arm u belongs to the set $\mathcal{A}_n = \{p_3, p_4, \ldots, p_{n-2}\}$.*

Lemma 10 (Lemma 8 in [27]). *For $n \geq 4$, all the borders of p_n form the set $\mathcal{B}_n = \{p_3, p_4, \ldots, p_{n-1}\}$.*

Lemma 11. *For $n \geq 5$, the border p_{n-1} of p_n is the longest proper cover of p_n.*

Proof. By Lemma 10, we know that p_{n-1} is the longest border of p_n. $F_n = F_{n-1} + F_{n-2}$. For $n \geq 5$, $F_n - 4 = F_{n-1} + F_{n-2} - 4$, and so $P_n - 2 = P_{n-1} + P_{n-2}$. Hence, $P_n \leq 2 \cdot P_{n-1}$. Clearly, p_{n-1} is either an adjacent or an overlapping border of p_n. Therefore, p_{n-1} is the longest proper cover of p_n. □

Lemma 12 (Lemma 2 in [21]). *Let u be a proper cover of x and let $v \neq u$ be a substring of x such that $|v| \leq |u|$. Then, v is a cover of $x \Leftrightarrow v$ is a cover of u.*

Lemma 13. *For $n \geq 5$, all covers of p_n form the set $C_n = \{p_4, p_5, \ldots, p_n\}$.*

Proof. By Lemma 11, for $n \geq 5$, p_{n-1} is the longest proper cover of p_n. Similarly, p_{n-2} is the longest proper cover of p_{n-1}, p_{n-3} is the longest proper cover of p_{n-2}, and so on. Therefore, by Lemma 12, it follows that all covers of p_n belong to the set $C_n = \{p_4, p_5, \ldots, p_n\}$ for all $n \geq 5$. □

Theorem 5. *For $n \geq 5$, every gapped-MCS in a Fibonacci word f_n corresponds to an occurrence of the substring 101 that is also a maximal gapped repeat.*

Proof. All gapped-MCSs are maximal gapped repeats. By Lemma 8, for $n \geq 5$, all maximal gapped repeats that are also suffixes of f_n are not gapped-MCSs.

By Lemma 9, for $n \geq 3$, if uvu is a maximal gapped repeat in f_n that is not a suffix of f_n, then the arm u belongs to the set $A_n = \{p_3, p_4, \ldots, p_{n-2}\}$. By Lemma 13, all covers of p_n belong to the set $C_n = \{p_4, p_5, \ldots, p_n\}$.

Suppose there exists a maximal gapped $w = uvu$, where $u \in A_n \setminus p_3$, that is a gapped-MCS. Assume n is odd. Then, $f_n = p_n \delta_n$, where $\delta_n = 01$, and $f_n[F_n - 1] = 0$. Since $u \in A_n \setminus p_3$, it ends with 1. Clearly, u does not occur ending at position $F_n - 1$. Now suppose n is even. By a similar reasoning, u does not occur ending at $F_n - 1$ since it ends with 01 and $f_n[F_n - 2] = f_n[F_n - 1] = 1$ (as $f_4 = 10110$ is a suffix of f_n). Hence, u ends within p_n. Since u is a cover of p_n, $w = uvu$ will contain at least three occurrences of u making it an open string. Therefore, there is no maximal gapped repeat of the form $w = uvu$, where $u \in A_n \setminus p_3$, that is a gapped MCS.

Now we consider the final case, where $u = p_3 = 1$. In this case, we get 101 as the only gapped-MCS. Any longer maximal gapped repeat will have at least three occurrences of $u = p_3 = 1$, making it an open string—a contradiction. □

Lemma 14. *The no. of gapped-MCSs in a Fibonacci word f_n, for $n \geq 5$, is:*

$$GM(f_n) = \begin{cases} F_{n-5}, & \text{if } n \text{ is odd,} \\ F_{n-5} + 1, & \text{if } n \text{ is even.} \end{cases} \tag{7}$$

Proof. We prove the Lemma by induction on n. By direct counting for the base cases, we verify that $GM(f_5) = 1$ and $GM(f_6) = 2$. For the inductive step, assume that Eq. (7) holds for $k \geq 7$. For $n \geq 3$, $f_3 = 101$ is a prefix of every f_n, and by Theorem 5 every gapped-MCS in a Fibonacci word f_n, for $n \geq 5$, corresponds to an occurrence of the substring 101 that is also a maximal gapped repeat. In $f_{k+1} = f_k f_{k-1}$, any new occurrences of 101 in f_{k+1} can only occur beginning in f_k and ending in f_{k-1}. Moreover, an occurrence of 101 as a suffix of f_k or a prefix of f_{k-1} may be lost by extension. Below we examine all such cases.

Suppose $k + 1$ is odd. By inductive hypothesis, $GM(f_k) = F_{k-5} + 1$ and $GM(f_{k-1}) = F_{k-6}$. Moreover, f_k ends with the suffix $f_2 = 10$ and f_{k-1} begins with $f_3 = 101$, resulting in the formation of the string 101 crossing the boundary. Clearly, this string is not a gapped-MCS as it is right-extendible. The gapped-MCS 101 which is the prefix of f_{k-1} is no longer a gapped-MCS since f_k ends with a 0 making it left-extendible. Thus, we get $GM(f_{k+1}) = GM(f_k) + GM(f_{k-1}) - 1 = (F_{k-5} + 1) + F_{k-6} - 1 = F_{k-4}$.

Suppose $k + 1$ is even. By inductive hypothesis, $GM(f_k) = F_{k-5}$ and $GM(f_{k-1}) = F_{k-6} + 1$. f_k ends with a 1, so the prefix 101 of f_{k-1} is retained as a gapped-MCS. By Lemma 8 the suffix $f_3 = 101$ of f_k is not a gapped-MCS. Thus, we get $GM(f_{k+1}) = GM(f_k) + GM(f_{k-1}) = F_{k-4} + 1$. □

Theorem 6. *The number of MCSs in a Fibonacci word f_n, denoted by $M(f_n)$, for $n \geq 5$, is given below in Eq. (8). Furthermore, $M(f_n) \approx 1.382F_n$.*

$$M(f_n) = \begin{cases} F_n + F_{n-2} - 1, & \text{if } n \text{ is odd,} \\ F_n + F_{n-2} - 2, & \text{if } n \text{ is even.} \end{cases} \tag{8}$$

Proof. Adding Eqs. (6), (7) and the Eq. in Lemma 6 and simplifying, we get Eq. (8). Next, we have $M(f_n) < F_n + F_{n-2} = \left(1 + \frac{F_{n-2}}{F_n}\right) F_n \approx \left(1 + \frac{1}{\phi^2}\right) F_n \approx 1.382F_n$ (since $\lim_{n \to \infty} \frac{F_{n-2}}{F_n} = \frac{1}{\phi^2}$). □

6   Experimental Analysis

All experiments were carried out on a server with an Intel Xeon Gold 6426Y CPU (64 cores, 128 threads, 75 MiB L3 cache), 250 GiB RAM, running RHEL 9.5 (kernel version 5.14.0-503.26.1). The code was implemented in C++ and compiled using GCC 11.5.0. The implementations for computing the suffix array (SA) and the longest common prefix array (LCP) were sourced from the *libsais* library [13], which is based on contributions from [15,22,25,26]. The implementation is available at https://github.com/neerjamhaskar/closedStrings.

For $\sigma = 2, 3$ and 4, we generate all strings of length n for $1 \leq n \leq 20$ and compute their MCSs. The results are plotted in Fig. 2 in the Appendix. As observed, the maximum number of MCSs increases with the size of the alphabet σ for strings of the same length. In total, the number of strings analyzed is equal to $\sum_{i=1}^{20} (2^i + 3^i + 4^i) \approx 1.47$ trillion. These exhaustive computations raise a natural question: Given n and σ, can we algorithmically construct a string that has the maximum possible number of MCSs?

Acknowledgments. We thank Simon J. Puglisi for introducing us to the MCS problem using SA and LCP, which led to Theorem 4. We are grateful to the reviewers for their valuable feedback.

Appendix

Table 1. The $\mathcal{MRC}$ array for the string $w[1..11] = mississippi$.

Index (i)	$\mathcal{MRC}[i]$
1	$(1,0)$
2	$(7,4),(1,0)$
3	$(6,3),(2,1)$
4	$(5,2),(3,1),(1,0)$
5	$(4,1),(1,0)$
6	$(2,1)$
7	$(1,0)$
8	$(4,1),(1,0)$
9	$(2,1)$
10	$(1,0)$
11	$(1,0)$

Table 2. Compact representation $\mathcal{C}(w)$ of $w[1..11] = mississippi$ with corresponding closed factors, maximal right-closed factors, and MCSs.

$\mathcal{C}(w)$	All Closed Factors	Maximal Right Closed Factors	MCS?
$(1,1,1)$	m	m	Yes
$(2,4,7)$	$issi, issis, ississ, ississi$	$ississi$	Yes
$(2,1,1)$	i	i	Yes
$(3,5,6)$	$ssiss, ssissi$	$ssissi$	No
$(3,1,2)$	s, ss	ss	Yes
$(4,5,5)$	$sissi$	$sissi$	No
$(4,3,3)$	sis	sis	Yes
$(4,1,1)$	s	s	No
$(5,4,4)$	$issi$	$issi$	No
$(5,1,1)$	i	i	Yes
$(6,1,2)$	s, ss	ss	Yes
$(7,1,1)$	s	s	No
$(8,4,4)$	$ippi$	$ippi$	Yes
$(8,1,1)$	i	i	Yes
$(9,1,2)$	p, pp	pp	Yes
$(10,1,1)$	p	p	No
$(11,1,1)$	i	i	Yes

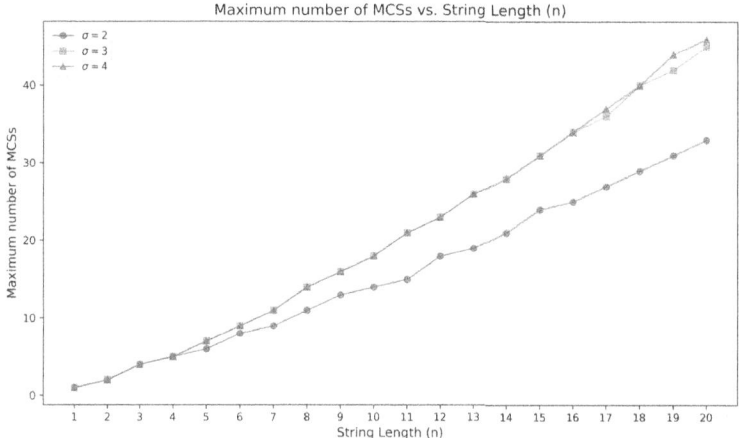

Fig. 2. Maximum number of MCSs in strings of length n, where $1 \leq n \leq 20$, for alphabets of size $\sigma = 2, 3$ and 4.

References

1. Alamro, H., Alzamel, M., Iliopoulos, C.S., Pissis, S.P., Watts, S., Sung, W.-K.: Efficient identification of k-closed strings. In: Boracchi, G., Iliadis, L., Jayne, C., Likas, A. (eds.) EANN 2017. CCIS, vol. 744, pp. 583–595. Springer, Cham (2017). https://doi.org/10.1007/978-3-319-65172-9_49
2. Badkobeh, G., et al.: Closed factorization. In: Holub, J., Zdǎrek, J. (eds.) Proceedings of PSC 2014, pp. 162–168. Czech Technical University in Prague (2014). https://www.stringology.org/papers/PSC2014.pdf
3. Badkobeh, G., et al.: Closed factorization. Disc. Appl. Math. **212**, 23–29 (2016). https://doi.org/10.1016/j.dam.2016.04.009
4. Badkobeh, G., De Luca, A., Fici, G., Puglisi, S.J.: Maximal closed substrings. In: Arroyuelo, D., Poblete, B. (eds.) String Processing and Information Retrieval, pp. 16–23. Springer, Cham (2022). https://doi.org/10.1007/978-3-031-20643-6_2
5. Badkobeh, G., Fici, G., Lipták, Z.: On the number of closed factors in a word. In: Dediu, A.-H., Formenti, E., Martín-Vide, C., Truthe, B. (eds.) LATA 2015. LNCS, vol. 8977, pp. 381–390. Springer, Cham (2015). https://doi.org/10.1007/978-3-319-15579-1_29
6. Badkobeh, G., Luca, A.D., Fici, G., Puglisi, S.: Maximal closed substrings (2022). https://doi.org/10.48550/arXiv.2209.00271
7. Bannai, H., et al.: Efficient algorithms for longest closed factor array. In: Iliopoulos, C., Puglisi, S., Yilmaz, E. (eds.) SPIRE 2015. LNCS, vol. 9309, pp. 95–102. Springer, Cham (2015). https://doi.org/10.1007/978-3-319-23826-5_10
8. Brodal, G.S., Pedersen, C.N.S.: Finding maximal quasiperiodicities in strings. In: Giancarlo, R., Sankoff, D. (eds.) CPM 2000. LNCS, vol. 1848, pp. 397–411. Springer, Heidelberg (2000). https://doi.org/10.1007/3-540-45123-4_33

9. Brown, M.R., Tarjan, R.E.: A fast merging algorithm. J. ACM **26**(2), 211–226 (1979). https://doi.org/10.1145/322123.322127
10. Bucci, M., De Luca, A., Fici, G.: Enumeration and structure of Trapezoidal words. Theor. Comput. Sci. **468**, 12–22 (2013). https://doi.org/10.1016/j.tcs.2012.11.007
11. De Luca, A., Fici, G., Zamboni, L.Q.: The sequence of open and closed prefixes of a Sturmian word. Adv. Appl. Math. **90**, 27–45 (2017). https://doi.org/10.1016/j.aam.2017.04.007
12. Fici, G.: A classification of Trapezoidal words. In: Ambrož, P., Štěpán Holub, Masáková, Z. (eds.) Proceedings 8th International Conference Words 2011, Prague, Czech Republic, 12–16 September 2011. Electronic Proceedings in Theoretical Computer Science, vol. 63, pp. 129–137. Open Publishing Association (2011). https://doi.org/10.4204/EPTCS.63.18
13. Grebnov, I.: libsais: a library for linear time suffix array, longest common prefix array and burrows wheeler transform construction based on induced sorting algorithm. GitHub repository (2021–2025). https://github.com/IlyaGrebnov/libsais, version 2.10.2. Accessed 12 June 2025
14. Jahannia, M., Mohammad-Noori, M., Rampersad, N., Stipulanti, M.: Closed Ziv–Lempel factorization of the m-bonacci words. Theor. Comput. Sci. **918**, 32–47 (2022). https://doi.org/10.1016/j.tcs.2022.03.019
15. Kärkkäinen, J., Manzini, G., Puglisi, S.J.: Permuted longest-common-prefix array. In: Kucherov, G., Ukkonen, E. (eds.) CPM 2009. LNCS, vol. 5577, pp. 181–192. Springer, Heidelberg (2009). https://doi.org/10.1007/978-3-642-02441-2_17
16. Kociumaka, T., Pissis, S.P., Radoszewski, J., Rytter, W., Waleń, T.: Fast algorithm for partial covers in words. Algorithmica **73**(1), 217–233 (2015). https://doi.org/10.1007/s00453-014-9915-3
17. Kolpakov, R., Kucherov, G.: On maximal repetitions in words. In: Ciobanu, G., Păun, G. (eds.) FCT 1999. LNCS, vol. 1684, pp. 374–385. Springer, Heidelberg (1999). https://doi.org/10.1007/3-540-48321-7_31
18. Kosolobov, D.: Closed repeats (2024). https://doi.org/10.48550/arXiv.2410.00209
19. Mhaskar, N., Smyth, W.: String covering: a survey. Fund. Inform. **190**(1), 17–45 (2023). https://doi.org/10.3233/FI-222164
20. Mieno, T., Takahashi, S., Seto, K., Horiyama, T.: Online and offline algorithms for counting distinct closed factors via sliding suffix trees. In: Královič, R., Kůrková, V. (eds.) SOFSEM 2025: Theory and Practice of Computer Science, pp. 172–183. Springer, Cham (2025). https://doi.org/10.1007/978-3-031-82697-9_13
21. Moore, D., Smyth, W.: An optimal algorithm to compute all the covers of a string. Inf. Process. Lett. **50**(5), 239–246 (1994). https://doi.org/10.1016/0020-0190(94)00045-X
22. Nong, G., Zhang, S., Chan, W.H.: Two efficient algorithms for linear time suffix array construction. IEEE Trans. Comput. **60**(10), 1471–1484 (2011). https://doi.org/10.1109/TC.2010.188
23. Parshina, O., Puzynina, S.: Finite and infinite closed-rich words. Theor. Comput. Sci. **984**, 114315 (2024). https://doi.org/10.1016/j.tcs.2023.114315
24. Parshina, O., Zamboni, L.Q.: Open and closed factors in Arnoux-Rauzy words. Adv. Appl. Math. **107**, 22–31 (2019). https://doi.org/10.1016/j.aam.2019.02.007
25. Timoshevskaya, N., Feng, W.c.: Sais-opt: on the characterization and optimization of the sa-is algorithm for suffix array construction. In: 2014 IEEE 4th International Conference on Computational Advances in Bio and Medical Sciences (ICCABS), pp. 1–6 (2014). https://doi.org/10.1109/ICCABS.2014.6863917

26. Xie, J.Y., Nong, G., Lao, B., Xu, W.: Scalable suffix sorting on a multicore machine. IEEE Trans. Comput. **69**(9), 1364–1375 (2020). https://doi.org/10.1109/TC.2020.2972546

27. Yamane, K., Nakashima, Y., Seto, K., Horiyama, T.: Maximal α-gapped repeats in a Fibonacci string. In: Královič, R., Kůrková, V. (eds.) SOFSEM 2025: Theory and Practice of Computer Science, pp. 337–350. Springer, Cham (2025). https://doi.org/10.1007/978-3-031-82697-9_25

Nyldon Factorization of Thue-Morse Words and Fibonacci Words

Kaisei Kishi[1(✉)], Kazuki Kai[2], Yuto Nakashima[2] (ID), Shunsuke Inenaga[2] (ID), and Hideo Bannai[3] (ID)

[1] Department of Information Science and Technology, Kyushu University, Fukuoka, Japan
kishi.kaisei.216@s.kyushu-u.ac.jp
[2] Department of Informatics, Kyushu University, Fukuoka, Japan
{nakashima.yuto.003,inenaga.shunsuke.380}@m.kyushu-u.ac.jp
[3] M&D Data Science Center, Institute of Integrated Research, Institute of Science Tokyo, Tokyo, Japan
hdbn.dsc@tmd.ac.jp

Abstract. The Nyldon factorization is a string factorization that is a non-decreasing product of Nyldon words. Nyldon words and Nyldon factorizations are recently defined combinatorial objects inspired by the well-known Lyndon words and Lyndon factorizations. In this paper, we investigate the Nyldon factorization of several words. First, we fully characterize the Nyldon factorizations of the (finite) Fibonacci and the (finite) Thue-Morse words. Moreover, we show that there exists a non-decreasing product of Nyldon words that is a factorization of the infinite Thue-Morse word.

1 Introduction

The Lyndon factorization [6] of a string is the unique factorization such that each factor is a Lyndon word and the sequence is a non-increasing product of them. More precisely, $w_1, \ldots, w_k$ is the Lyndon factorization of a string w if the sequence satisfies the following three conditions:

1. $w = w_1 \cdots w_k$,
2. $w_i \in \Sigma^+$ is a Lyndon word for all i satisfying $1 \leq i \leq k$,
3. $w_i \succeq w_{i+1}$ for all i satisfying $1 \leq i < k$.

A string x is said to be a Lyndon word [14] if x is lexicographically smaller than any of its cyclic rotations (or suffixes). This is a simple and fundamental definition of Lyndon words. Lyndon words and Lyndon factorizations are well-known combinatorial objects, and there are lots of studies on them and its variants (e.g., [2–4,9,17]). On the other hand, Lyndon words can be defined in a recursive way: strings of length 1 are Lyndon words, strings of length more than 1 are Lyndon words if there is no factorization into a non-increasing product of multiple Lyndon words.

G. Badkobeh et al. (Eds.): SPIRE 2025, LNCS 16073, pp. 188–201, 2026.
https://doi.org/10.1007/978-3-032-05228-5_16

Nyldon words and Nyldon factorizations are symmetric structures which were introduced by Grinberg's natural question [7]: How do the (Lyndon) structures and properties change when changing from "non-increase" to "non-decrease"? Charlier et al. [5] answered several questions on Nyldon words, for instance, the Nyldon factorization is also unique for any string and there is a unique Nyldon word for each conjugacy class of primitive words. Later, Garg [8] resolved several questions introduced by Charlier et al. [5]. Thanks to their studies, a lot of nice structures and properties of Nyldon words and relations between Lyndon words and Nyldon words were revealed, but there are still many open questions on Nyldon words and Nyldon factorizations. One interesting question is given as follows: Can we define Nyldon words by a simpler way such as definitions of Lyndon words by cyclic rotations or suffixes.[1] Such a property may accelerate not only combinatorial studies but also algorithmic studies on strings by using Nyldon words.

In this paper, to further understand Nyldon words and Nyldon factorizations, we consider the Nyldon factorizations of Fibonacci and Thue-Morse words. Our main contributions are three-fold.

- We fully characterize the Nyldon factorization of the finite Fibonacci words (Sect. 3).
- We fully characterize the Nyldon factorization of the finite Thue-Morse words (Sect. 4).
- We show that there exists a non-decreasing sequence of Nyldon words that is a factorization of the infinite Thue-Morse words (Sect. 5).

Siromoney et al. [18] generalized the notion of Lyndon words to infinite words, and they also showed that any infinite word can be factorized into a unique non-increasing sequence of Lyndon words, finite and/or infinite: either (1) the infinite sequence of finite Lyndon words or (2) the concatenation of a finite sequence of finite Lyndon words and a single infinite Lyndon word. Melançon fully characterized the Lyndon factorization of the infinite Fibonacci words [15, 16] and Ido and Melançon characterized the Lyndon factorization of the infinite Thue-Morse words [10].

Our basic idea for the infinite Thue-Morse words is from [10]. Our result gives a factorization in type-(1). However, we have not shown its uniqueness yet. The difficulty of showing its uniqueness may come from the fact that a natural generalization of Nyldon words for infinite words is not known. We believe that there is a nice generalization of Nyldon words for infinite cases as well as Lyndon words and the generalization may implies the uniqueness of Nyldon factorization for infinite words.

---

[1] Note that a class of words defined as the set of largest cyclic rotations is the anti-Lyndon words and as the set of largest suffixes is the inverse Lyndon words.

2   Preliminaries

Strings

Let Σ be an *alphabet*. An element of Σ^* is called a *string*. The length of a string w is denoted by $|w|$. The empty string ε is the string of length 0. Let Σ^+ be the set of non-empty strings, i.e., $\Sigma^+ = \Sigma^* \setminus \{\varepsilon\}$. For any strings x and y, let $x \cdot y$ (or sometimes xy) denote the concatenation of the two strings. For a string $w = xyz$, x, y and z are called a *prefix*, *substring*, and *suffix* of w, respectively. They are called a *proper prefix*, a *proper substring*, and a *proper suffix* of w if $x \neq w$, $y \neq w$, and $z \neq w$, respectively. The i-th symbol of a string w is denoted by $w[i]$, where $1 \leq i \leq |w|$. For a string w and two integers $1 \leq i \leq j \leq |w|$, let $w[i..j]$ denote the substring of w that begins at position i and ends at position j. For convenience, let $w[i..j] = \varepsilon$ when $i > j$. Also, let $w[..i] = w[1..i]$ and $w[i..] = w[i..|w|]$ and $w' = w[1..|w| - 1]$. For a string w, let $w^1 = w$ and let $w^k = ww^{k-1}$ for any integer $k \geq 2$. Also, for a string w and a prefix x of w, let $x^{-1}w = w[|x| + 1..]$ and for a string w a suffix z of w, let $wz^{-1} = w[..|w| - |z|]$. A sequence of k strings $w_1, \ldots, w_k$ is called a *factorization* of a string w if $w = w_1 \cdots w_k$. We call k the size of a *factorization* of a string w. For a binary string w, $\overline{w}$ denotes the bit-wise flipped string of w (e.g., $\overline{aab} = bba$ over $\{a, b\}$). Let $\prec$ denote a (strict) total order on an alphabet Σ. A total order $\prec$ on the alphabet induces a total order on the set of strings called the *lexicographic order* w.r.t. $\prec$, also denoted as $\prec$, i.e., for any two strings $x, y \in \Sigma^*$, We write $x \prec y$ if and only if x is a proper prefix of y, or, there exists $1 \leq i \leq \min\{|x|, |y|\}$ s.t. $x[1..i-1] = y[1..i-1]$ and $x[i] \prec y[i]$. We also write $x \preceq y$ if and only if $x \prec y$ or $x = y$. A mapping $\phi : \Sigma^* \to \Sigma^*$ is called *morphism* on an alphabet Σ if $\phi(xy) = \phi(x)\phi(y)$ for any $x, y \in \Sigma^*$. Let $\phi^1(x) = \phi(x)$ and let $\phi^k(x) = \phi(\phi^{k-1}(x))$ for any integer $k \geq 2$.

Lyndon word and Lyndon factorizations

A string w is a *Lyndon word* [14] w.r.t. a lexicographic order $\prec$, if and only if $w \prec w[i..]$ for all $1 < i \leq |w|$, i.e., w is lexicographically smaller than all its proper suffixes with respect to $\prec$. The *Lyndon factorization* [6] of a string w, denoted by $LF(w)$, is a unique factorization $\lambda_1, \ldots, \lambda_m$ of w, such that each $\lambda_i \in \Sigma^+$ is a Lyndon word and $\lambda_i \succeq \lambda_{i+1}$ for $1 \leq i < m$. Let $w = abaabbaabbaab$. The Lyndon factorization of w is ab, $aabb$, $aabb$, aab. Since $LF(w)$ is unique for any w, Lyndon word can be defined recursively as follows: every string w whose length is 1 is a Lyndon word, and every string w whose length is longer than 1 is a Lyndon word if and only if w has no factorization that is a lexicographically non-increasing sequence of Lyndon words of length shorter than $|w|$.

Nyldon word and Nyldon factorizations

Nyldon words [7] are defined recursively in a similar way to Lyndon words: every string w whose length is 1 is a Nyldon word, and every string w whose length is longer than 1 is a Nyldon word if and only if w cannot be factorized into shorter Nyldon words. More precisely, there is no factorization $\gamma_1, \ldots, \gamma_m$ of w that satisfies the following three conditions:

1. each $\gamma_i \in \Sigma^+$ is a Nyldon word,
2. $\gamma_i \preceq \gamma_{i+1}$ for all $1 \leq i < m$,
3. $m \geq 2$.

Notice that there is a lexicographical reversal: $\gamma_i \preceq \gamma_{i+1}$ while $\lambda_i \succeq \lambda_{i+1}$. It is known that any string w has a unique factorization that satisfies the first two conditions [5]. We call such factorization of a string w the *Nyldon factorization* of w and denote it by $NF(w)$.

We call a string x a Nyldon proper suffix of w if x is a Nyldon word and x is a proper suffix of w, and a string y the longest Nyldon proper suffix of w if y is a Nyldon proper suffix of w and y is longer than any other Nyldon proper suffixes of w. $\mathsf{lnps}(w)$ denotes the longest Nyldon proper suffix of w. For $w = aabba$, $NF(w) = a, a, b, ba$ and the Nyldon proper suffixes of w are a and ba. Hence, $\mathsf{lnps}(w) = ba$. Notice that ba is a Nyldon word since the possible factorization b, a is not a Nyldon factorization.

Fibonacci words and Thue-Morse words

The k-th (finite) Fibonacci word F_k over a binary alphabet $\{a, b\}$ is defined as follows: $F_0 = b$, $F_1 = a$, $F_k = F_{k-1} \cdot F_{k-2}$ for any $k \geq 2$. Let f_k be the length of the k-th Fibonacci word (i.e., $f_k = |F_k|$). We also define the infinite Fibonacci words over an alphabet $\{a, b\}$ as $\mathcal{F} = \lim_{k \to \infty} F_k$.

The k-th (finite) Thue-Morse word TM_k over a binary alphabet $\{a, b\}$ is defined as follows: $TM_0 = a$, $TM_k = TM_{k-1} \cdot \overline{TM_{k-1}}$ for any $k \geq 1$. It is clear from the definition that $|TM_k| = 2^k$ holds. We also define the infinite Thue-Morse words over an alphabet $\{a, b\}$ as $\mathcal{TM} = \lim_{k \to \infty} TM_k$. The Thue-Morse morphism τ over a binary alphabet $\{a, b\}$ is defined as follows: $\tau(a) = ab$, $\tau(b) = ba$. We call τ Thue-Morse morphism since $\mathcal{TM} = \lim_{n \to \infty} \tau^n(a)$.

3 Nyldon Factorization of Fibonacci Words

In this section, we fully characterize the Nyldon factorization of the k-th Fibonacci word F_k. We start with known properties on Nyldon words.

Lemma 1 (Theorem 20 of [5]). *Let $w \in \Sigma^+$, $|w| \geq 2$ and $w = ps$ where $s = \mathsf{lnps}(w)$. Then w is a Nyldon word iff p is a Nyldon word and $p \succ s$.*

Lemma 2 (Theorem 13 of [5]). *Let w be a Nyldon word. For each Nyldon proper suffix s of w, we have $s \prec w$.*

Our main result in this section is given by Theorem 1. It explains that the k-th Fibonacci word is always factorized into the sequence of two Nyldon words.

Theorem 1. *For every $k \geq 2$, the Nyldon factorization of F_k is $a, F_k[2..f_k]$.*

It is easy to see that the product of strings $a \cdot F_k[2..f_k]$ represents F_k, and $a \prec F_k[2..f_k]$ also holds since $F_k[2] = b$. Hence, we only have to show $F_k[2..f_k]$ to be a Nyldon word in order to prove Theorem 1. We prove it in Lemma 3. For convenience, let $H_k = F_k[2..f_k]$ for every $k \geq 2$, and h_k denote the length of H_k (i.e., $h_k = f_k - 1$).

Lemma 3. *(i) For every $k \geq 2$, H_k and $H_k \cdot a$ are Nyldon words. (ii) For every $k \geq 4$, $\mathsf{Inps}(H_k) = H_{k-2}$ and $\mathsf{Inps}(H_k \cdot a) = H_{k-2} \cdot a$.*

Proof. We prove the two statements by induction on k.

Base case. For $k = 2, 3$ of (i), we can simply check that $H_2 = b$, $H_2 \cdot a = ba$, $H_3 = ba$, and $H_3 \cdot a = baa$ are Nyldon word. For $k = 4$, we can also check $\mathsf{Inps}(H_4) = \mathsf{Inps}(baab) = b = H_2$. Hence, H_4 is a Nyldon word by the fact that $H_4 \cdot (\mathsf{Inps}(H_4))^{-1} = baa = H_3$ is a Nyldon word and Lemma 1. Moreover, we can also check that $\mathsf{Inps}(H_4 \cdot a) = \mathsf{Inps}(baaba) = ba = H_2 \cdot a$ holds and $H_4 \cdot a = baaba$ is a Nyldon word. Thus, the two statements also hold for $k = 4$.

Induction Step. Suppose that the statement holds for any k satisfying $k < j$ for some $j \geq 5$. We show that the statements hold for $k = j$ (see also Fig. 1). Firstly, we show that $\mathsf{Inps}(H_j) = H_{j-2}$. By the definitions of F_k and H_k, we can write $H_j = H_{j-1} \cdot a \cdot H_{j-2}$ and H_{j-2} is a Nyldon word by induction hypothesis. Assume on the contrary that there exists a longer Nyldon suffix $x \cdot H_{j-2}$ of H_j for some non-empty string x. Here, x is also assumed to be the shortest such string. Namely, H_{j-2} is the longest Nyldon proper suffix of $x \cdot H_{j-2}$. By Lemma 1, x must be a Nyldon word that satisfies $x \succ H_{j-2}$. Moreover, x is a suffix of $H_{j-1} \cdot a$. Since $\mathsf{Inps}(H_{j-1} \cdot a) = H_{j-3} \cdot a$ by induction hypothesis, x is also a suffix of $H_{j-3} \cdot a$. We can also see that $H_{j-3} \cdot a$ is a prefix H_{j-2} by the definitions of F_k and H_k. These facts and Lemma 2 imply that $x \preceq H_{j-3} \cdot a \prec H_{j-2}$. This contradicts the inequality $x \succ H_{j-2}$. Hence, $\mathsf{Inps}(H_j) = H_{j-2}$ holds. Moreover, we can see that H_j is a Nyldon word by Lemma 1 since $\mathsf{Inps}(H_j) = H_{j-2}$, $H_{j-1} \cdot a$ is a Nyldon (by induction hypothesis), and H_{j-2} is a prefix of $H_{j-1} \cdot a$ (i.e., $H_{j-1} \cdot a \succ H_{j-2}$). In a similar way, we can also show that $\mathsf{Inps}(H_j \cdot a) = H_{j-2} \cdot a$ holds and $H_j \cdot a$ is a Nyldon word. Therefore, the statement holds for every k.

4   Nyldon Factorization of Thue-Morse Words

In this section, we fully characterize the Nyldon factorization of the k-th Thue-Morse word TM_k. We show that the factorization can be characterized recursively. Note that $w' = w[1..|w| - 1]$ for a string w.

Theorem 2. *Let $NF(TM_{2k-1}) = x_1, x_2, \ldots, x_{c_k}$. For every $n \geq 4$,*

$$NF(TM_n) = \begin{cases} x_1, \ldots, x_{c_k-1}, x_{c_k} \cdot \overline{TM'_{2k-2}}, b \cdot TM_{2k-2} & \text{if } n = 2k, \\ x_1, \ldots, x_{c_k-1}, x_{c_k} \cdot \overline{TM'_{2k-2}}, b \cdot TM_{2k-2} \cdot \overline{TM_{2k}} & \text{if } n = 2k + 1. \end{cases}$$

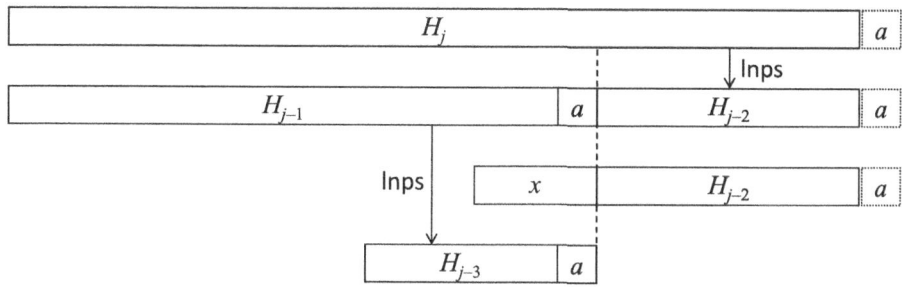

Fig. 1. Illustration for proof of Lemma 3.

Proof. We show this statement by induction on k. First, we can check the statement holds for $k = 2$ by the definition of the Nyldon factorization as follows.

- $NF(TM_3) = a, b, ba, baab$
- $NF(TM_4) = a, b, ba, baab \cdot baa, b \cdot abba$
- $NF(TM_5) = a, b, ba, baab \cdot baa, b \cdot abba \cdot baababbaabbabaab$

For convenience, we define

$$X_k = x_{c_k} \cdot \overline{TM'_{2k-2}}, Y_k = b \cdot TM_{2k-2}, \text{ and } Z_k = b \cdot TM_{2k-2} \cdot \overline{TM_{2k}}.$$

Suppose that the statement holds for $k = j - 1$ for some $j \geq 3$. Now we prove the statement also holds for $k = j$. To prove the factorization is the Nyldon factorization of the string, we will prove the following three conditions.

1. $TM_{2j} = x_1 \cdot x_2 \cdots x_{c_{j-1}} \cdot X_j \cdot Y_j$ and $TM_{2j+1} = x_1 \cdot x_2 \cdots x_{c_{j-1}} \cdot X_j \cdot Z_j$.
2. $x_1 \prec x_2 \prec \ldots \prec x_{c_{j-1}} \prec X_j \prec Y_j$ and $x_1 \prec x_2 \prec \ldots \prec x_{c_{j-1}} \prec X_j \prec Z_j$.
3. $x_i \ (1 \leq i \leq c_j)$, X_j, Y_j, Z_j are Nyldon words.

The first property can be shown by induction hypothesis as follows.

$$x_1 \cdot x_2 \cdots x_{c_{j-1}} \cdot X_j \cdot Y_j = x_1 \cdot x_2 \cdots x_{c_{j-1}} \cdot x_{c_j} \cdot \overline{TM'_{2j-2}} \cdot b \cdot TM_{2j-2}$$
$$= TM_{2j-1} \cdot \overline{TM'_{2j-2}} \cdot b \cdot TM_{2j-2}$$
$$= TM_{2j-1} \cdot \overline{TM_{2j-1}} = TM_{2j}.$$

$$x_1 \cdot x_2 \cdots x_{c_{j-1}} \cdot X_j \cdot Z_j = x_1 \cdot x_2 \cdots x_{c_{j-1}} \cdot x_{c_j} \cdot \overline{TM'_{2j-2}} \cdot b \cdot TM_{2j-2} \cdot \overline{TM_{2j}}$$
$$= TM_{2j} \cdot \overline{TM_{2j}} = TM_{2j+1}.$$

Next, we consider the second condition. By induction hypothesis, $x_1 \prec x_2 \prec \cdots \prec x_{c_j}$ holds. We can see that $x_{c_j} \prec X_j$ holds since x_{c_j} is a prefix of X_j. In the rest of this part, we show $X_j \prec Y_j \prec Z_j$ instead of proving $X_j \prec Y_j$ and $X_j \prec Z_j$. Since Y_j is a prefix of Z_j, $Y_j \prec Z_j$ holds. When $j = 2$, we have

$X_j = b \cdot aab \cdot baa \prec b \cdot abba = Y_j$. When $j > 2$, we have $X_j = x_{c_j} \cdot \overline{TM'_{2j-2}} = b \cdot TM_{2j-4} \cdot \overline{TM_{2j-2}} \cdot \overline{TM'_{2j-2}}$ by induction hypothesis. To obtain $X_j \prec Y_j$ for every $j > 2$, we can use a part of results in the next section (e.g., $X_j = t_{12}(j-2)$ and $Y_j = t_4(j-1)$ in Lemma 7). Hence $X_j \prec Y_j \prec Z_j$ holds.

Finally, we consider the third condition. By induction hypothesis, $x_1, x_2, \ldots, x_{c_{j-1}}$ are Nyldon words. In this part, we also use a part of results in the next section. We can see that $X_j = t_{12}(j-2)$, $Y_j = t_4(j-1)$, and $Z_j = t_{10}(j-1)$. These words are Nyldon words by Lemma 9 which will be shown in the next section.

Therefore, the given factorizations are Nyldon factorizations of TM_{2k} and TM_{2k+1} for every $k \geq 3$.   □

5   Factorization for Infinite Thue-Morse Word

In this section, we show that there exists an infinite non-decreasing sequence of Nyldon words that is a factorization of the infinite Thue-Morse words. First, we define the sequence in two ways and show that they are equivalent (Lemma 4).

Definition 1. Let $w_1 = a$, $w_2 = b$, $w_3 = ba$, $w_4 = baabbaa$, and $w_n = b \cdot TM_{2n-8} \cdot \overline{TM_{2n-6}} \cdot \overline{TM'_{2n-6}}$ for every $n \geq 5$.

Definition 2. Let $\tilde{w}_1 = a$, $\tilde{w}_2 = b$, $\tilde{w}_3 = ba$, $\tilde{w}_4 = baabbaa$, $\tilde{w}_5 = b \cdot TM_2 \cdot \overline{TM_4} \cdot \overline{TM'_4}$, and $\tilde{w}_{n+1} = (baa)^{-1}\tau^2(\tilde{w}_n) \cdot baa$ for every $n \geq 5$.

Lemma 4. For every $n \geq 1$, $w_n = \tilde{w}_n$.

Proof. We prove it by induction on n. We can check that the statement holds for every $1 \leq n \leq 5$ by the definitions. Suppose that $w_k = \tilde{w}_k$ holds for some $k \geq 5$. Then, we have

$$\tilde{w}_{k+1} = (baa)^{-1}\tau^2(\tilde{w}_k) \cdot baa = (baa)^{-1}\tau^2(w_k) \cdot baa$$
$$= (baa)^{-1}\tau^2(b \cdot TM_{2k-8} \cdot \overline{TM_{2k-6}} \cdot \overline{TM'_{2k-6}}) \cdot baa$$
$$= b \cdot TM_{2k-6} \cdot \overline{TM_{2k-4}} \cdot \overline{TM'_{2k-4}} = w_{k+1}.$$

Thus, $w_n = \tilde{w}_n$ holds for every n.   □

In the rest of this section, we use w_n (also for representing $\tilde{w}_n$) to represent the sequence. The next theorem is a main result of this section that explains the sequence $(w_n)_{n \geq 1}$ is an infinite non-decreasing sequence of Nyldon words representing the infinite Thue-Morse words $\mathcal{TM}$.

Theorem 3. *The sequence of words $(w_n)_{n \geq 1}$ satisfies the following properties;*

1. $\mathcal{TM} = \lim_{n \to \infty}(w_1 \cdots w_n)$ *(Lemma 5).*
2. $w_n \preceq w_{n+1}$ *for every $n \geq 1$ (Lemma 6).*
3. w_n *is a Nyldon word for every $n \geq 1$ (Lemma 9).*

5.1 Factorization Property

We show the first property of Theorem 3. The property explains that the sequence $(w_n)_{n \geq 1}$ represents a factorization of $\mathcal{TM}$. By the definition of the morphism τ of the Thue-Morse word, we prove the following equation.

$$\tau^2 \left(\prod_{n \geq 1} w_n \right) = \prod_{n \geq 1} w_n. \tag{1}$$

To prove the equation, we observe relations between some factors and the morphism τ^2.

Observation 1. *The following conditions hold.*

1. $\tau^2(w_1) = \tau^2(a) = abba = w_1 w_2 w_3$
2. $\tau^2(w_2) = \tau^2(b) = baab = (baabbaa)(baa)^{-1} = w_4(baa)^{-1}$
3. $(baa)^{-1}\tau^2(w_3)\tau^2(w_4)(baa) = w_5$
4. *For every $n \geq 5$, w_n has b as a prefix, and $\tau^2(w_n)$ has baa as a prefix.*

Now we can prove the equation by using the observation as follows.

Lemma 5. $\mathcal{TM} = \lim_{n \to \infty}(w_1 \cdots w_n)$ *holds.*

Proof. We show that Equation (1) holds.

$$\tau^2 \left(\prod_{n \geq 1} w_n \right) = \tau^2(w_1)\tau^2(w_2)\tau^2(w_3)\tau^2(w_4)(baa) \prod_{n \geq 5}(baa)^{-1}\tau^2(w_n)(baa)$$

$$= w_1 w_2 w_3 \cdot w_4 \cdot w_5 \cdot \prod_{n \geq 5}(baa)^{-1}\tau^2(w_n)(baa)$$

$$= w_1 w_2 w_3 w_4 w_5 \prod_{n \geq 6} w_n = \prod_{n \geq 1} w_n.$$

Hence, $\mathcal{TM} = \lim_{n \to \infty}(w_1 \cdots w_n)$ holds. □

5.2 Ordering Property

We show the second property of Theorem 3. The property explains that the words in the sequence $(w_n)_{n \geq 1}$ are sorted in lexicographically non-decreasing order.

Lemma 6. *For every $n \geq 1$, $w_n \preceq w_{n+1}$ holds.*

Proof. We show that $w_n = x \cdot a \cdot y$ and $w_{n+1} = x \cdot b \cdot z$ for some strings x, y, and z for every $n \geq 5$. Firstly, we can check the statement holds for $n = 5$ with $x = babbabaab$. We prove this statement by induction on n. Suppose that $w_k = x \cdot a \cdot y$ and $w_{k+1} = x \cdot b \cdot z$ for some strings x, y, and z for some $k \geq 5$. By the definition of w_k and the induction hypothesis, $w_{k+1} = (baa)^{-1}\tau^2(w_k) \cdot baa = (baa)^{-1}\tau^2(x \cdot a \cdot y) \cdot baa$ and $w_{k+2} = (baa)^{-1}\tau^2(w_{k+1}) \cdot baa = (baa)^{-1}\tau^2(x \cdot b \cdot z) \cdot baa$ hold. Hence, $w_{k+1} = XaY$ and $w_{k+2} = XbZ$ hold where $X = (baa)^{-1}\tau^2(x)$, $Y = bba\tau^2(y) \cdot baa$, and $Z = aab\tau^2(z) \cdot baa$. Thus, the statement holds for every $n \geq 5$. This fact also implies that $w_n \preceq w_{n+1}$ for every $n \geq 5$. On the other hand, we can also check that $w_1 \preceq w_2 \preceq \ldots \preceq w_5$. Therefore, $w_n \preceq w_{n+1}$ holds for every $n \geq 1$. □

5.3 Nyldon Property

We show the third property of Theorem 3. Our proof is based on an important combinatorial property of Nyldon words which was explained in Lemma 1. We consider the specific Nyldon substrings in w_n that are produced by recursively applying the property (i.e., $\mathsf{lnps}(\cdot)$) to Nyldon substrings. Let $\Pi_k = \{t_i(k) \mid i \in [1, 12]\}$ be the set of twelve types of substrings that are Nyldon words for every $k \geq 1$, where $[l, r]$ represents the set of integers i such that $l \leq i \leq r$. Every element in Π_k is defined as follows. For convenience, we also represent the substrings by using $A = TM_{2k}$ and $B = \overline{TM}_{2k}$.

$$
\begin{aligned}
t_1(k) &= b \cdot TM'_{2k+1} & &= b \cdot A \cdot B' \\
t_2(k) &= b \cdot TM_{2k+1} & &= b \cdot A \cdot B \\
t_3(k) &= b \cdot TM_{2k+1} \cdot \overline{TM}'_{2k} & &= b \cdot A \cdot B \cdot B' \\
t_4(k) &= b \cdot TM_{2k} & &= b \cdot A \\
t_5(k) &= b \cdot TM_{2k} \cdot TM_{2k-1} \cdot \overline{TM}'_{2k-2} & &= b \cdot A \cdot TM_{2k-1} \cdot \overline{TM}'_{2k-2} \\
t_6(k) &= b \cdot TM_{2k} \cdot TM'_{2k+1} & &= b \cdot A \cdot A \cdot B' \\
t_7(k) &= b \cdot TM_{2k} \cdot TM_{2k+1} & &= b \cdot A \cdot A \cdot B \\
t_8(k) &= b \cdot TM_{2k} \cdot TM_{2k+1} \cdot \overline{TM}'_{2k} & &= b \cdot A \cdot A \cdot B \cdot B' \\
t_9(k) &= b \cdot TM_{2k} \cdot \overline{TM}'_{2k+2} & &= b \cdot A \cdot B \cdot A \cdot A \cdot B' \\
t_{10}(k) &= b \cdot TM_{2k} \cdot \overline{TM}_{2k+2} & &= b \cdot A \cdot B \cdot A \cdot A \cdot B \\
t_{11}(k) &= b \cdot TM_{2k} \cdot TM_{2k+1} \cdot \overline{TM}'_{2k+2} & &= b \cdot A \cdot A \cdot B \cdot B \cdot A \cdot A \cdot B' \\
t_{12}(k) &= b \cdot TM_{2k} \cdot \overline{TM}_{2k+2} \cdot \overline{TM}'_{2k+2} & &= b \cdot A \cdot B \cdot A \cdot A \cdot B \cdot B \cdot A \cdot A \cdot B'
\end{aligned}
$$

Notice that $t_{12}(k - 4) = w_k$ for every $k \geq 5$. Hence, our goal of this part is to show that every string in Π_k is a Nyldon word. This implies the third property of Theorem 3 given by Lemma 9. Firstly, we introduce properties of strings in Π_k.

Observation 2. *For every $k \geq 1$, $t_4(k) \prec t_5(k) \prec t_6(k) \prec t_7(k) \prec t_8(k) \prec t_{11}(k) \prec t_1(k) \prec t_2(k) \prec t_9(k) \prec t_{10}(k) \prec t_{12}(k) \prec t_3(k)$ holds.*

Lemma 7. *For every $k \geq 1$ and $\ell_1, \ell_2 \in [1, 12]$, $t_{\ell_1}(k) \prec t_{\ell_2}(k+1)$ holds.*

Proof. We first prove $t_3(k) \prec t_4(k+1)$ for $k \geq 1$. It is because

$$
\begin{aligned}
t_3(k) &= b \cdot TM_{2k+1} \cdot \overline{TM'_{2k}} \\
&\prec b \cdot TM_{2k+1} \cdot \overline{TM_{2k+1}} \\
&= b \cdot TM_{2k+2} \\
&= t_4(k+1).
\end{aligned}
$$

By this fact and Observation 2, Lemma 7 holds.   □

Now we can show that any string in Π_k is a Nyldon word.

Lemma 8. *For every $k \geq 1$ and $\ell \in [1, 12]$, $t_\ell(k) \in \Pi_k$ is a Nyldon word.*

Proof. We prove $t_\ell(k)$ is a Nyldon word for every $k \geq 1$ and each $\mathsf{Inps}(t_\ell(k))$ is a string given in Table 1 for every $k \geq 2$. We prove these by induction on k and ℓ.

Base Case. We can check the statements hold for $k = 1, 2$ based on the definitions.

Induction Step. Let i, j be integers satisfying $i > 2$ and $j \in [1, 12]$. Suppose that the statement holds for every (k, ℓ) such that $1 \leq k < i$, or $k = i$ and $\ell < j$. We show that the statement also holds for $(k, \ell) = (i, j)$.

Table 1. List of $t_\ell(k) \cdot \mathsf{Inps}(t_\ell(k))^{-1}$ and $\mathsf{Inps}(t_\ell(k))$ ($k > 1$)

$t_\ell(k) \in \Pi_k$	$t_\ell(k) \cdot \mathsf{Inps}(t_\ell(k))^{-1}$	$\mathsf{Inps}(t_\ell(k))$
$t_1(k)$	$t_3(k-1)$	$t_9(k-1)$
$t_2(k)$	$t_3(k-1)$	$t_{10}(k-1)$
$t_3(k)$	$t_3(k-1)$	$t_{12}(k-1)$
$t_4(k)$	$t_3(k-1)$	$t_4(k-1)$
$t_5(k)$	$t_3(k-1)$	$t_8(k-1)$
$t_6(k)$	$t_5(k)$	$t_9(k-1)$
$t_7(k)$	$t_5(k)$	$t_{10}(k-1)$
$t_8(k)$	$t_5(k)$	$t_{12}(k-1)$
$t_9(k)$	$t_2(k)$	$t_6(k)$
$t_{10}(k)$	$t_2(k)$	$t_7(k)$
$t_{11}(k)$	$t_8(k)$	$t_6(k)$
$t_{12}(k)$	$t_1(k)$	$t_{11}(k)$

There are twelve cases, but we cannot give the proof for all cases. Here, we explain a common overview for all cases and also give a proof for the case when $j = 4, 6, 12$.

[Sketch of proof for each case] Let $t_j(i) = p_0 s_0$ where $p_0 \cdot s_0$ is a factorization given in Table 1. Firstly, we check that $t_j(i)$ is the concatenation of p and s in fact by the definition of $t_j(i)$. By induction hypothesis, we can see that s_0 is a Nyldon suffix of $t_j(i)$. Then, we want to prove $\mathsf{Inps}(t_j(i)) = s_0$. Assume on the contrary that there exists a longer Nyldon suffix $x \cdot s$ (x is the shortest such string). This assumption implies that s_0 is the longest Nyldon proper suffix of $x \cdot s_0$. By Lemma 1, x must be a Nyldon word such that $x \succ s_0$ holds. Moreover, x is a suffix of $s_1 = \mathsf{Inps}(p_0)$. Since $s_1 = t_{j'}(i - \ell)$ for some ℓ by induction hypothesis, x is also a Nyldon suffix of s_1. By Lemma 2, $x \preceq s_1$. Hence, we have $s_0 \prec x \preceq s_1$. However, we can also show that $s_0 \succ s_1$. Let $s_0 = t_{i_0}(j_0)$ and $s_1 = t_{i_1}(j_1)$. If $j_1 < j_0$ holds, we can use Lemma 7, otherwise, we can show it directly by using Observation 2. This contradiction implies that $\mathsf{Inps}(t_j(i)) = s_0$. Finally, we want to prove $t_j(i)$ is a Nyldon word. We can see that $p_0 \succ s_0$ holds by Lemma 7 or Observation 2. Hence, we can obtain $t_j(i) = p_0 s_0$ by applying Lemma 1.

[Proof for $t_4(i)$] Here, we consider the case when $j = 4$. Firstly, we show that $\mathsf{Inps}(t_4(i)) = t_4(i - 1)$. By the definitions of Thue-Morse words, we can write

$$
\begin{aligned}
t_4(i) &= b \cdot TM_{2i} \\
&= b \cdot TM_{2i-1} \cdot \overline{TM'_{2i-2}} \cdot b \cdot TM_{2i-2} \\
&= t_3(i-1) \cdot t_4(i-1)
\end{aligned}
$$

and the suffix $t_4(i-1)$ is a Nyldon word by induction hypothesis. Assume on the contrary that there exists a longer Nyldon suffix $x \cdot t_4(i-1)$ of $t_4(i)$ for some nonempty string x. Here, x is also assumed to be the shortest such string. Namely, $t_4(i-1)$ is the longest Nyldon proper suffix of $x \cdot t_4(i-1)$. By Lemma 1, x must be a Nyldon word such that $x \succ t_4(i-1)$ holds. Moreover, x is a suffix of $t_3(i-1)$. Since $\mathsf{Inps}(t_3(i-1)) = t_{12}(i-2)$ by induction hypothesis, x is also a Nyldon suffix of $t_{12}(i-2)$. By Lemma 2, $x \prec t_{12}(i-2)$. Hence, we have $t_4(i-1) \prec x \preceq t_{12}(i-2)$. However, this fact contradicts $t_4(i-1) \succ t_{12}(i-2)$ by Lemma 7. This implies that $\mathsf{Inps}(t_4(i)) = t_4(i-1)$. Finally, we prove $t_4(i)$ is a Nyldon word by Lemma 1. We know that $t_4(i) = t_3(i-1) \cdot t_4(i-1)$, $t_3(i-1) \succ t_4(i-1)$ by Observation 2, $\mathsf{Inps}(t_4(i)) = t_4(i-1)$, and $t_3(i-1)$ is a Nyldon word by induction hypothesis. Therefore, $t_4(i)$ is a Nyldon word by these facts and Lemma 1.

[Proof for $t_6(i)$] Here, we consider the case when $j = 6$. Firstly, we show that $\mathsf{Inps}(t_6(i)) = t_9(i-1)$. By the definitions of Thue-Morse words, we can write

$$
\begin{aligned}
t_6(i) &= b \cdot TM_{2i} \cdot TM'_{2i+1} \\
&= b \cdot TM_{2i} \cdot TM_{2i-1} \cdot \overline{TM'_{2i-2}} \cdot b \cdot TM_{2i-2} \cdot \overline{TM'_{2i}} \\
&= t_5(i) \cdot t_9(i-1)
\end{aligned}
$$

and the suffix $t_9(i-1)$ is a Nyldon word by induction hypothesis. Assume on the contrary that there exists a longer Nyldon suffix $x \cdot t_9(i-1)$ of $t_6(i)$ for some nonempty string x. Here, x is also assumed to be the shortest such string. Namely,

$t_9(i-1)$ is the longest Nyldon proper suffix of $x \cdot t_9(i-1)$. By Lemma 1, x must be a Nyldon word such that $x \succ t_9(i-1)$ holds. Moreover, x is a suffix of $t_5(i)$. Since $\mathsf{lnps}(t_5(i)) = t_8(i-1)$ by induction hypothesis, x is also a Nyldon suffix of $t_8(i-1)$. By Lemma 2, $x \prec t_8(i-1)$. Hence, we have $t_9(i-1) \prec x \preceq t_8(i-1)$. However, this fact contradicts $t_9(i-1) \prec x \prec t_8(i-1)$ by Observation 2. This implies that $\mathsf{lnps}(t_6(i)) = t_9(i-1)$. Finally, we prove $t_6(i)$ is a Nyldon word by Lemma 1. We know that $t_6(i) = t_5(i) \cdot t_9(i-1)$, $t_5(i) \succ t_9(i-1)$ by Lemma 7, $\mathsf{lnps}(t_6(i)) = t_9(i-1)$, and $t_5(i)$ is a Nyldon word by induction hypothesis. Therefore, $t_6(i)$ is a Nyldon word by these facts and Lemma 1.

[Proof for $t_{12}(i)$] Here, we consider the case when $j = 12$. Firstly, we show that $\mathsf{lnps}(t_{12}(i)) = t_{11}(i)$. By the definitions of Thue-Morse words, we can write

$$t_{12}(i) = b \cdot TM_{2i} \cdot \overline{TM_{2i+2}} \cdot \overline{TM'_{2i+2}}$$
$$= b \cdot TM'_{2i+1} \cdot b \cdot TM_{2i} \cdot TM_{2i+1} \cdot \overline{TM'_{2i+2}}$$
$$= t_1(i) \cdot t_{11}(i)$$

and the suffix $t_{11}(i)$ is a Nyldon word by induction hypothesis. Assume on the contrary that there exists a longer Nyldon suffix $x \cdot t_{11}(i)$ of $t_{12}(i)$ for some non-empty string x. Here, x is also assumed to be the shortest such string. Namely, $t_{11}(i)$ is the longest Nyldon proper suffix of $x \cdot t_{11}(i)$. By Lemma 1, x must be a Nyldon word such that $x \succ t_{11}(i)$ holds. Moreover, x is a suffix of $t_1(i)$. Since $\mathsf{lnps}(t_1(i)) = t_9(i-1)$ by induction hypothesis, x is also a Nyldon suffix of $t_9(i-1)$. By Lemma 2, $x \prec t_9(i-1)$. Hence, we have $t_{11}(i) \prec x \preceq t_9(i-1)$. However, this fact contradicts $t_{11}(i) \succ t_9(i-1)$ by Lemma 7. This implies that $\mathsf{lnps}(t_{12}(i)) = t_{11}(i)$. Finally, we prove $t_{12}(i)$ is a Nyldon word by Lemma 1. We know that $t_{12}(i) = t_1(i) \cdot t_{11}(i)$, $t_1(i) \succ t_{11}(i)$ by Observation 2, $\mathsf{lnps}(t_{12}(i)) = t_{11}(i)$, and $t_1(i)$ is a Nyldon word by induction hypothesis. Therefore, $t_{12}(i)$ is a Nyldon word by these facts and Lemma 1.

Finally, we obtain this lemma. □

Now, we are ready to prove the following (third) property.

Lemma 9. *For every* $n \geq 1$, w_n *is a Nyldon word.*

Proof. We can see that $w_1 = a$, $w_2 = b$, $w_3 = ba$, and $w_4 = baabbaa$ are Nyldon words. Recall that $t_{12}(k-4) = w_k$ for every $k \geq 5$ by the definitions. By combining with Lemma 8, w_n is a Nyldon word for every $n \geq 1$. □

6   Conclusions

We considered the Nyldon factorizations of Fibonacci and Thue-Morse words. We gave the Nyldon factorization of the finite Fibonacci words, and fully characterize the Nyldon factorization of the finite Thue-Morse words. We also showed that there exists a non-decreasing sequence of Nyldon words that is a factorization of the infinite Thue-Morse words. There are numerous studies on factorizations of certain well-known string families, such as Fibonacci words and Thue-Morse words. For instance, the Lyndon factorization of the infinite Fibonacci words and Sturmian words [15,16], the Lyndon factorization of the infinite Thue-Morse words [10], the Crochemore factorization of the Sturmian words [1], repetition factorizations of the finite Fibonacci words [11,13], palindromic Ziv-Lempel and Crochemore factorizations of n-bonacci infinite words [12] have been considered. Studies on factorizations for specific strings like these can be useful not only for combinatorics on words, but also for algorithmic studies (e.g., analyzing the complexity of algorithms). We hope our new results also contribute to algorithms that use the Nyldon words and the Nyldon factorizations.

Acknowledgments. This work was supported by JSPS KAKENHI Grant Numbers JP25K00136 (YN), JP23K24808, JP23K18466 (SI), JP24K02899 (HB), and JST BOOST, Japan Grant Number JPMJBS2406 (KK).

References

1. Berstel, J., Savelli, A.: Crochemore factorization of Sturmian and other infinite words. In: International Symposium on Mathematical Foundations of Computer Science, pp. 157–166. Springer (2006)
2. Bonizzoni, P., De Felice, C., Zaccagnino, R., Zizza, R.: Inverse Lyndon words and inverse Lyndon factorizations of words. Adv. Appl. Math. **101**, 281–319 (2018). https://doi.org/10.1016/j.aam.2018.08.005, https://www.sciencedirect.com/science/article/pii/S0196885818300964
3. Bonizzoni, P., De Felice, C., Riccardi, B., Zaccagnino, R., Zizza, R.: Unveiling the connection between the inverse Lyndon factorization and the canonical inverse Lyndon factorization via a border property. In: Královic, R., Kucera, A. (eds.) 49th International Symposium on Mathematical Foundations of Computer Science, MFCS 2024, August 26-30, 2024, Bratislava, Slovakia. LIPIcs, vol. 306, pp. 31:1–31:14. Schloss Dagstuhl - Leibniz-Zentrum für Informatik (2024). https://doi.org/10.4230/LIPICS.MFCS.2024.31, https://doi.org/10.4230/LIPIcs.MFCS.2024.31
4. Brlek, S., Lachaud, J.O., Provençal, X., Reutenauer, C.: Lyndon + Christoffel = digitally convex. Pattern Recogn. **42**(10), 2239–2246 (2009). https://doi.org/10.1016/j.patcog.2008.11.010, https://www.sciencedirect.com/science/article/pii/S0031320308004706, selected papers from the 14th IAPR International Conference on Discrete Geometry for Computer Imagery 2008
5. Émilie Charlier, Philibert, M., Stipulanti, M.: Nyldon words. J. Combinatorial Theory, Series A **167**, 60–90 (2019). https://doi.org/10.1016/j.jcta.2019.04.002
6. Chen, K.T., Fox, R.H., Lyndon, R.C.: Free differential calculus. IV. The quotient groups of the lower central series. Ann. Math. **68**(1), 81–95 (1958)
7. Grinberg, D.: "Nyldon words":understanding a class of words factorizing the free monoid increasingly. https://mathoverflow.net/questions/187451/ (2014)
8. Garg, S.: New results on Nyldon words and Nyldon-like sets. Adv. Appl. Math. **131**, 102249 (2021). https://doi.org/10.1016/j.aam.2021.102249, https://www.sciencedirect.com/science/article/pii/S0196885821000877
9. Hirakawa, R., Nakashima, Y., Inenaga, S., Takeda, M.: Counting Lyndon subsequences. In: Holub, J., Zdárek, J. (eds.) Prague Stringology Conference 2021, Prague, Czech Republic, August 30-31, 2021, pp. 53–60. Czech Technical University in Prague, Faculty of Information Technology, Department of Theoretical Computer Science (2021). http://www.stringology.org/event/2021/p05.html

10. Ido, A., Melançon, G.: Lyndon factorization of the Thue-Morse word and its relatives. Discrete Math. Theoretical Comput. Sci. DMTCS **1**, 43–52 (1997). https://api.semanticscholar.org/CorpusID:2228097

11. Inoue, H., et al.: Factorizing strings into repetitions. Theory Comput. Syst. **66**(2), 484–501 (2022). https://doi.org/10.1007/S00224-022-10070-3

12. Jahannia, M., Mohammad-noori, M., Rampersad, N., Stipulanti, M.: Palindromic Ziv-Lempel and Crochemore factorizations of m-bonacci infinite words. Theoret. Comput. Sci. **790**, 16–40 (2019)

13. Kishi, K., Nakashima, Y., Inenaga, S.: Largest repetition factorization of Fibonacci words. In: Nardini, F.M., Pisanti, N., Venturini, R. (eds.) String Processing and Information Retrieval - 30th International Symposium, SPIRE 2023, Pisa, Italy, September 26-28, 2023, Proceedings. Lecture Notes in Computer Science, vol. 14240, pp. 284–296. Springer (2023). https://doi.org/10.1007/978-3-031-43980-3_23, https://doi.org/10.1007/978-3-031-43980-3_23

14. Lyndon, R.C.: On Burnside's problem. Trans. Am. Math. Soc. **77**, 202–215 (1954)

15. Melançon, G.: Lyndon factorization of infinite words. In: Puech, C., Reischuk, R. (eds.) STACS 1996. LNCS, vol. 1046, pp. 147–154. Springer, Heidelberg (1996). https://doi.org/10.1007/3-540-60922-9_13

16. Melançon, G.: Lyndon factorization of Sturmian words. Discret. Math. **210**(1), 137–149 (2000). https://doi.org/10.1016/S0012-365X(99)00123-5

17. Mucha, M.: Lyndon Words and Short Superstrings. In: Proceedings of the 2013 Annual ACM-SIAM Symposium on Discrete Algorithms (SODA), pp. 958–972 (2013). https://doi.org/10.1137/1.9781611973105.69, https://epubs.siam.org/doi/abs/10.1137/1.9781611973105.69

18. Siromoney, R., Mathew, L., Dare, V., Subramanian, K.: Infinite Lyndon words. Inf. Process. Lett. **50**(2), 101–104 (1994). https://doi.org/10.1016/0020-0190(94)90016-7, https://www.sciencedirect.com/science/article/pii/0020019094900167

String Matching with a Dynamic Pattern

Bruno Monteiro$^{(\boxtimes)}$ ⓘ and Vinicius dos Santos ⓘ

Computer Science Department, Federal University of Minas Gerais,
Belo Horizonte, Brazil
malettabruno@gmail.com, viniciussantos@dcc.ufmg.br

Abstract. In this work, we tackle a natural variation of the String
Matching Problem on the case of a dynamic pattern, that is, given a
static text T and a pattern P, we want to support character insertions
and deletions to the pattern, and after each operation compute how many
times it occurs in the text. We show a simple and practical algorithm
using Suffix Arrays that achieves $\mathcal{O}(\log |T|)$ update time, after $\mathcal{O}(|T|)$
preprocessing time. We show how to extend our solution to support sub-
string deletion, transposition (moving a substring to another position
of the pattern), and copy (copying a substring and pasting it in a specific position), in the same time complexities. Our solution can also be
extended to support an online text (inserting characters to one end of
the text), maintaining the same amortized bounds.

Keywords: Strings · Algorithms · Suffix array · String matching

1 Introduction

The String Matching Problem consists of, given two strings usually called the *text*
and the *pattern*, computing the indices where the pattern occurs in the text. After
Knuth, Morris, and Pratt [25,30] settled it with a linear-time solution in 1970,
work was done on variations of the problem. Fischer and Paterson [16] introduced
the problem of *Pattern Matching with Wildcards*, in which we can have "don't-
care" symbols in the text or the pattern, that match with any other symbol.
Another studied variation is the *Approximate Pattern Matching*, which consists
of finding occurrences of the pattern subject to at most k mismatches [20,27].
Due to its wide applicability, pattern matching and related problems remain an
active and well-studied research topic [12,26].

Another problem of interest is the Indexing Problem. This problem can be
solved *offline* (if all patterns are known in advance) in linear time [1], for a
constant size alphabet. In the *online* scenario, we want to preprocess only the
text and answer queries efficiently. Classical solutions for this problem use suffix
trees or suffix arrays. Both data structures can be computed directly in linear
time and space [14,21], assuming the alphabet symbols are integers bounded
by $|T|^c$, for text T and $c \in \mathcal{O}(1)$. Of course, if the alphabet is unbounded, in
the comparison model there is a lower bound of $\Omega(|T| \log |T|)$ (from sorting)

G. Badkobeh et al. (Eds.): SPIRE 2025, LNCS 16073, pp. 202–216, 2026.
https://doi.org/10.1007/978-3-032-05228-5_17

Table 1. Deterministic worst-case query time for the Indexing Problem (for pattern P and text T) across different data structures that can be built in linear time and space

Data Structure	Query Time	Source
Suffix Tree	$\mathcal{O}(\|P\|\log\|\Sigma\|)$ or $\mathcal{O}(\|P\|+\log\|T\|)$	[32] [10]
Suffix Array	$\mathcal{O}(\|P\|+\log\|T\|)$	[28]
Suffix Tray	$\mathcal{O}(\|P\|+\log\|\Sigma\|)$	[9]

for building these data structures [11]. When it comes to query time, the data structures achieve different bounds (see Table 1).

It is also possible to maintain the $\mathcal{O}(\|P\|+\log\|T\|)$ query time while supporting updates of inserting a symbol to the end of the text (also called *online text*) in $\mathcal{O}(\log\|T\|)$ worst-case time [4]. Studies have been made in the case of a *dynamic text*, in which we can insert or delete symbols in arbitrary positions of the text while supporting efficient queries [2,15,18]. Further work has explored maintaining a collection of strings subject to splits and concatenates while supporting pattern queries [17].

The case of a dynamic pattern has been given less attention. Amir and Kondratovsky [3] were the first to give a sub-linear solution for updating the pattern and querying for the number of occurrences of the pattern in the text, naming it the *modified pattern reporting problem*. They show how to maintain a pattern subject to symbol change (insertion and deletion) in $\mathcal{O}(\log\|T\|)$ time, and substring copy-pasting and deletion in $\mathcal{O}(\log\|T\| + \ell)$ time, with ℓ being the size of the modified substring. This is done after $\mathcal{O}(\|T\|\sqrt{\log\|T\|})$ preprocessing time and memory.

We can maintain $T\$P$ (the text concatenated with a separator and the pattern) using dynamic suffix arrays [23], answering how many times P occurs in T in $\mathcal{O}(\log^4\|T\|)$ time. A generalization of this approach can support answering whether a substring of a dynamic pattern occurs in a substring of a dynamic text in $\tilde{\mathcal{O}}(1)$ time [7].

Using several insights regarding the suffix array, we achieve better time bounds for the aforementioned problem (see Table 2), with a simpler algorithm. We show how we can maintain a dynamic pattern, subject to character insertion or deletion and substring deletion, transposition (moving a substring to another position of the pattern), and copy (copying a substring and pasting it in a specific position). After every update to the pattern, we can output the number of times the pattern occurs in the text.

In addition, we apply our algorithm for an online text, that is, supporting insertion of a symbol to one end of the text. We achieve amortized logarithmic bounds (see Table 3). While Amir and Kondratovsky [3] outline a suffix tree based approach claiming logarithmic worst-case bounds, this work provides a

Table 2. Overview of the time bounds we achieve for the *modified pattern reporting problem*. Here, an edit is an insertion or deletion, and ℓ is the size of the modified substring

Operation	[3]	This work
Preprocessing Time and Space	$\mathcal{O}(|T|\sqrt{\log|T|})$	$\mathcal{O}(|T|)$
Pattern Symbol Edit	$\mathcal{O}(\log|T|)$	$\mathcal{O}(\log|T|)$
Pattern Substring Edit	$\mathcal{O}(\log|T| + \ell)$	$\mathcal{O}(\log|T|)$

simpler suffix array based solution with clear logarithmic amortized bounds for all operations.

Table 3. Overview of the time bounds we achieve for the *modified pattern reporting problem* with an online text. Here, an edit is an insertion or deletion, the $\star$ symbol represents amortized bounds, and ℓ is the size of the modified substring

Operation	[3]	This work
Preprocessing Time and Space	$\mathcal{O}(|T|\sqrt{\log|T|})$	$\mathcal{O}(|T|)$
Pattern Symbol Edit	$\mathcal{O}(\log|T|)$	$\mathcal{O}(\log|T|)^{\star}$
Pattern Substring Edit	$\mathcal{O}(\log|T| + \ell)$	$\mathcal{O}(\log|T|)^{\star}$
Text Symbol Extension	$\mathcal{O}(\log|T|)$	$\mathcal{O}(\log|T|)^{\star}$

For convenience of presentation, we define Problem 1, which can be shown to be equivalent to the *modified pattern reporting problem*.

Problem 1. (**Dynamic Pattern and Static Text Matching**)
Input: A string T that represents the text, several updates to the pattern string P that can be one of the following.

1. Pattern search: set the current pattern to a new pattern provided as input;
2. Pattern symbol edit: insertion or deletion of some character of the current pattern;
3. Pattern substring edit: substring deletion, transposition (moving the substring to another position), or copy on the current pattern.

Output: After every operation, the number of times the current pattern occurs in the text.

The solution from the literature [3] uses suffix trees. Our solution with suffix arrays is considerably simpler, reduces the preprocessing time and space requirement, and improves upon the time required for the substring operations.

Note that in this work we do not assume that the text T is necessarily much larger than the pattern P ($|T| \gg |P|$). Even though we assume $|T| > |P|$, factors of, for example, $\log |T|$ and $|P|$ are both taken into account in the time complexity analysis. That is, we are interested in the cases when the pattern can be large relative to the text.

2   Suffix Arrays and Suffix Ranges

The suffix array was first developed by Manber and Myers [28] as a more space-efficient alternative to Suffix Trees, a data structure developed for solving the Indexing Problem [28,32]. The lemmas presented in this section are standard, but we prove them here for completeness.

Definition 1 (Suffix Array of a String). The *suffix array* of a string S, denoted as $\mathcal{SA}(S)$, is an array of length $|S|$ such that $\mathcal{SA}(S)[i]$ stores the starting index of the i-th smallest suffix of S, in lexicographical order.

Note that there are no ties, so the suffix array is unique. It is very common to use, along with the suffix array, the *lcp* array.

Definition 2 (Longest Common Prefix Array of a String). The *longest common prefix array* (or simply *lcp* array) of a string S, denoted as $\mathcal{LCP}(S)$, is an array of size $|S| - 1$ such that $\mathcal{LCP}(S)[i] = lcp(S_{\mathcal{SA}[i]}, S_{\mathcal{SA}[i+1]})$, that is, $\mathcal{LCP}(S)[i]$ is the *lcp* between the i-th and $(i + 1)$-th smallest suffixes of S.

Both the suffix array and the *lcp* array can be computed in linear time [21, 22,24], assuming the alphabet symbols are integers bounded by $|T|^c$, for text T and $c \in \mathcal{O}(1)$.

Lemma 1. *Let S be a string and $\mathcal{SA}$ its suffix array. For $0 \leq i < j < |S|$, if there is a common prefix between $S_{\mathcal{SA}[i]}$ and $S_{\mathcal{SA}[j]}$ of length ℓ, then $lcp(S_{\mathcal{SA}[i]}, S_{\mathcal{SA}[k]}) \geq \ell$, for all $i < k < j$.*

Proof. From definition of suffix array, we know that $S_{\mathcal{SA}[i]} \prec S_{\mathcal{SA}[k]} \prec S_{\mathcal{SA}[j]}$. By contradiction, assume that the statement is false. This means that, for some $i < k < j$, there is an integer $x < \ell$ such that $lcp(S_{\mathcal{SA}[i]}, S_{\mathcal{SA}[k]}) = x$ and $S_{\mathcal{SA}[i]}[x] < S_{\mathcal{SA}[k]}[x]$. But since $lcp(S_{\mathcal{SA}[i]}, S_{\mathcal{SA}[j]}) \geq \ell$, we have that $S_{\mathcal{SA}[i]}[x] = S_{\mathcal{SA}[j]}[x]$, implying $S_{\mathcal{SA}[j]} \prec S_{\mathcal{SA}[k]}$, a contradiction.

Lemma 2. *The lcp between any two suffixes of S, S_i and S_j, is equal to the minimum over the range in the lcp array (assuming $\mathcal{ISA}[i] < \mathcal{ISA}[j]$):*

$$lcp(S_i, S_j) = \min_{\mathcal{ISA}[i] \leq k < \mathcal{ISA}[j]} \mathcal{LCP}[k], \tag{1}$$

This also holds for any set of strings sorted lexicographically.

Proof. Let $\ell = lcp(S_i, S_j)$. From Lemma 1, we know that the first ℓ characters of S_i are equal to those of $S_{\mathcal{SA}[\mathcal{ISA}[i]+1]}$ (the next suffix in the suffix array order), and are also equal to those of $S_{\mathcal{SA}[\mathcal{ISA}[i]+2]}$, and so on until S_j. Therefore, the $\mathcal{LCP}$ values of these positions are at least ℓ. And they can not be all greater than ℓ, otherwise we would have $lcp(S_i, S_j) > \ell$. So the minimum of the $\mathcal{LCP}$ values of the range is exactly ℓ. All these arguments are also valid for any set of strings sorted lexicographically.

From Equation 2 we can solve *lcp* queries of arbitrary suffixes in constant time, since range minimum queries can be answered in constant time after a linear-time preprocessing [6].

2.1   Suffix Range

The reason why suffix arrays are useful for the Indexing Problem is because the occurrence of any pattern defines a *range* in the suffix array, which we call *suffix range* (see Definition 3).

Definition 3 (Suffix Range). Let P, T be strings, and let S be the set of indices where P occurs in T. The *suffix range* of P with respect to T, denoted by $\mathcal{SR}(P, T)$ is the set $\{i : \mathcal{SA}(T)[i] \in S\}$. From Lemma 3, this set is a range of indices, so we can represent it by a range $[\ell, r)$, meaning that the suffix range is $\ell, \ell + 1, \ldots, r - 1$. If P does not occur in T, we represent the suffix range by the empty range $[0, 0)$.

Lemma 3. *If P occurs in T, then $\mathcal{SR}(P, T)$ is a range of indices.*

Proof. Let $\mathcal{SA}$ be the suffix array of T and let S be the set of indices where P occurs in T. By contradiction, assume that there are indices $i < k < j$ such that $\mathcal{SA}[i], \mathcal{SA}[j] \in S$ and $\mathcal{SA}[k] \notin S$. We know that $lcp(T_{\mathcal{SA}[i]}, T_{\mathcal{SA}[j]}) \geq |P|$, so from Lemma 1 we have that $lcp(T_{\mathcal{SA}[i]}, T_{\mathcal{SA}[k]}) \geq |P|$, so $\mathcal{SA}[k]$ is an occurrence of P, a contradiction.

It turns out that $\mathcal{SR}(P, T)$ can be computed in $\mathcal{O}(|P| + \log |T|)$ time using binary search and some clever *lcp* insights [28].

3   Suffix Range Updates

The key contribution of this section is an efficient method for updating the pattern's suffix range following pattern edits.

We begin by tackling concatenation: given $|A|, |B|, \mathcal{SR}(A, T)$, and $\mathcal{SR}(B, T)$, we can compute $\mathcal{SR}(A \circ B, T)$ efficiently, as detailed in Theorem 1 and Algorithm 1. Here, $A \circ B$ denotes the concatenation of strings A and B. While this result appears in prior work [19], we include the proof for completeness and consistency of notation.

Lemma 4. *If $A \circ B$ occurs in T, then $\mathcal{SR}(A \circ B, T)$ is a sub-range of $\mathcal{SR}(A, T)$.*

Proof. All of the suffixes from $\mathcal{SR}(A \circ B, T)$ have A as a prefix, therefore they are also in $\mathcal{SR}(A, T)$.

Theorem 1 *([19], cf. Lemma 2). Given $|A|$, $|B|$, $\mathcal{SR}(A, T)$, and $\mathcal{SR}(B, T)$, we can compute $\mathcal{SR}(A \circ B, T)$ in $\mathcal{O}(\log |T|)$ time.*

Proof. From Lemma 4, we want to find a sub-range of $\mathcal{SR}(A, T)$, so it suffices to find its first and last positions. Let $[\ell, r)$ be the range $\mathcal{SR}(A, T)$. We want to find the first $i \in [\ell, r)$ such that $T_{\mathcal{SA}[i]+|A|}$ has B as a prefix. But note that, since $T_{\mathcal{SA}[\ell]}, T_{\mathcal{SA}[\ell+1]}, \ldots, T_{\mathcal{SA}[r-1]}$ all have an *lcp* of at least $|A|$, then $T_{\mathcal{SA}[\ell]+|A|}, T_{\mathcal{SA}[\ell+1]+|A|}, \ldots, T_{\mathcal{SA}[r-1]+|A|}$ appear in this order in the suffix array, because their lexicographical comparison is not decided by their first $|A|$ characters, so skipping them maintain their order. Therefore, we can do a binary search on the range $[\ell, r)$, and when checking some suffix, we look at the position of the suffix that skips $|A|$ characters from that suffix, and check if it is in $\mathcal{SR}(B, T)$, using the $\mathcal{ISA}$. We can therefore find the first and last position in $[\ell, r)$ that correspond to $\mathcal{SR}(A \circ B, T)$, with two binary searches.

A visual representation of Lemma 4 is shown in Fig. 1. The algorithm is implemented in Algorithm 1, using the type of binary search that maintains that the answer is in the half-open range $[\ell, r)$.

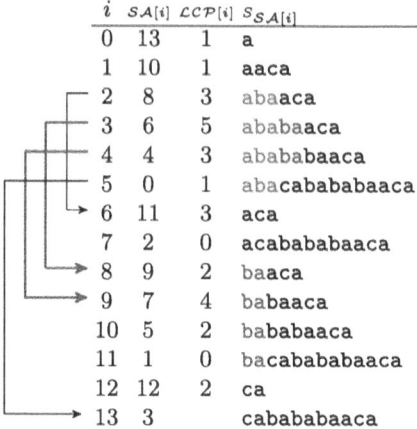

i	$S\!A[i]$	$LCP[i]$	$S_{S\!A[i]}$
0	13	1	a
1	10	1	aaca
2	8	3	abaaca
3	6	5	ababaaca
4	4	3	abababaaca
5	0	1	abacabababaaca
6	11	3	aca
7	2	0	acabababaaca
8	9	2	baaca
9	7	4	babaaca
10	5	2	bababaaca
11	1	0	bacabababaaca
12	12	2	ca
13	3		cabababaaca

Fig. 1. For $S =$ abacabababaaca, in red, $\mathcal{SR}(\text{aba}, S)$, and, in blue, $\mathcal{SR}(\text{ba}, S)$. The arrows illustrate the fact that the suffixes from $\mathcal{SR}(\text{aba}, S)$ skipping $|\text{aba}| = 3$ characters occur in increasing order, so we can use binary search to find the ones that end up in $\mathcal{SR}(\text{ba}, S)$, as in Theorem 1

We now describe how to update the suffix range when deleting characters from either end of the pattern. For this, we need the following lemma.

Lemma 5. *Given any index $i \in \mathcal{SR}(P, T)$, we can find $\mathcal{SR}(P, T)$ in $\mathcal{O}(\log |T|)$ time.*

Proof. We just need to find $\ell = \min_{j \leq i}\{j : lcp(T_{\mathcal{SA}[j]}, T_{\mathcal{SA}[i]}) \geq |P|\}$ and $r = \min_{i < j}\{j : lcp(T_{\mathcal{SA}[j]}, T_{\mathcal{SA}[i]}) < |P|\}$, and this gives us $\mathcal{SR}(P,T) = [\ell, r)$. Both these indices can be found with a binary search using lcp queries, since by Lemma 2 the lcp value is monotonic when we increase the range: if $k < j \leq i$, $lcp(T_{\mathcal{SA}[k]}, T_{\mathcal{SA}[i]}) \leq lcp(T_{\mathcal{SA}[j]}, T_{\mathcal{SA}[i]})$, and symmetrically for the other direction.

Algorithm 1 Suffix Range Concatenation.

   Input: $|A|, |B|, \mathcal{SR}(A,T), \mathcal{SR}(B,T)$.
   Output: $\mathcal{SR}(A \circ B, T)$.
   Time complexity: $\mathcal{O}(\log |T|)$.

$(lb, rb) \leftarrow \mathcal{SR}(B, T)$

$(\ell, r) \leftarrow \mathcal{SR}(A, T)$ ▷ Compute first index of the answer
while $\ell < r$ **do**
$\quad m \leftarrow \lfloor \frac{\ell + r}{2} \rfloor$
$\quad$ **if** $\mathcal{SA}[m] + |A| = |T|$ or $\mathcal{ISA}[\mathcal{SA}[m] + |A|] < lb$ **then**
$\quad\quad \ell \leftarrow m + 1$
$\quad$ **else**
$\quad\quad r \leftarrow m$
$first \leftarrow \ell$

$(\ell, r) \leftarrow \mathcal{SR}(A, T)$ ▷ Compute last index of the answer
while $\ell < r$ **do**
$\quad m \leftarrow \lfloor \frac{\ell + r}{2} \rfloor$
$\quad$ **if** $\mathcal{SA}[m] + |A| = |T|$ or $\mathcal{ISA}[\mathcal{SA}[m] + |A|] < rb$ **then**
$\quad\quad \ell \leftarrow m + 1$
$\quad$ **else**
$\quad\quad r \leftarrow m$
$last \leftarrow \ell$

if $l = r$ **then**
$\quad$ **return** $[0, 0)$ ▷ Empty range: no occurrences
else
$\quad$ **return** $[first, last)$

Lemma 6. *If P occurs in T, given $\mathcal{SR}(P,T)$ we can compute the suffix range of P with $k < |P|$ characters deleted from the beginning or the end in $\mathcal{O}(\log |T|)$ time.*

Proof. Using Lemma 5 it is easy to delete some amount of characters from the end of the pattern: from Lemma 4, we know that the suffix range contains the suffix range we had. So we can take any index of the suffix range we had and extend it to find the new suffix range.

   To delete some amount of characters from the beginning, we apply a similar strategy. Let P' be the pattern P with the first k characters deleted. If

$i \in \mathcal{SR}(P, T)$, then $\mathcal{ISA}(\mathcal{SA}[i] + k) \in \mathcal{SR}(P', T)$. Therefore, we can again use Lemma 5 and compute the new suffix range.

Now we are ready to tackle arbitrary character edits in the pattern.

Theorem 2. *Let P be a pattern that occurs in the text T. Let P' be P with the i-th character edited (inserted or deleted). Given $\mathcal{SR}(P, T)$, we can compute $\mathcal{SR}(P', T)$ in $\mathcal{O}(\log |T|)$ time.*

Proof. Assume we want to perform a character insertion. From Lemma 6, we can compute $\mathcal{SR}(P[0, i], T)$ and $\mathcal{SR}(P[i, |P|), T)$ in $\mathcal{O}(\log |T|)$ time. That is, we split the pattern at the index we want to change. To insert some character $c \in \Sigma$, we can then compute $\mathcal{SR}(c, T)$ in $\mathcal{O}(\log |T|)$ time using binary search, and then concatenate the suffix ranges as outlined in Theorem 1. In the case of a deletion, we just need to concatenate $\mathcal{SR}(P[0, i], T)$ and $\mathcal{SR}(P[i + 1, |P|), T)$.

So far, we assumed the P pattern always occurs in the text T. However, if an edit causes P to no longer appear in T, the suffix range becomes empty $(\mathcal{SR}(P, T) = [0, 0))$, and we lose the matching information. To handle this, we adopt the idea of an occurrence partition (or cover) found in previous work [5, 8].

4   Dealing with Patterns that Do Not Occur

Now we look at the case when the pattern might not occur in the text. What we can do is represent the pattern as a *concatenation* of strings that occur in the text, which we call an *occurrence partition*, see Definition 4.

Definition 4 (Occurrence Partition). Let P be a pattern and T be a text. An *occurrence partition* of P with respect to T is a sequence of strings $(P_1, P_2, \ldots, P_k)$ that partition P, that is, $P = P_1 \circ P_2 \circ \cdots \circ P_k$, and satisfies two properties:

1. Occurrence: P_i either occurs as a substring of T or $|P_i| = 1$ and P_i represents a character that does not occur in T.
2. Maximality: $P_i \circ P_{i+1}$ does not occur in T.

We also denote the minimum size of such partition as $|P|_T$.

Note that not all occurrences partitions are minimum, for example, for $T = $ aabcaba and $P = $ abaaba, (ab, aab, a) is an occurrence partition, but there is another of size two: (aba, aba).

Lemma 7. *Assuming all characters of P occur in T, a pattern P occurs in a text T if, and only if, the occurrence partition of P with respect to T has size one.*

Proof. Follows directly from the definition.

The idea is then to represent an occurrence partition of the pattern, and, for every string in the partition, maintain its suffix range. When asked to return the number of occurrences of the pattern, we return zero if the size of the partition is greater than one, or the length of the suffix range otherwise, as per Lemma 7 (taking care of characters that do not occur in the text).

We can represent the occurrence partition in a balanced binary search tree, since we want to apply modifications, and this will allow us to do that in $\mathcal{O}(\log |P|)$ time.

We are left with the task of maintaining the occurrence partition properties when modifying some character. Turns out that this is easy: given some index to edit, we can find the string from the partition that should be edited (by traversing the binary search tree). After that, we can apply Theorem 2, but not merge the suffix ranges when they would result in an empty range (occurrence property). This might break the maximality property, so we might need to apply some concatenations. It turns out that at most 4 concatenations are needed to ensure the maximality property, as per Lemma 8.

Lemma 8. *Let P be a pattern and $(P_1, P_2, \ldots, P_k)$ some occurrence partition of P with respect to the text T. After a character edit in P, we can find an updated occurrence partition using at most 4 concatenations.*

Proof. Assume our edit was in the i-th string of the partition, so we had the partition $(\ldots, P_{i-2}, P_{i-1}, P_i, P_{i+1}, P_{i+2}, \ldots)$. Following our strategy, P_i might be split into 3 parts (in the case of an insertion, the middle part represents the new character), say Q_1, Q_2, Q_3. It might be the case that $P_{i-1} \circ Q_1 \circ Q_2 \circ Q_3 \circ P_{i+1}$ occurs in the text (4 concatenations are necessary in this case). But this is the worst-case, since, from the maximality property of the initial partition, we know that $P_{i-2} \circ P_{i-1}$ does not occur in the text, so it can not be the case that $P_{i-2} \circ P_{i-1} \circ S$ occurs, for any string S. The same applies with P_{i+1} and P_{i+2}.

Consider the following example: assume that we have the text $T = \text{cababaa}$ and the pattern $P = \text{abcaabb}$. An occurrence partition of P with respect to T is (ab, c, aa, b, b). Now assume we want to insert a character b in P at index 4, that is, between the two a's. This will turn P into abcababb. To do this, we first need to split the string aa, so our representation becomes (ab, c, a, a, b, b); then we find the suffix range of the new character b, and insert it to our representation: (ab, c, a, b, a, b, b). Now we are left with concatenating adjacent strings in the representation (starting from the b we inserted, in both directions), to fix the maximality property. After doing that, the final representation will become (ab, cabab, b), so we performed 4 concatenations. We can see that, since ab $\circ$ c (concatenation of first and second strings in the initial representation) does not occur in T, then ab $\circ$ c $\circ$ abab also does not occur in T.

Theorem 3. *Let P be a pattern and let T be the text. Let P' be P with the i-th character edited (inserted or deleted). If we have the occurrence partition of P, we can find an occurrence partition of P' in $\mathcal{O}(\log |T|)$ time.*

Proof. From Lemma 8, we only need to use Theorem 1 $\mathcal{O}(1)$ times. Since we can do all necessary operations in the binary search tree used to represent the occurrence partition in $\mathcal{O}(\log |P|) \subseteq \mathcal{O}(\log |T|)$ time, we can edit the pattern in $\mathcal{O}(\log |T|)$ time.

5   Improving Pattern Search

We now discuss the pattern search operation, that is, set the current pattern as some pattern P given in the input. We assume the current pattern before the operation is empty. Of course, this can be viewed as repeated pattern edit, and therefore can be done in $\mathcal{O}(|P| \log |T|)$ time. But we can also compute it in $\mathcal{O}(|P| + |P|_T \log |T|)$ time.

Lemma 9. *Greedily taking each time the largest prefix of the pattern that occurs in the text produces an occurrence partition of size $|P|_T$.*

Proof. Any minimum occurrence partition can be transformed to the partition that the greedy algorithm produces, by repeatedly shifting the positions of the divisions between the strings to the right (from left to right).

Lemma 10. *The algorithm from [28] to compute $\mathcal{SR}(P,T)$ can be implemented in $\mathcal{O}(\log |T| + \ell)$ time, if ℓ is the size of the largest prefix of P that occurs in T.*

Proof. The only operation inside the binary search that has nontrivial cost is the *lcp* computation between the pattern and some suffix of the text. But note that the total time that takes is bounded by ℓ, since every time it runs another iteration it is increasing the size of a prefix of P that occurs in T.

From Lemma 10, we can repeatedly run the algorithm from [28], and each time from its output we can figure out the size of the string to use for our partition. This is running the greedy algorithm, and from Lemma 9 we know that this produces a partition of size $|P|_T$, and adding the total cost we get $\mathcal{O}(|P|)$ plus $\mathcal{O}(\log |T|)$ times the size of the partition. Therefore, we can find the occurrence partition in $\mathcal{O}(|P| + |P|_T \log |T|)$ time.

It turns out we can, given $\mathcal{SR}(P,T)$, compute $\mathcal{SR}(P \circ \mathsf{c}, T)$ for some character $\mathsf{c} \in \Sigma$ in $\mathcal{O}(\log |\Sigma|)$ time after $\mathcal{O}(|T|)$ preprocessing time, by essentially using the suffix tree (storing, for each of the $\mathcal{O}(|T|)$ possible ranges, its child ranges, and doing binary search over those – there can be at most $|\Sigma|$ children).

6   Substring Edits and Running Time Bounds

Substring modifications can be achieved by operations in the binary search tree that represents the occurrence partition. Deletion and transposition of a substring of the pattern can be done by split and join operations in the binary search tree. By similar arguments as before, we see that only a constant number of concatenations of suffix ranges are needed after each operation.

Substring copying can be achieved using persistent binary search trees in a similar way, since the persistence of nodes and subtrees allows us to "copy" a subtree [13,31].

The time bounds of our solution for Problem 1 are described in Table 4, along with the time bounds achieved by Amir and Kondratovsky [3] using suffix trees. Our solution for the static text case is considerably simpler, and we improve on the preprocessing time and space and on the substring edits.

Table 4. Time bounds for Problem 1. ℓ is the size of the modified substring

Operation	[3]	This work
Preprocessing Time and Space	$\mathcal{O}(|T|\sqrt{\log|T|})$	$\mathcal{O}(|T|)$
Pattern Search	$\mathcal{O}(|P|\log|T|)$	$\mathcal{O}(|P|+|P|_T\log|T|)$ or $\mathcal{O}(|P|\log|\Sigma|)$
Pattern Symbol Edit	$\mathcal{O}(\log|T|)$	$\mathcal{O}(\log|T|)$
Pattern Substring Edit	$\mathcal{O}(\log|T|+\ell)$	$\mathcal{O}(\log|T|)$

7   Online Text

We can extend our solutions for Problem 1 to maintain an *online* text, that is, also support text extensions (on one end). We will assume we want to extend the text on the left.

Problem 2. (**Dynamic Pattern and Online Text Matching**)
Input: A string T that represents the text, several updates to the pattern string P that can be one of the following.

1. Pattern search: computing the representation of a pattern;
2. Pattern symbol edit: insertion or deletion of some character of the pattern;
3. Pattern substring edit: substring deletion, transposition (moving the substring to another position), or copy.

Also, we support inserting a new symbol to the left of the text.
Output: After every update to the pattern, the number of times it occurs in the text.

To solve Problem 2, we use the online suffix array described in [29], which maintains the suffix array structure in a balanced binary search tree, allowing for text extensions in amortized logarithmic time. By storing pointers to the relevant tree nodes, we can maintain the suffix ranges. The problem is that, after a symbol extension in the text, the occurrence partition might lose the maximality property.

But note that we don't need to enforce the maximality property: it is enough to try to concatenate the first and second elements of the occurrence partition after every operation.

This obviously maintains correctness, because we just need to know if the pattern will occur or not, and this will be the case if, and only if, we are able to concatenate all elements of the occurrence partition into one (Lemma 7).

Also note that this does not add to the time complexities of the operations, in an amortized sense. The idea is that the number of concatenations we make is bounded by the number of splits, and we do a constant number of splits per operation. This is formalized in Theorem 4.

Theorem 4. *For a text T, Problem 2 can be solved with $\mathcal{O}(|T|)$ time construction and $\mathcal{O}(\log |T|)$ amortized time per operation.*

Proof. We will use a potential function to define our amortized complexities [11]. Let D_i be the state of our data structure after operation i, and let $E(D_i)$ be the number of elements of the occurrence partition. Define our potential function $\Phi(D_i) = E(D_i) \cdot \log |T|$. Every operation increases the number of elements by a constant, so this adds an amortized cost of $\mathcal{O}(\log |T|)$ to each operation. If we assume that we then perform k concatenations, we have $\Phi(D_i) - \Phi(D_{i-1}) = c_1 \log |T| - k \cdot \log |T|$, for some constant c_1.

Finally, summing the real cost with the change of our amortized function, we get the amortized cost, for some constant c_2:

$$c_2 \log |T| + k \log |T| + (\Phi(D_i) - \Phi(D_{i-1})) \tag{2}$$
$$= c_2 \log |T| + k \log |T| + c_1 \log |T| - k \cdot \log |T| \tag{3}$$
$$= c_2 \log |T| + c_1 \log |T| \tag{4}$$
$$\in \mathcal{O}(\log |T|) \tag{5}$$

The time complexities for Problem 2 are summarized in Table 5.

Table 5. Time bounds for Problem 2. The $\star$ symbol represents amortized bounds, and ℓ is the size of the modified substring

Operation	[3]	This work
Preprocessing Time and Space	$\mathcal{O}(|T|\sqrt{\log |T|})$	$\mathcal{O}(|T|)$
Pattern Symbol Edit	$\mathcal{O}(\log |T|)$	$\mathcal{O}(\log |T|)^\star$
Pattern Substring Edit	$\mathcal{O}(\log |T| + \ell)$	$\mathcal{O}(\log |T|)^\star$
Text Symbol Extension	$\mathcal{O}(\log |T|)$	$\mathcal{O}(\log |T|)^\star$

8   Conclusion

In this work, we tackled the *modified pattern reporting problem* defined by Amir and Kondratovsky [3]. We improved upon their algorithm, with a simpler and faster algorithm – our preprocessing time is linear, and we can support substring edits in sub-linear time (see Tabel 4).

In addition, we apply our algorithm for an online text, achieving amortized logarithmic bounds (see Table 5). The paper by Amir and Kondratovsky [3] claim to achieve worst-case logarithmic bounds, though their analysis omits key details that prevent independent verification.

To illustrate the feasibility of the proposed solution, it was implemented in C++ (available in a public repository[1]) and tested with random test data against a naive solution. Our algorithm performed orders of magnitude faster than the naive algorithm (in the worst case) even for strings with size in the order of one million, suggesting its practicability.

References

1. Aho, A.V., Corasick, M.J.: Efficient string matching: an aid to bibliographic search. Commun. ACM **18**(6), 333–340 (1975)
2. Alstrup, S., Brodal, G.S., Rauhe, T.: Pattern matching in dynamic texts. In: Proceedings of the Eleventh Annual ACM-SIAM Symposium on Discrete Algorithms, pp. 819–828 (2000)
3. Amir, A., Kondratovsky, E.: Searching for a modified pattern in a changing text. In: International Symposium on String Processing and Information Retrieval, pp. 241–253. Springer (2018)
4. Amir, A., Kopelowitz, T., Lewenstein, M., Lewenstein, N.: Towards real-time suffix tree construction. In: International Symposium on String Processing and Information Retrieval, pp. 67–78. Springer (2005)
5. Amir, A., Landau, G.M., Lewenstein, M., Sokol, D.: Dynamic text and static pattern matching. ACM Trans. Algorithms (TALG) **3**(2), 19–es (2007)
6. Bender, M.A., Farach-Colton, M.: The LCA Problem Revisited. In: Gonnet, G.H., Viola, A. (eds.) LATIN 2000. LNCS, vol. 1776, pp. 88–94. Springer, Heidelberg (2000). https://doi.org/10.1007/10719839_9
7. Boneh, I., Golan, S., Kraus, M.: \ optimal algorithm for fully dynamic lz77. arXiv preprint arXiv:2502.12000 (2025)
8. Charalampopoulos, P., Gawrychowski, P., Pokorski, K.: Dynamic longest common substring in polylogarithmic time. arXiv preprint arXiv:2006.02408 (2020)
9. Cole, R., Kopelowitz, T., Lewenstein, M.: Suffix trays and suffix Trists: structures for faster text indexing. In: International Colloquium on Automata, Languages, and Programming, pp. 358–369. Springer (2006)
10. Cole, R., Lewenstein, M.: Multidimensional matching and fast search in suffix trees. In: Proceedings of the Fourteenth Annual ACM-SIAM Symposium on Discrete Algorithms, pp. 851–852 (2003)

---

[1] https://github.com/brunomaletta/DynamicPatternMatching.

11. Cormen, T.H., Leiserson, C.E., Rivest, R.L., Stein, C.: Introduction to algorithms. MIT Press (2022)
12. Das, R., He, M., Kondratovsky, E., Munro, J.I., Wu, K.: Internal masked prefix sums and its connection to fully internal measurement queries. In: International Symposium on String Processing and Information Retrieval, pp. 217–232. Springer (2022)
13. Driscoll, J.R., Sarnak, N., Sleator, D.D., Tarjan, R.E.: Making data structures persistent. J. Comput. Syst. Sci. **38**(1), 86–124 (1989)
14. Farach, M.: Optimal suffix tree construction with large alphabets. In: Proceedings 38th Annual Symposium on Foundations of Computer Science, pp. 137–143. IEEE (1997)
15. Ferragina, P., Grossi, R.: Optimal on-line search and sublinear time update in string matching. SIAM J. Comput. **27**(3), 713–736 (1998)
16. Fischer, M.J., Paterson, M.S.: String matching and other products. In: Complexity of Computation, RM Karp (editor), SIAM-AMS Proceedings. vol. 7, pp. 113–125 (1974)
17. Gawrychowski, P., Karczmarz, A., Kociumaka, T., Łcki, J., Sankowski, P.: Optimal dynamic strings. In: Proceedings of the Twenty-Ninth Annual ACM-SIAM Symposium on Discrete Algorithms, pp. 1509–1528. SIAM (2018)
18. Gu, M., Farach, M., Beigel, R.: An efficient algorithm for dynamic text indexing. In: Proceedings of the Fifth Annual ACM-SIAM Symposium on Discrete Algorithms, pp. 697–704. SODA '94, Society for Industrial and Applied Mathematics, USA (1994)
19. Huynh, T.N., Hon, W.K., Lam, T.W., Sung, W.K.: Approximate string matching using compressed suffix arrays. Theoret. Comput. Sci. **352**(1–3), 240–249 (2006)
20. Ivanov, A.: Distinguishing an approximate word s inclusion on turing machine in real time. Izv. Acad. Nauk USSR Ser. Mat **48**, 520–568 (1984)
21. Kärkkäinen, J., Sanders, P.: Simple linear work suffix array construction. In: International Colloquium on Automata, Languages, and Programming, pp. 943–955. Springer (2003)
22. Kasai, T., Lee, G., Arimura, H., Arikawa, S., Park, K.: Linear-time longest-common-prefix computation in suffix arrays and its applications. In: CPM. vol. 2089, pp. 181–192. Springer (2001)
23. Kempa, D., Kociumaka, T.: Dynamic suffix array with polylogarithmic queries and updates. In: Proceedings of the 54th Annual ACM SIGACT Symposium on Theory of Computing, pp. 1657–1670 (2022)
24. Kim, D.K., Sim, J.S., Park, H., Park, K.: Constructing suffix arrays in linear time. J. Discrete Algorithms **3**(2–4), 126–142 (2005)
25. Knuth, D.E., Morris, J.H., Jr., Pratt, V.R.: Fast pattern matching in strings. SIAM J. Comput. **6**(2), 323–350 (1977)
26. Kociumaka, T., Radoszewski, J., Rytter, W., Waleń, T.: Internal pattern matching queries in a text and applications. SIAM J. Comput. **53**(5), 1524–1577 (2024)
27. Landau, G.M., Vishkin, U.: Efficient string matching with k mismatches. Theoret. Comput. Sci. **43**, 239–249 (1986)
28. Manber, U., Myers, G.: Suffix arrays: a new method for on-line string searches. SIAM J. Comput. **22**(5), 935–948 (1993)
29. Monteiro, B.: String Matching with a Dynamic Pattern. Master's thesis, Universidade Federal de Minas Gerais (2024)

30. Morris, J., Jr., Pratt, V.: A linear pattern-matching algorithm. University of California, Berkeley (1970)
31. Sarnak, N., Tarjan, R.E.: Planar point location using persistent search trees. Commun. ACM **29**(7), 669–679 (1986)
32. Weiner, P.: Linear pattern matching algorithms. In: 14th Annual Symposium on Switching and Automata Theory (swat 1973), pp. 1–11. IEEE (1973)

Smallest Suffixient Sets
as a Repetitiveness Measure

Gonzalo Navarro[1,2](✉) (iD), Giuseppe Romana[3](✉) (iD),
and Cristian Urbina[1,2](✉) (iD)

[1] Department of Computer Science, University of Chile, Santiago, Chile
{gnavarro,crurbina}@dcc.uchile.cl
[2] Center for Biotechnology and Bioengineering (CeBiB), Santiago, Chile
[3] Department of Mathematics and Computer Science, University of Palermo,
Palermo, Italy
giuseppe.romana01@unipa.it

Abstract. A suffixient set is a novel combinatorial object that captures
the essential information of repetitive strings in a way that, provided with
a random access mechanism, supports various forms of pattern matching.
In this paper, we study the size χ of the smallest suffixient set as a
repetitiveness measure: we place it between known measures and study
its sensitivity to various string operations.

Keywords: Repetitive sequences · Burrows-Wheeler Transform · Text
compressibility

1 Introduction

The study of repetitive string collections has recently attracted considerable
interest from the stringology community, triggered by practical challenges such
as representing huge collections of similar strings in a way that they can be
searched and mined directly in highly compressed form [25, 26]. An example is
the *European '1+ Million Genomes' Initiative*[1] which aims at sequencing over a
million human genomes: while this data requires around 750TB of storage in raw
form (using 2 bits per base), the high similarity between human genomes would
allow storing it in querieable form using two orders of magnitude less space.

An important aspect of this research is to understand how to measure repetitiveness, especially when those measures reflect the size of compressed representations that offer different access and search functionalities on the collection.
Various repetitiveness measures have been proposed, from abstract lower bounds
to those related to specific text compressors and indices; a relatively up-to-date
survey is maintained [27]. Understanding how those measures relate to each other
sheds light on what search functionality is obtained at what space cost.

---

[1] https://digital-strategy.ec.europa.eu/en/policies/1-million-genomes.

G. Badkobeh et al. (Eds.): SPIRE 2025, LNCS 16073, pp. 217–232, 2026.
https://doi.org/10.1007/978-3-032-05228-5_18

A relevant measure proposed recently is the size χ of the smallest *suffixient set* of the text collection [6], whose precise definition will be given later. Within $O(\chi)$ size, plus a random-access mechanism on the string, it is possible to support some text search functionalities, such as finding one occurrence of a pattern, or finding its maximal exact matches (MEMs), which is of central use on various bioinformatic applications [4].

While there has been some work already on how to build minimal suffixient sets and how to index and search a string within their size, less is known about that size, χ, as a measure of repetitiveness. It is only known [6] that $\gamma = O(\chi)$ and $\chi = O(\bar{r})$ on every string family, where γ is the size of the smallest *string attractor* of the collection (a measure that lower bounds most repetitiveness measures) [18] and $\bar{r}$ is the number of equal-letter runs of the Burrows-Wheeler Transform (BWT) [3] of the reversed string.

In this paper we better characterize χ as a repetitiveness measure. First, we study how it behaves when the string undergoes updates, showing in particular that it grows by $O(1)$ when appending or prepending symbols, but that it can grow additively by $\Omega(\log n)$ upon arbitrary edit operations or rotations, and by $\Omega(\sqrt{n})$ when reversing the string. Second, we show that $\chi = O(r)$ on every string family, where r is the number of equal-letter runs of the BWT of the string. We also show that there are string families where $\chi = o(v)$, where v is the size of the smallest lexicographic parse [28] (an alternative to the size of the Lempel-Ziv parse [20], which behaves similarly). In particular, this holds on the Fibonacci strings, where we fully characterize the only 2 smallest suffixient sets of size 4, and further prove that $\chi \le \sigma + 2$ on all substrings of episturmian words over an alphabet of size σ. Since $v = O(r)$ on all string families, this settles χ as a strictly smaller measure than r, which is a more natural characterization than in terms of the reverse string. We also show that χ is incomparable with most "copy-paste" based measures [25], as there are families where it is strictly smaller and others where it is strictly larger than any of those measures.

This result relates to the important question of whether a measure μ is *reachable* (i.e., one can represent the string within $O(\mu)$ space), *accessible* (i.e., one can access any string position from an $O(\mu)$-size representation, in sublinear time), or *searchable* (i.e., one can search for patterns in sublinear time within space $O(\mu)$). Measure r is, curiously, the only one to date being reachable and searchable, but not known to be accessible. Now χ emerges as a measure smaller than r, which can search if provided with a mechanism to efficiently access substrings (r does not need access to support searches). Unlike r, χ is yet not known to be reachable (as its relation to the smallest known reachable measure, the size b of the smallest bidirectional macro scheme [31], remains unknown). As said, it is known that $\gamma = O(\chi)$, but it is unknown whether γ is reachable or not.

2   Preliminaries

An *ordered alphabet* $\Sigma = \{a_1, \ldots, a_\sigma\}$ is a finite set of symbols equipped with a total order $<$ such that $a_1 < a_2 < \cdots < a_\sigma$. When $\sigma = 2$, we assume $\Sigma = \{\mathtt{a}, \mathtt{b}\}$

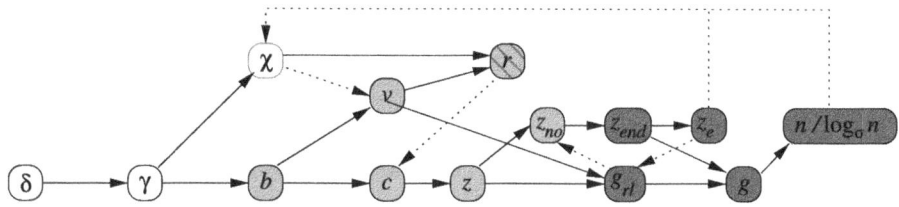

Fig. 1. Relations between relevant repetitiveness measures and how our results place χ among them. An arrow $\mu_1 \rightarrow \mu_2$ means that $\mu_1 = O(\mu_2)$ for all strings and, save for $c \rightarrow z$, $z_{no} \rightarrow z_{end}$, and $z_{end} \rightarrow z_e$, there is a string family where $\mu_1 = o(\mu_2)$. The dotted arrows mark only this last condition, so they are not transitive. Measures in light gray nodes are known to be reachable; those in dark gray are accessible and searchable; and r is hatched because it is searchable but not known to be accessible.

with $\mathbf{a} < \mathbf{b}$. The special symbol $\$$, if it appears, is always assumed to be the smallest of the alphabet.

A *string* $w[1 . . n]$ (or simply w if it is clear from the context) of *length* $|w| = n$ over the alphabet Σ is a sequence $w[1]w[2] \cdots w[n]$ of symbols where $w[i] \in \Sigma$ for all $i \in [1, n]$. The *empty string* of length 0 is denoted ϵ. We denote by Σ^* the set of all strings over Σ. Additionally, we let $\Sigma^+ = \Sigma^* \setminus \{\epsilon\}$ and $\Sigma^k = \{w \in \Sigma^* \mid |w| = k\}$. We denote by $w[i . . j]$ the substring $w[i]w[i + 1] \cdots w[j]$. If $x = x[1 . . n]$ and $y = y[1 . . m]$ are strings, we define the *concatenation operation* applied on x and y, as the string obtained by juxtaposing these two strings, that is, $x \cdot y = x[1]x[2] \cdots x[n]y[1] \cdots y[m] = xy$. A string x is a *substring* of w if $w = yxz$ for some $y, z \in \Sigma^*$. A string x is a *prefix* of w if $w = xy$ for some $y \in \Sigma^*$. Analogously, x is a *suffix* of w if $w = yx$ for some $y \in \Sigma^*$. We say that substrings, prefixes, and suffixes are *non-trivial* if they are different from w and ϵ. The set of substrings of w is denoted by $\mathcal{F}_w$. We also let $\mathcal{F}_w(k) = \mathcal{F}_w \cap \Sigma^k$. The *reverse* of a finite string w is the string $w^R = w[n] \cdot w[n-1] \cdots w[1]$. We denote by $\mathcal{R}(w)$ the multiset of rotations of $w[1 . . n]$, that is, $\mathcal{R}(w) = \{w[i + 1 . . n]w[1 . . i] \mid i \in [1 . . n]\}$. The *Burrows-Wheeler transform* (BWT) of a string w, denoted $\mathtt{BWT}(w)$, is the transformation of w obtained by collecting the last symbol of all rotations in $\mathcal{R}(w)$ in lexicographic order. The *BWT matrix* $B(w)$ of w is the $(n \times n)$-matrix where the i-th row is the i-th rotation of w in lexicographic order.

A *right-infinite string* $\mathbf{w}$ —we use **boldface** to emphasize its infinite length— over Σ is any infinite sequence $\mathbb{Z}^+ \rightarrow \Sigma$. The set of all infinite strings over Σ is denoted Σ^ω. A substring of $\mathbf{w}$ is the finite string $\mathbf{w}[i . . j]$ for any $1 \le i \le j$. A prefix of $\mathbf{w}$ is a finite substring of the form $\mathbf{w}[1 . . n]$ for some $n \ge 0$. The *substring complexity function* $P_{\mathbf{w}}(k) : \mathbb{Z}^+ \cup \{0\} \rightarrow \mathbb{Z}^+$ counts the number of distinct substrings of length k in $\mathbf{w}$, for any $k \in \mathbb{Z}^+ \cup \{0\}$, that is, $P_{\mathbf{w}}(k) = |\mathcal{F}_{\mathbf{w}}(k)|$. For a finite string $w[1 . . n]$, the domain of P_w is restricted to $[0 . . n]$.

2.1   Measures of Repetitiveness

In this work, we will relate χ, in asymptotic terms, with several well-established measures of repetitiveness [25, 27]: $\delta = \max_{k \in [0..n]}(\mathcal{F}_w(k)/k)$ (a measure of substring complexity), γ (the smallest string attractor), b (the size of the smallest bidirectional macro scheme), z (the size of the Lempel-Ziv parse), z_{no} (the same without allowing phrases to overlap their sources), z_e (the size of the greedy LZ-End parse), z_{end} (the size of the minimal LZ-End parse), v (the size of the smallest lexicographic parse), r (the number of equal-letter runs in the BWT of the string), g (the size of the smallest context-free grammar generating only the string), g_{rl} (the same allowing run-length rules), and c (the size of the smallest collage system generating only the string). Except for δ, γ and r, these measures are said to be *copy-paste* because they refer to a way of cutting the sequence into chunks that can be copied from elsewhere in the same sequence. Indeed, δ and γ are lower-bound measures, the former known to be unreachable and the latter not known to date to be reachable; all the others are. The smallest measures known to be accessible (and searchable) are z_{end} and g_{rl}, and r is searchable but not known to be accessible.

The known relations between those measures are summarized in Fig. 1, where we have added the results we obtain in this paper with respect to χ.

2.2   Edit Operations and Sensitivity Functions

The so-called *edit operations* are *insertion*, *substitution* and *deletion* of a single character on a string. We denote $\text{ins}_\Sigma(w)$, $\text{sub}_\Sigma(w)$, $\text{del}_\Sigma(w)$ the sets of strings that can be obtained by applying an edit operation to w. In addition, we let $\text{prepend}_\Sigma(w)$ and $\text{append}_\Sigma(w)$ be $\text{ins}_\Sigma(w)$ restricted to the insertion being made at the beginning and the end of the string, respectively.

A repetitiveness measure μ is *monotone* or *non-decreasing* to the insertion of a single character if $\mu(w') - \mu(w) \geq 0$ for any w and $w' \in \text{ins}_\Sigma(w)$. More generally, the *additive sensitivity* and *multiplicative sensitivity* functions of a repetitiveness measure μ to the insertion of a single character are the maximum possible values of $\mu(w') - \mu(w)$ and $\mu(w')/\mu(w)$, respectively. We define the concept of monotonicity and sensitivity functions for the remaining string operations analogously.

3   Suffixient Sets and the Measure χ

In this section we define the central combinatorial objects and measures we analyse on this work. Note that some of our definitions are slightly different from their original formulation [4, 5], because we do not always assume that all strings are $-terminated.

Definition 1 (Right-maximal Substrings and Right-extensions [4, 5]).
Let $w \in \Sigma^*$. A substring x of w is right-maximal *if there exist at least two distinct symbols $a, b \in \Sigma$ such that both xa and xb are substrings of w. For any*

right-maximal substring x *of* w, *the substrings* xa *with* $a \in \Sigma$ *are called* right-extensions. *We denote the set of right-extensions in* w *by* $E_r(w) = \{xa \mid \exists b : b \neq a, xa \in \mathcal{F}_w, xb \in \mathcal{F}_w\}$.

We distinguish a special class of right-extensions that are not suffixes of any other right-extension.

Definition 2 (Super-maximal Extensions *[4, 5]*). *The set of* super-maximal extensions *of* w *is* $\mathcal{S}_r(w) = \{x \in E_r(w) \mid \forall y \in E_r, y = zx \Rightarrow z = \varepsilon\}$. *Moreover, we let* $\mathtt{sre}(w) = |\mathcal{S}_r(w)|$.

We now define suffixient sets for strings not necessarily $-terminated; we introduce later the special terminator $.

Definition 3 (Suffixient Set *[4, 5]*). *Let* $w[1 .. n] \in \Sigma^*$. *A set* $S \subseteq [1 .. n]$ *is a* suffixient set *for* w *if for every right-extension* $x \in E_r(w)$ *there exists* $j \in S$ *such that* x *is a suffix of* $w[1 .. j]$.

Intuitively, a suffixient set is a collection of positions of $[1 .. |w|]$ capturing all the right-extensions appearing in w. The smallest suffixient sets, which are suffixient sets of minimum size, have also been characterized in terms of super-maximal right-extensions. The next definition simplifies the original one [4,5].

Definition 4 (Smallest Suffixient Set). *Let* $w[1 .. n] \in \Sigma^*$. *A suffixient set* $S \subseteq [1 .. n]$ *is a* smallest suffixient set *for* w *if there is a bijection* $pos : \mathcal{S}_r \to S$ *such that every* $x \in \mathcal{S}_r$ *is a suffix of* $w[1 .. pos(x)]$.

In its original formulation, the measure χ is defined over $-terminated strings. Here, we define $\chi(w)$ with the $ being implicit, not being part of w.

Definition 5 (Measure χ *[4, 5]*). *Let* $w \in \Sigma^*$ *and assume* $ \notin \mathcal{F}_w$. *Then,* $\chi(w) = |\mathcal{S}|$, *where* $\mathcal{S}$ *is a smallest suffixient set for* w.

One can see from the above definitions that χ is well-defined because $\chi(w) = \mathtt{sre}(w$)$. We will use this relation to prove results on χ via $\mathtt{sre}$.

4   Sensitivity of χ to String Operations

The sensitivity to string operations has been studied for many repetitiveness measures [1, 9, 10, 14, 15, 24, 29, 30]. It is desirable for a repetitiveness measure to not change much upon small changes in the sequence. Some repetitiveness measures are resistant to edit operations. For instance, b, z and g can only increase by a multiplicative constant after an edit operation [1], though they can increase only by a $O(1)$ additive factor when prepending or appending a character. On the other hand, r can increase by a $\Theta(\log n)$ factor when appending a character [15, Prop. 37]. Other results have been obtained concerning more complex string operations, like reversing a string [14], or applying a string morphism [9,10].

In this section we study how $\mathtt{sre}$ and χ behave in this respect. We start by proving the following useful lemma.

Lemma 1. *If $E_r(w_1) \subseteq E_r(w_2)$, then $\mathtt{sre}(w_1) \leq \mathtt{sre}(w_2)$.*

Proof. Let $x, y \in \mathcal{S}_r(w_1)$ with $x \neq y$. Because $x \in E_r(w_2)$, there exists $z \in \mathcal{S}_r(w_2)$ with x a suffix of z. Because y is not a suffix of x and vice versa, y cannot be a suffix of z. Therefore, the map $x \mapsto z$ with $x \in \mathcal{S}_r(w_1)$, $z \in \mathcal{S}_r(w_2)$, and $z = z'x$ for some $z' \in \Sigma^*$ is injective and then $\mathtt{sre}(w_1) \leq \mathtt{sre}(w_2)$. □

We now prove that $\mathtt{sre}(w)$ grows only by $O(1)$ when prepending or appending characters.

Lemma 2. *Let $w \in \Sigma^*$, and $c \in \Sigma$. It holds $\mathtt{sre}(w) \leq \mathtt{sre}(wc) \leq \mathtt{sre}(w) + 2$.*

Proof. The lower bound follows from Lemma 1. For the upper bound, we analyse the new right-extensions that may arise due to appending c to w. For any fixed suffix xc of wc:

1. if xa does not appear in w for any $a \neq c$, then xc induces no new right-extensions in wc;
2. if for some $a \neq b$, xa and xb were both substrings of w, and $c \neq a$ and $c \neq b$, then xc is a new right-extension of wc;
3. if x is always followed by $a \neq c$ in w (hence, xa is not a right-extension of w), then both xa and xc are new right-extensions of wc.

Cases 1 and 2 induce at most one new super-maximal right-extension in total for all possible xc, namely the longest right-extension in wc that is a suffix of wc. For Case 3, consider a fixed $a \in \Sigma$. For all the increasing-length suffixes $x_1 c, x_2 c, \ldots, x_t c$ of wc that became right-extensions together with $x_1 a, x_2 a, \ldots, x_t a$, one can see that the latter form a chain of suffixes of $x_t a$. Hence, we only have one possible new super-maximal right-extension ending with a, namely $x_t a$. Observe that the chain of suffixes $x_1 a, x_2 a, \ldots, x_t a$ is unique: if the suffix x is always followed by a, any suffix y of x is either right-maximal in w (and y falls within Case 2), or it is always followed by an a (because x is always followed by an a), i.e. $y = x_i$ for some $i \in [1 .. t]$. □

Lemma 3. *Let $w \in \Sigma^*$ and $c \in \Sigma$. It holds $\mathtt{sre}(w) \leq \mathtt{sre}(cw) \leq \mathtt{sre}(w) + 2$.*

Proof. The lower bound follows from Lemma 1. For the upper bound, let cxa be the smallest prefix of cw that is not a right-extension of w, but is a right-extension of cw (if it exists). This means that cxa does not appear in w (otherwise, it would be a right-extension of w), so no prefix of cw of length $|cxa|$ or more is right-maximal. Hence, all prefixes of cw shorter than cxa were already right-extensions, and all prefixes longer than cxa cannot be right-extensions. Therefore, cxa together with some cxb appearing in w are the only possible new right-extensions in cw with respect to w. □

By letting $c = \$ \notin \mathcal{F}_w$ in Lemma 2, we relate χ to $\mathtt{sre}$ (note that χ is always at least $\mathtt{sre} + 1$ because of the new super-maximal extension ending with $\$$). This makes clear the relation between Combinatorics on words [21] with suffixient sets, via the common notion of *right-special factors* (what we call here right-maximal substrings).

Corollary 1. *Let $w \in \Sigma^*$. It holds $\mathtt{sre}(w) + 1 \leq \chi(w) \leq \mathtt{sre}(w) + 2$.*

Note that, while the value $\mathtt{sre}(w)$ is non-decreasing after appending a character, this is not the case for the measure χ.

Lemma 4. *The measure χ is not monotone to appending a character.*

Proof. Let $w = \mathtt{abaab}$. It holds $\mathcal{S}_r(w\$) = \{\mathtt{aa}, \mathtt{ab}, \mathtt{ab}\$, \mathtt{aba}\}$ and $\mathcal{S}_r(wa\$) = \mathcal{S}_r(\mathtt{abaaba}\$) = \{\mathtt{ab}, \mathtt{aba}\$, \mathtt{abaa}\}$. Hence, $\chi(w) = 4$ and $\chi(wa) = 3$. $\qquad\square$

Now we study how much $\mathtt{sre}(w)$ can vary upon edit operations in arbitrary positions, rotations, and reversals. We will use the following famous string family.

Definition 6. *A binary de Bruijn sequence of order $k > 0$ [2] contains every binary string in $\{\mathtt{a}, \mathtt{b}\}^k$ as a substring exactly once. The length of these strings is $n = 2^k + (k-1)$. The set of binary de Bruijn sequences of order k is $\mathtt{dB}(k)$.*

Lemma 5. *It holds $\mathtt{sre}(w) = 2^k = \Omega(n)$ for any $w[1 \mathinner{\ldotp\ldotp} n] \in \mathtt{dB}(k)$.*

Proof. Let $w[1 \mathinner{\ldotp\ldotp} n]$ be a binary de Bruijn string of order k. By definition, w contains every binary string of length k as a substring exactly once. As all the possible pairs of strings $x\mathtt{a}$ and $x\mathtt{b}$ of length k appear in w, it follows that all the strings in $\mathcal{F}_w(k)$ are right-extensions. Moreover, each $x\mathtt{a}$ and $x\mathtt{b}$ of length k are super-maximal right-extensions: otherwise, there would exist some $c \in \{\mathtt{a}, \mathtt{b}\}$ such that $cx\mathtt{a}$ and $cx\mathtt{b}$ are both substrings of w, which raises a contradiction since the k-length string cx cannot appear twice in w. Moreover, there are no right-maximal strings of length k or greater; hence, there are no right-extensions of length greater than k. It follows that $\mathtt{sre}(w) = |\mathcal{F}_w(k)| = 2^k = \Omega(n)$. $\qquad\square$

The following lemma uses the de Bruijn family to show that $\mathtt{sre}$ can grow by $\Omega(\log n)$ upon arbitrary edit operations and rotations.

Lemma 6. *Let $w = \mathtt{a}^k \mathtt{b} \mathtt{a}^{k-2} \mathtt{b} x \mathtt{a} \mathtt{b}^k \mathtt{a}^{k-1} \in \mathtt{dB}(k)$ be the lexicographically smallest binary de Bruijn sequence of order k [11,12]. It holds:*

1. *(Ins) $\mathtt{sre}(w) - \mathtt{sre}(w') = 2k - 2$ if $w' = \mathtt{a}^{2k-2} \mathtt{b} x \mathtt{a} \mathtt{b}^k \mathtt{a}^{k-1}$,*
2. *(Sub) $\mathtt{sre}(w) - \mathtt{sre}(w') = 2k - 3$ if $w' = \mathtt{a}^k \mathtt{b} \mathtt{a}^{k-2} \mathtt{b} x \mathtt{a} \mathtt{b}^{k-1} c \mathtt{a}^{k-1}$,*
3. *(Del) $\mathtt{sre}(w) - \mathtt{sre}(w') = 2k - 4$ if $w' = \mathtt{a}^k \mathtt{b} \mathtt{a}^{k-2} \mathtt{b} x \mathtt{a} \mathtt{b}^k c \mathtt{a}^{k-1}$,*
4. *(Rot) $\mathtt{sre}(w) - \mathtt{sre}(w') = 2k - 2$ if $w' = \mathtt{b} \mathtt{a}^{k-2} \mathtt{b} x \mathtt{a} \mathtt{b}^k \mathtt{a}^{2k-1}$.*

Proof. Observe that in each claim, w is obtained after performing a string operation on the corresponding w': in Claim 1, $w \in \mathtt{ins}_\Sigma(w')$; in Claim 2, $w \in \mathtt{sub}_\Sigma(w')$; in Claim 3, $w \in \mathtt{del}_\Sigma(w')$; in Claim 4, $w \in \mathcal{R}(w')$. We prove each claim separately by comparing the super-maximal extensions of w' before and after performing the string operation on w' that yields w, for which $\mathtt{sre}(w) = 2^k$ by Lemma 5.

For Claim 1, note that $\mathtt{sre}(w')$ is the same as $\mathtt{sre}(\mathtt{a}^k \mathtt{b} x \mathtt{a} \mathtt{b}^k \mathtt{a}^{k-1})$, as prepending the character $\mathtt{a}$ multiple times to this string to obtain w' never increases $\mathtt{sre}$; it only updates the super-maximal extension $\mathtt{a}^k$ to $\mathtt{a}^{k+1}$ and $\mathtt{a}^{k-1}\mathtt{b}$ to $\mathtt{a}^k\mathtt{b}$, and so on. For simplicity, we let $w' = \mathtt{a}^k \mathtt{b} x \mathtt{ab}^k \mathtt{a}^{k-1}$. The string w' does not contain substrings of length k of the form $\mathtt{a}^i \mathtt{ba}^{k-i-1}$ for $i \in [1 .. k-2]$, nor the substring $\mathtt{ba}^{k-2}\mathtt{b}$. Note that for each of these substrings $y \in \mathcal{F}_w(k)$ with $y \notin \mathcal{F}_{w'}(k)$, the other corresponding right-extension y' in w sharing a length $k-1$ prefix with y is not a right-extension in w'. Moreover, note that all the suffixes of length $k-1$ of these y are not suffixes of one another, nor of the length $k-1$ suffixes of any of the substrings y' in w'. Hence, all $k-1$ length binary strings still appear in w' as the suffix of some length k substring that remains a right-extension in w', and hence, super-maximal extensions of w' have to be of length at least k. As each string of length k appearing in w' is unique, there are no super-maximal extensions of length greater than k. Thus, $\mathtt{sre}(w') = 2^k - 2(k-1)$ because we are losing $k-1$ pairs of super-maximal extensions of length k with respect to w. It follows that by inserting the $\mathtt{b}$ in w' to yield w, $\mathtt{sre}$ increases by $2k-2$.

For Claim 2, note that exactly k substrings of length k are lost when substituting the last $\mathtt{b}$ of w by $\mathtt{c}$: those of the form $\mathtt{b}^i \mathtt{a}^{k-i}$ with $i > 0$. This means that substrings ending in $\mathtt{b}^i \mathtt{a}^{k-i-1}$ with $0 < i < k$ are not right-maximal in w', hence, $2(k-1)$ super-maximal extensions are lost. Moreover, $\mathtt{b}^{k-2}$ is still a right-maximal substring, since $\mathtt{b}^{k-1}$ and $\mathtt{b}^{k-2}\mathtt{c}$ occur in w'. Observe that only $\mathtt{b}^{k-2}\mathtt{c}$ is a super-maximal extension, while $\mathtt{b}^{k-1}$ is a suffix of $\mathtt{ab}^{k-1}$. Thus, $\mathtt{sre}(w') = 2^k - 2(k-1) + 1$ and $\mathtt{sre}(w) - \mathtt{sre}(w') = 2k-3$.

For Claim 3, the analysis is similar to Claim 2, but in w', $\mathtt{b}^{k-1}$ remains as a super-maximal extension. Thus, $\mathtt{sre}(w') = 2^k - 2(k-1) + 2$ and $\mathtt{sre}(w) - \mathtt{sre}(w') = 2k-4$.

For Claim 4, the analysis is similar to Claim 1, but in w', $\mathtt{ba}^{k-2}\mathtt{b}$ appears, while $\mathtt{a}^{k-1}\mathtt{b}$ does not. Thus, $\mathtt{sre}(w') = 2^k - 2(k-1)$ and $\mathtt{sre}(w) - \mathtt{sre}(w') = 2k-2$. $\qquad\square$

We now show that $\mathtt{sre}$ can grow by $\Omega(\sqrt{n})$ upon string reversals.

Lemma 7. *Let $k > 0$. Let $w_k = \prod_{i=0}^{k-1} \mathtt{ca}^i \mathtt{ba}^{k-i-1} \#_i \mathtt{a}^i \mathtt{ba}^{k-i-1} \$_i$ on the alphabet $\Sigma = \{\mathtt{a}, \mathtt{b}, \mathtt{c}\} \cup \bigcup_{i \in [0 .. k-1]} \{\#_i, \$_i\}$. It holds $\mathtt{sre}(w_k) - \mathtt{sre}(w_k^R) = k-1$.*

Proof. Observe that by construction, any substring of w_k containing $\#_i$ or $\$_i$ is not right-maximal, as these symbols are unique. Hence, the right-extensions of w_k cannot cross from one side to the other side of those special delimiters. Moreover, substrings of the form $\mathtt{a}^i \mathtt{ba}^{k-i-1}$ for $i \in [0 .. k-1]$ appear exactly twice in w_k and their right-extensions are super-maximal. By looking at the structure of the string w_k and carefully analyzing its right-extensions, one can verify that the super-maximal right-extensions of w_k are the following:

1. $\mathtt{ba}^{k-1}$ and $\mathtt{c}$
2. $\mathtt{a}^i \mathtt{ba}^{k-i-1}\#_i$ and $\mathtt{a}^i \mathtt{ba}^{k-i-1}\$_i$ for $i \in [0 .. k-1]$,
3. $\mathtt{ca}^i$ and $\mathtt{ca}^{i-1}\mathtt{b}$ for $i \in [1 .. k-1]$,
4. $\mathtt{a}^i \mathtt{ba}^{k-i-1}$ for $i \in [1 .. k-1]$.

This sums to a total of $5k - 1$ super-maximal extensions in w_k. In the reversed string $w_k^R = \prod_{i=0}^{k-1} \$_{k-i-1} \mathtt{a}^i \mathtt{ba}^{k-i-1} \#_{k-i-1} \mathtt{a}^i \mathtt{ba}^{k-i-1}\mathtt{c}$, we have instead:

1. $\mathtt{ba}^{k-1}$ and $\$_{k-1}$,
2. $\mathtt{a}^i\mathtt{ba}^{k-i-1}\#_{k-i-1}$ and $\mathtt{a}^i\mathtt{ba}^{k-i-1}\mathtt{c}$ for $i \in [0 .. k-1]$,
3. $\mathtt{a}^{k-i-1}\mathtt{c}\$_{k-i-2}$ for $i \in [1 .. k-2]$, and $\mathtt{a}^{k-2}\mathtt{c}\$_{k-2}$,
4. $\mathtt{a}^i\mathtt{ba}^{k-i-1}$ for $i \in [1 .. k-1]$.

This sums to a total of $4k$ super-maximal extensions in w_k^R. Thus, $\mathtt{sre}(w_k) - \mathtt{sre}(w_k^R) = (5k-1) - 4k = k-1$. □

We give an example of the words w_k and w_k^R of Lemma 7, and their super-maximal right-extensions.

Example 1. Let $w_3 = \mathtt{cbaa}\#_0\mathtt{baa}\$_0\mathtt{caba}\#_1\mathtt{aba}\$_1\mathtt{caab}\#_2\mathtt{aab}\$_2$. It can be verified that the super-maximal right-extensions of w_3 are:

1. baa and c;
2. baa$\#_0$ and baa$\$_0$; aba$\#_1$ and aba$\$_1$; aab$\#_2$ and aab$\$_2$;
3. ca and cb; caa and cab;
4. aba and aab.

Similarly, let $w_3^R = \$_2\mathtt{baa}\#_2\mathtt{baac}\$_1\mathtt{aba}\#_1\mathtt{abac}\$_0\mathtt{aab}\#_0\mathtt{aabc}$. The super-maximal right-extensions of w_3^R are:

1. baa and $\$_2$;
2. baa$\#_2$ and baac; aba$\#_1$ and abac; aab$\#_0$ and aabc;
3. ac$\$_0$; ac$\$_1$;
4. aba and aab.

One can see that $\mathtt{sre}(w_3) = 14$, $\mathtt{sre}(w_3^R) = 12$, and hence, $\mathtt{sre}(w_3) - \mathtt{sre}(w_3^R) = 2$, as stated in Lemma 7.

Formally, the additive sensitivity of a measure of repetitiveness μ to a string operation ρ can be defined as a function $AS_{\mu,\rho} : \mathbb{Z}^+ \to \mathbb{R}$, where $AS_{\mu,\rho}(n) = \max_{w \in \Sigma^n}(\max_{w' \in \rho(w)}(\mu(w')) - \mu(w))$, that is the maximum achievable difference among all the strings. Overall, we obtain the following result on the additive sensitivity of sre, which, by Corollary 1, can be written in terms of χ.

Corollary 2. *The following bounds on the additive sensitivity of the measure χ to string operations hold:*

1. $AS_{\chi,\rho}(n) = \Omega(\log n)$ for $\rho \in \{\mathtt{ins}, \mathtt{del}, \mathtt{sub}, \mathcal{R}(\cdot)\}$;
2. $AS_{\chi,\mathtt{rev}}(n) = \Omega(\sqrt{n})$, where $\mathtt{rev}(w) = \{w^R\}$.

Proof. Claim 1 follows by Lemma 6, where $n = |w| = \Theta(2^k)$ and $AS_{\chi,\rho}(n) = \Omega(k) = \Omega(\log n)$, for all $\rho \in \{\mathtt{ins}, \mathtt{del}, \mathtt{sub}, \mathcal{R}(\cdot)\}$. Claim 2 follows by Lemma 7, where $n = |w_k| = \Theta(k^2)$ and $AS_{\chi,\mathtt{rev}}(n) = \Omega(k) = \Omega(\sqrt{n})$. □

Finally, we show upper bounds on the sensitivity of χ to string operations.

Lemma 8. *Let $w \in \Sigma^*$ and $w' \in \text{ins}_\Sigma(w) \cup \text{del}_\Sigma(w) \cup \text{sub}_\Sigma(w) \cup \mathcal{R}(w) \cup \{w^R\}$.*
It holds

$$\chi(w') - \chi(w) = O\left(\delta \max\left(1, \log(n/\delta \log \delta)\right) \log \delta\right) \text{ and}$$
$$\chi(w') \,/\, \chi(w) = O\left(\max\left(1, \log(n/\delta \log \delta)\right) \log \delta\right).$$

Proof. To prove our thesis, we rely on the relations $\delta \leq \chi \leq 2\bar{r}$ [4] and
$r = O(\delta \max(1, \log(n/\delta \log \delta)) \log \delta)$ [17]. Moreover, since the multiplicative sensitivity of the measure δ to any of the string operations is $O(1)$ [1], for any
$w \in \Sigma^*$ it holds $\bar{r}(w) = r(w^R) = O(\delta \max(1, \log(n/\delta \log \delta)) \log \delta)$. The the-
sis follows by considering the worst case, that is $\chi(w) = \Theta(\delta)$ and $\chi(w') = \Theta(\delta \max(1, \log(n/\delta \log \delta)) \log \delta)$. $\qquad\square$

5   Relating χ to Other Repetitiveness Measures

Previous work [4] established that $\gamma = O(\chi)$ and $\chi = O(\bar{r})$ on every string
family. In this section we obtain the more natural result that χ is always $O(r)$,
and that it can be asymptotically strictly smaller, $\chi = o(r)$, on some string
families (we actually prove $\chi = o(v)$). We also show that χ is incomparable with
all the copy-paste measures except b, in the sense that there are string families
where χ is asymptotically strictly smaller than each other, and vice versa.

5.1   Proving $\chi = O(r)$

We first prove that χ is asymptotically upper-bounded by the number r of runs
in the BWT of the sequence. As for the measure χ, we assume that the BWT is
computed after appending the $\$$ symbol.

Lemma 9. *It always holds that $\chi \leq 2r$.*

Proof. Let x_i denotes the ith rotation of $w\$$ in lexicographic order, for each
$i \in [1 .. |w| + 1]$, and let u_i be the longest common prefix between the rotations
x_i, x_{i+1}, for each $i \in [1 .. |w|]$. We further define $s : [1 .. n + 1] \rightarrow [0 .. n]$ as
$s(i) = j$ if $x_i = w[j + 1 .. |w|]\$w[1 .. j]$, i.e., the number of cyclic shift to the
right required to transform x_i into $w\$.$[2] As the symbol $\$$ occurs only once in
$w\$$, the function s is bijective.

Note that each right-extension of $w\$$ can be written as $u_i c$, for some $i \in [1 .. |w|]$ and $c \in \Sigma$. Consider now the set

$$S = \bigcup_{i \in [1 .. |w|]} \{s(i) + |u_i| + 1, s(i + 1) + |u_i| + 1\},$$

that is the set of positions where the occurrences of the right-extensions $u_i c_1$ and
$u_i c_2$ end in $w\$$, where $u_i c_1$ and $u_i c_2$ are the prefix of x_i and x_{i+1} respectively,

---

[2] The function s mimics the well-known Suffix Array [23], here omitted for simplicity
of exposition.

for some $c_1, c_2 \in \Sigma$ such that $c_1 < c_2$. It follows by construction that the set S is a suffixient set of $w\$$.

We now show that $|S| \leq 2r$. Let us factorize each pair of consecutive rotations in the BWT-matrix as $x_i = u_i v_i c_i$ and $x_{i+1} = u_i v_i' c_{i+1}$. Observe that $v_i, v_i' \neq \varepsilon$ [9, Corollary 8], $v_i[1] \neq v_i'[1]$, and $c_i = \text{BWT}(w\$)[i]$ for all $i \in [1 \mathinner{.\,.} |w| + 1]$. A well-known property of the BWT-matrix is that if $c_i = c_{i+1} = c \in \Sigma$, then there exists $j \in [1 \mathinner{.\,.} |w|]$ such that $x_j = c u_i v_i$ and $x_{j+1} = c u_i v_i'$ [3]. As a consequence, one has that $s(j) + |u_j| + 1 = (s(i) - 1) + (|u_i| + 1) + 1 = s(i) + |u_i| + 1$ and $s(j + 1) + |u_j| + 1 = (s(i + 1) - 1) + (|u_i| + 1) + 1 = s(i + 1) + |u_i| + 1$, and the procedure can be reiterated as long as x_j and x_{j+1} end with the same symbol. It follows that the same set can be written as

$$S = \{s(i) + |u_i| + 1, s(i + 1) + |u_i| + 1 \mid i \in [1 \mathinner{.\,.} |w|] \land \text{BWT}[i] \neq \text{BWT}[i + 1]\},$$

i.e., the size of S is at most twice the number of equal-letter runs in $\text{BWT}(w\$)$, and the thesis follows. $\qquad\square$

5.2   A Family with $\chi = o(v)$ (and Thus $o(r)$)

We will now show that $\chi = o(v)$ on the so-called Fibonacci words, which also implies $\chi = o(r)$ in that string family because $v = O(r)$ [28]. Combined with Lemma 9, this implies that χ is a strictly smaller measure than r. In contrast, χ is incomparable with v, as we show later. On our way, we obtain some relevant byproducts about the structure of suffixient sets on Fibonacci, and more generally, episturmian words.

Definition 7 *([8, 16]). An infinite string $\boldsymbol{w}$ is* episturmian *if it has at most one right-maximal substring of each length and its set of substrings is closed under reversal, that is, $\mathcal{F}_{\boldsymbol{w}} = \mathcal{F}_{\boldsymbol{w}}^R$. It is* standard episturmian *(or* epistandard*) if, in addition, all the right-maximal substrings of $\boldsymbol{w}$ are of the form $\boldsymbol{w}[1 \mathinner{.\,.} i]^R$ with $i \geq 0$, i.e., they are the reverse of some prefix of $\boldsymbol{w}$.*

Lemma 10. *Let $\boldsymbol{w} \in \Sigma^\omega$ be an episturmian word with $\sigma \geq 2$. Then, $\mathtt{sre}(\boldsymbol{w}[i \mathinner{.\,.} j]) \leq \sigma$ for $i, j \geq 0$.*

Proof. Let $\boldsymbol{w}$ be an epistandard word. The right-extensions $x_1, x_2, \ldots$ ending with $a \in \Sigma$ form a *suffix-chain* where each x_i is a suffix of x_{i+1}. There is one of those suffix-chains for each character $a \in \Sigma$.

Let $\boldsymbol{w}$ be episturmian but not necessarily epistandard. There exists some epistandard word $\boldsymbol{s}$ with the same set of substrings, i.e., $\mathcal{F}_{\boldsymbol{w}} = \mathcal{F}_{\boldsymbol{s}}$ [8]. Therefore, for any episturmian word $\boldsymbol{w}$, there exist exactly σ suffix-chains of right-extensions.

When considering substrings of $\boldsymbol{w}$, the super-maximal right-extension in $\boldsymbol{w}[i \mathinner{.\,.} j]$ ending with $a \in \Sigma$ is the longest right-extension of $\boldsymbol{w}$ ending with a that remains a right-extension in $\boldsymbol{w}[i \mathinner{.\,.} j]$. It follows that for any substring $\boldsymbol{w}[i \mathinner{.\,.} j]$ of any episturmian word $\boldsymbol{w}$, $\mathtt{sre} \leq \sigma$. $\qquad\square$

Combining this result with Corollary 1, we obtain the following bound.

Corollary 3. *For any episturmian word $\boldsymbol{w} \in \Sigma^\omega$ it holds $\chi(\boldsymbol{w}[i \mathinner{.\,.} j]) \leq \sigma + 2$.*

The next lemma precisely characterizes the suffixient sets of Fibonacci words, a particular case of epistandard words that will be useful to relate χ with v.

Definition 8. *Let $F_1 = b$, $F_2 = a$, and $F_k = F_{k-1}F_{k-2}$ for $k \geq 3$ be the Fibonacci family of strings. Their lengths, $f_k = |F_k|$, form the Fibonacci sequence.*

Lemma 11. *Every Fibonacci word $F_k\$$ has a suffixient set of size at most 4. For $k \geq 6$, the only smallest suffixient sets for $F_k\$$ are $\{f_k+1, f_k-1, f_{k-1}-1, p\}$, where $p \in \{f_{k-2}+1, 2f_{k-2}+1\}$.*

Proof. The upper bound of 4 stems directly from Corollary 3, because the infinite Fibonacci word is binary epistandard. For $k \geq 3$, there exist strings H_k such that $F_k = F_{k-1}F_{k-2} = H_k cd$ and $F_{k-2}F_{k-1} = H_k dc$, for $cd = $ ab or $cd = $ ba depending on the parity of k [22]. Let us call $F_k' = H_k dc = F_{k-2}F_{k-1}$, that is, F_k with the last two letters exchanged; thus $F_k = F_{k-1}F_{k-2} = F_{k-2}F_{k-1}'$.

Note that $F_{k-1} = H_{k-1}dc$ prefixes F_k. On the other hand, we can write $F_k = F_{k-1}F_{k-2} = F_{k-2}F_{k-3}F_{k-2} = F_{k-2}F_{k-1}' = F_{k-2}H_{k-1}cd$. Therefore, string H_{k-1} is right-maximal in F_k. Its extensions, $H_{k-1}d$ and $H_{k-1}c$, are super-maximal because there are no other occurrences of H_{k-1} in F_k: (i) H_{k-1} cannot occur starting at positions $f_{k-2}+2$ or $f_{k-2}+3$ because it occurs at $f_{k-2}+1$, so H_{k-1} should match itself with an offset of 1 or 2, which is impossible because it prefixes F_{k-1} and all F_{k-1} for $k-1 \geq 5$ start with abaab; (ii) H_{k-1} cannot occur starting at positions 2 to f_{k-2} because its prefix F_{k-2} should occur inside the prefix $F_{k-2}F_{k-2}$ of $F_k = F_{k-2}F_{k-1}' = F_{k-2}F_{k-2}F_{k-3}'$, and so F_{k-2} should equal a rotation of itself, which is impossible [7, Cor. 3.2]. The two positions following H_{k-1}, $f_{k-1}-1$ and f_k-1, then appear in any suffixient set.

On the other hand, F_{k-2} is followed by \$ in $F_k\$$, and it also prefixes $F_k = F_{k-2}F_{k-1}'$, therefore F_{k-2} is right-maximal. The first occurrence is preceded by F_{k-1}, and hence by c, and the second by no symbol. F_{k-2} also occurs in F_k at position $f_{k-2}+1$, as seen above, preceded by F_{k-2} and thus by d. There are no other occurrences of F_{k-2} in F_k because (i) it cannot occur starting at positions 2 to f_{k-2} by the same reason as point (ii) of the previous paragraph; (ii) it cannot appear starting at positions $f_{k-2}+2$ to $f_{k-1}-2$ because $F_k = F_{k-2}F_{k-2}F_{k-3}'$ and $F_{k-3}'[1, f_{k-3}-2] = F_{k-3}[1, f_{k-3}-2] = F_{k-2}[1..f_{k-3}-2]$, thus such an occurrence would also match a rotation of F_{k-2}, which is impossible as noted above; (iii) it cannot appear starting at positions $f_{k-1}-1$ or f_{k-1} because, since it matches at position $f_{k-1}+1$, F_{k-2} would match itself with an offset of 1 or 2, which is impossible as noted in point (i) of the previous paragraph. The right-extensions of F_{k-2} are then super-maximal. The one followed by \$ occurs ending at position $f_k + 1$. The other two are followed by a because they are followed by F_{k-2} and by F_{k-3}' and all F_k for $k \geq 2$ start with a. We can then choose either ending position for a suffixient set, $f_{k-2}+1$ or $2f_{k-2}+1$. □

Corollary 4. *There exist string families where $\chi = o(v)$.*

Proof. It follows from Lemma 11 and the fact that $v = \Omega(\log n)$ on the odd Fibonacci words [28, Thm. 28]. □

5.3   Uncomparability of χ with Copy-Paste Measures

Finally, we show that χ is incomparable with most copy-paste measures. This follows from χ being $\Theta(n)$ on de Bruijn sequences and $O(1)$ on Fibonacci strings. Because $g = O(n/\log n)$ on de Bruijn sequences [28] and by Lemma 5, we have:

Corollary 5. *There exists a string family with* $\chi = \Omega(g \log n)$.

   This result is particularly relevant because all the copy-paste based measures μ, with the exception of z_e, are $O(g)$. Corollary 5 then implies $\mu = o(\chi)$ on de Bruijn sequences for all these measures μ.
   While it has been said that $z_e = O(n/\log n)$ on binary sequences as well [19], this referred to the version that adds to each phrase the next nonmatching character. Because z_e is not an optimal parse, it is not obvious that this also holds for the version studied later in the literature, which does not add the next character. We then prove next that $z_e = o(\chi)$ holds on de Bruijn words.

Lemma 12. *There exists a string family with* $\chi = \Omega\left(z_e \frac{\log n \log \log \log n}{(\log \log n)^2}\right)$.

Proof. It always holds that $z_e = O\left(z \frac{\log^2(n/z)}{\log \log(n/z)}\right)$ [13]. In de Bruijn sequences it holds that $z = \Theta(n/\log n)$, so $n/z = \Theta(\log n)$. Therefore, $z_e = O\left(z \frac{(\log \log n)^2}{\log \log \log n}\right)$, and replacing $z = \Theta(n/\log n)$ we get $z_e = O\left(n \frac{(\log \log n)^2}{\log n \log \log \log n}\right)$. By Lemma 5, this yields $\chi = \Omega\left(z_e \frac{\log n \log \log \log n}{(\log \log n)^2}\right) = \omega(z_e)$ on de Bruijn sequences.   □

Corollary 6. *The measure* χ *is uncomparable to* $\mu \in \{z, z_{no}, z_e, z_{end}, v, g, g_{rl}, c\}$.

Proof. From Corollary 5 and Lemma 12, and that z, z_{no}, z_{end}, v, g_{rl} and c are always $O(g)$, it follows that there are string families where $\mu = o(\chi)$, for any $\mu \in \{z, z_{no}, z_e, z_{end}, v, g, g_{rl}, c\}$. On the other hand, from Lemma 11 and Corollary 4, and that $c = \Omega(\log n)$ on Fibonacci words [28, Thm. 32] and $c = O(\mu)$ for any $\mu \in \{z, z_{no}, z_e, z_{end}, g_{rl}, g\}$ [28, Thm. 30], it follows that there are string families where $\chi = o(\mu)$, for any $\mu \in \{z, z_{no}, z_e, z_{end}, v, g, g_{rl}, c\}$.   □

6   Conclusions and Open Questions

We have contributed to the understanding of χ as a new measure of repetitiveness, better finding its place among more studied ones. Figure 1 shows the (now) known relations around χ (cf. [27]).
   There are still many interesting open questions about χ. One of the most important is whether χ is reachable. Proving $b = O(\chi)$ would settle this question on the affirmative, and at the same time give the first copy-paste measure that is comparable with χ. We conjecture, instead, that χ is not reachable, proving which would imply that γ is also unreachable, a long-time open question.

One consequence of Corollary 5 is that $\chi \notin O(g \log^k(n/g))$ for any $k > 0$. It could be the case, though, that $\chi = O(\delta \log n)$, because the separation of χ and δ on de Bruijn sequences is a $\Theta(\log n)$ factor.

Regarding edit operations, it seems that $\mathtt{sre}(w')/\mathtt{sre}(w)$ is $O(1)$ for all the string operations we considered. Showing a multiplicative constant for insertion would imply the existence of a constant for rotation and vice versa. It is also open whether $r = O(\chi \log \chi)$. If this were true —and provided that χ has $O(1)$ multiplicative sensitivity to string operations— it would imply that r has $O(\log n)$ multiplicative sensitivity to these operations, making the already known lower bounds on multiplicative sensitivity [1, 14, 15] tight. If the conjecture were false, then χ could be considerably smaller than r in some string families.

Acknowledgements. We thank Davide Cenzato, Nicola Prezza, and Francisco Olivares for their code to compute smallest suffixient sets https://github.com/regindex/suffixient, which was helpful to propose and discard hypotheses on the behavior of χ, and for useful discussions on suffixient sets.

Funding Information. G.N. and C.U. were partially funded by Basal Funds FB0001 and AFB240001, ANID, Chile; and FONDECYT Project 1-230755, ANID, Chile. G.R. was partially funded by the MUR PRIN Project "PINC, Pangenome INformatiCs: from Theory to Applications" (Grant No. 2022YRB97K), funded by Next Generation EU PNRR M4 C2, Inv. 1.1 and by the INdAM - GNCS Project CUP_E53C24001950001. C.U. was partially funded by ANID-Subdirección de Capital Humano/Doctorado Nacional/2021-21210580, ANID, Chile; and NIC Chile Doctoral Scholarship, NIC, Chile.

Disclosure of Interests. The authors have no competing interests to declare that are relevant to the content of this article.

References

1. Akagi, T., Funakoshi, M., Inenaga, S.: Sensitivity of string compressors and repetitiveness measures. Inf. Comput. **291**, 104999 (2023). https://doi.org/10.1016/j.ic.2022.104999
2. Bruijn, de, N.: A combinatorial problem. Proceedings of the Section of Sciences of the Koninklijke Nederlandse Akademie van Wetenschappen te Amsterdam **49**(7), 758–764 (1946)
3. Burrows, M., Wheeler, D.: A block sorting lossless data compression algorithm. Tech. Rep. 124, Digital Equipment Corporation (1994)
4. Cenzato, D., et al.: Suffixient arrays: a new efficient suffix array compression technique. CoRR 2407.18753 (2025). https://doi.org/10.48550/arXiv.2407.18753
5. Cenzato, D., Olivares, F., Prezza, N.: On computing the smallest suffixient set. In: Proceedings of 31st International Symposium on String Processing and Information Retrieval (SPIRE 2024). Lecture Notes in Computer Science, vol. 14899, pp. 73–87. Springer (2024). https://doi.org/10.1007/978-3-031-72200-4_6
6. Depuydt, L., Gagie, T., Langmead, B., Manzini, G., Prezza, N.: Suffixient sets. CoRR 2312.01359 (2023). https://doi.org/10.48550/arXiv.2312.01359

7. Droubay, X.: Palindromes in the Fibonacci word. Inf. Process. Lett. **55**(4), 217–221 (1995). https://doi.org/10.1016/0020-0190(95)00080-V
8. Droubay, X., Justin, J., Pirillo, G.: Episturmian words and some constructions of de Luca and Rauzy. Theoret. Comput. Sci. **255**(1), 539–553 (2001). https://doi.org/10.1016/S0304-3975(99)00320-5
9. Fici, G., Romana, G., Sciortino, M., Urbina, C.: On the impact of morphisms on BWT-runs. In: Proceedings of 34th Annual Symposium on Combinatorial Pattern Matching (CPM 2023). Leibniz International Proceedings in Informatics, vol. 259, pp. 10:1–10:18. Schloss Dagstuhl - Leibniz-Zentrum für Informatik (2023). https://doi.org/10.4230/LIPIcs.CPM.2023.10
10. Fici, G., Romana, G., Sciortino, M., Urbina, C.: Morphisms and BWT-run sensitivity. In: Proceedings of 50th International Symposium on Mathematical Foundations of Computer Science (MFCS 2025). To appear (2025)
11. Fredricksen, H.: A survey of full length nonlinear shift register cycle algorithms. SIAM Rev. **24**(2), 195–221 (1982). https://doi.org/10.1137/1024041
12. Gabric, D., Sawada, J., Williams, A., Wong, D.: A framework for constructing de Bruijn sequences via simple successor rules. Discret. Math. **341**(11), 2977–2987 (2018). https://doi.org/10.1016/j.disc.2018.07.010
13. Gawrychowski, P., Kosche, M., Manea, F.: On the number of factors in the LZ-end factorization. In: Proceedings of 30th International Symposium on String Processing and Information Retrieval (SPIRE 2023). Lecture Notes in Computer Science, vol. 14240, pp. 253–259. Springer (2023). https://doi.org/10.1007/978-3-031-43980-3_20
14. Giuliani, S., Inenaga, S., Lipták, Z., Prezza, N., Sciortino, M., Toffanello, A.: Novel results on the number of runs of the Burrows-Wheeler-Transform. In: Proc. 47th International Conference on Current Trends in Theory and Practice of Computer Science (SOFSEM 2021). Lecture Notes in Computer Science, vol. 12607, pp. 249–262. Springer (2021). https://doi.org/10.1007/978-3-030-67731-2_18
15. Giuliani, S., Inenaga, S., Lipták, Z., Romana, G., Sciortino, M., Urbina, C.: Bit catastrophes for the Burrows-Wheeler transform. Theory Comput. Syst. **69**(2), 19 (2025). https://doi.org/10.1007/s00224-024-10212-9
16. Glen, A., Justin, J.: Episturmian words: a survey. RAIRO - Theoret. Inf. Appl. **43**(3), 403–442 (2009). https://doi.org/10.1051/ita/2009003
17. Kempa, D., Kociumaka, T.: Resolution of the Burrows-Wheeler transform conjecture. Commun. ACM **65**(6), 91–98 (2022). https://doi.org/10.1145/3531445
18. Kempa, D., Prezza, N.: At the roots of dictionary compression: String attractors. In: Proceedings of 50th Annual ACM Symposium on the Theory of Computing (STOC 2018), pp. 827–840. ACM (2018). https://doi.org/10.1145/3188745.3188814
19. Kreft, S., Navarro, G.: On compressing and indexing repetitive sequences. Theoret. Comput. Sci. **483**, 115–133 (2013). https://doi.org/10.1016/j.tcs.2012.02.006
20. Lempel, A., Ziv, J.: On the complexity of finite sequences. IEEE Trans. Inf. Theory **22**(1), 75–81 (1976). https://doi.org/10.1109/TIT.1976.1055501
21. Lothaire, M.: Algebraic Combinatorics on Words. Encyclopedia of Mathematics and its Applications, Cambridge University Press, New York, NY, USA (2002). https://doi.org/10.1017/CBO9781107326019
22. de Luca, A.: A combinatorial property of the Fibonacci words. Inf. Process. Lett. **12**(4), 193–195 (1981). https://doi.org/10.1016/0020-0190(81)90099-5
23. Manber, U., Myers, E.W.: Suffix arrays: a new method for on-line string searches. SIAM J. Comput. **22**(5), 935–948 (1993). https://doi.org/10.1137/0222058

24. Mantaci, S., Restivo, A., Romana, G., Rosone, G., Sciortino, M.: A combinatorial view on string attractors. Theoret. Comput. Sci. **850**, 236–248 (2021). https://doi.org/10.1016/j.tcs.2020.11.006
25. Navarro, G.: Indexing highly repetitive string collections, part I: repetitiveness measures. ACM Comput. Surv. **54**(2), article 29 (2021). https://doi.org/10.1145/3434399
26. Navarro, G.: Indexing highly repetitive string collections, part II: Compressed indexes. ACM Comput. Surv. **54**(2), article 26 (2021). https://doi.org/10.1145/3432999
27. Navarro, G.: Indexing highly repetitive string collections. CoRR 2004.02781 (2022). https://doi.org/10.48550/arXiv.2004.02781
28. Navarro, G., Ochoa, C., Prezza, N.: On the approximation ratio of ordered parsings. IEEE Trans. Inf. Theory **67**(2), 1008–1026 (2021). https://doi.org/10.1109/TIT.2020.3042746
29. Navarro, G., Olivares, F., Urbina, C.: Generalized straight-line programs. Acta Informatica **62**(1), 14 (2025). https://doi.org/10.1007/s00236-025-00481-3
30. Navarro, G., Urbina, C.: Repetitiveness measures based on string morphisms. Theoret. Comput. Sci. **1043**, 115259 (2025). https://doi.org/10.1016/j.tcs.2025.115259
31. Storer, J.A., Szymanski, T.G.: Data compression via textual substitution. J. ACM **29**(4), 928–951 (1982). https://doi.org/10.1145/322344.322346

Longest Unbordered Factors
on Run-Length Encoded Strings

Shoma Sekizaki$^{(\boxtimes)}$ and Takuya Mieno(ORCID)

University of Electro-Communications, Chofu, Japan
s2431091@edu.cc.uec.ac.jp, tmieno@uec.ac.jp

Abstract. A border of a string is a non-empty proper prefix of the string that is also a suffix. A string is unbordered if it has no border. The longest unbordered factor is a fundamental notion in stringology, closely related to string periodicity. This paper addresses the longest unbordered factor problem: given a string of length n, the goal is to compute its longest factor that is unbordered. While recent work has achieved subquadratic and near-linear time algorithms for this problem, the best known worst-case time complexity remains $O(n \log n)$ [Kociumaka et al., ISAAC 2018]. In this paper, we investigate the problem in the context of compressed string processing, particularly focusing on run-length encoded (RLE) strings. We first present a simple yet crucial structural observation relating unbordered factors and RLE-compressed strings. Building on this, we propose an algorithm that solves the problem in $O(m^{1.5} \log^2 m)$ time and $O(m \log^2 m)$ space, where m is the size of the RLE-compressed input string. To achieve this, our approach simulates a key idea from the $O(n^{1.5})$-time algorithm by [Gawrychowski et al., SPIRE 2015], adapting it to the RLE setting through new combinatorial insights. When the RLE size m is sufficiently small compared to n, our algorithm may show linear-time behavior in n, potentially leading to improved performance over existing methods in such cases.

Keywords: string algorithms · unbordered factors · run-length encoding

1 Introduction

A non-empty string b is called a *border* of another string T if b is both a prefix and a suffix of T. A string is said to be *bordered* if it has a border, and *unbordered* otherwise. Unbordered factors are known to have a deep connection with the smallest period of the string. The length of a string uv is called a *period* of a string T if $T = (uv)^k u$ for some strings u, v and an integer $k \geq 1$. The concept of string periodicity is fundamental and has applications in various areas of string processing, including pattern matching, text compression, and sequence assembly in bioinformatics [5, 15, 19].

In 1979, Ehrenfeucht and Silberger [9] posed the problem of determining the conditions under which $\tau(T) = \pi(T)$ holds for a string T of length n, where

G. Badkobeh et al. (Eds.): SPIRE 2025, LNCS 16073, pp. 233–247, 2026.
https://doi.org/10.1007/978-3-032-05228-5_19

$\tau(T)$ denotes the length of the longest unbordered factor of T and $\pi(T)$ denotes the smallest period of T. They further conjectured that $\tau(T) \leq n/2$ implies $\tau(T) = \pi(T)$. However, this conjecture was disproved by Assous and Pouzet [2], who provided a counterexample. Subsequently, some progress was made toward the conjecture [6,7,11–13,20]. Finally, in 2012, Holub and Nowotka [14] solved this longstanding open problem, showing that $\tau(T) = \pi(T)$ holds if $\tau(T) \leq 3n/7$, and that this bound is tight due to the counterexample of [2].

This result led to increased research activity on algorithms for computing the longest unbordered factor [8,10,16,18]. As a special case, when a string of length n is periodic (i.e., its smallest period is at most $n/2$), its longest unbordered factor can be computed in $O(n)$ time [8]. Unfortunately, since many strings are non-periodic on average [18], this linear-time approach has limited applicability. For the general case, a straightforward $O(n^2)$-time algorithm can be designed by constructing *border arrays* [15], which can be computed in linear time, for all suffixes of a string T of length n. The resulting n border arrays indicate whether each factor of T is bordered or unbordered. The first non-trivial algorithm for computing the longest unbordered factor was proposed by Loptev et al. [18], who presented an algorithm with average-case running time $O(n^2/\sigma^4)$, where σ is the alphabet size. Furthermore, Cording et al. [4] proved that the expected length of the longest unbordered factor of a random string is $n - O(\sigma^{-1})$, and used this result to propose an average-case $O(n)$-time algorithm. In terms of worst-case time complexity, all of the above are quadratic-time algorithms. The first worst-case subquadratic-time algorithm was given by Gawrychowski et al. [10], whose algorithm runs in $O(n\sqrt{n})$ time. We will later review the basic strategy of their algorithm, which we simulate in our approach. The state-of-the-art algorithm for this problem is an $O(n \log n)$-time algorithm proposed by Kociumaka et al. [16,17], which exploits combinatorial properties of unbordered factors and sophisticated data structures, including the *prefix-suffix query* (PSQ) data structure. As of 2018 [16], their algorithm was reported to run in $O(n \log n \log^2 \log n)$ time in the worst case due to the cost of constructing the PSQ data structures. Later improvements [17] sped up the construction of the PSQ data structure to linear time, bringing the overall algorithm down to $O(n \log n)$ time. Whether the longest unbordered factor can be computed in $O(n)$ time remains open.

In this paper, instead of directly aiming to an $O(n)$-time algorithm, we propose an efficient solution in the context of *compressed string processing*. Especially, this work focuses on the *run-length encoding* (RLE) of a string. We first give a simple but important relationship between unbordered strings and RLE strings (Lemma 2). Using this relationship, we propose an RLE-based algorithm for computing all longest unbordered factors that runs in $O(m\sqrt{m} \log^2 m)$ time and uses $O(m \log^2 m)$ space, where m is the size of the RLE-compressed string. When m is sufficiently small (e.g., $m < n^{2/3-\varepsilon}$ for a small constant $\varepsilon > 0$), our approach achieves $O(n)$ time, thus improving the worst-case complexity over existing methods for such cases. On the one hand, the high-level idea of our approach is inspired by the algorithm of Gawrychowski et al. [10], which achieves subquadratic time via a non-trivial combination of fundamental string data structures and combinatorial techniques on strings. On the other hand, our algorithm differs in details and require techniques specially tailored to unbordered factors in RLE strings, particularly in Sects. 4.2 and 4.3.

Several proofs are omitted due to space limitations. All the omitted proofs can be found in the full version of this paper [22].

2 Preliminaries

Let Σ be an alphabet. An element in Σ is called a character. An element in $\Sigma^\star$ is called a string. The empty string is the string of length 0, which is denoted by ε. A string in which all characters are identical is called a unary string. The concatenation of strings S and T is written as $S \cdot T$, or simply ST when there is no confusion. Let T be a non-empty string. If $T = X \cdot Y \cdot Z$ holds for some strings X, Y, and Z, then X, Y, and Z are called a prefix, a factor, and a suffix of T, respectively. Further, they are called a proper prefix, a proper factor, and a proper suffix of T if $X \neq T, Y \neq T$, and $Z \neq T$, respectively. A non-empty string B is called a border of T if B is both a proper prefix and a proper suffix of T. We call the occurrence of B as a prefix (resp. suffix) of T the prefix-occurrence (resp. suffix-occurrence) of border B. We call the longest border of T the border of T. A string T is said to be bordered If T has a border, and is said to be unbordered otherwise. We denote by $|T|$ the length of T. For an integer i with $1 \leq i \leq |T|$, we denote by $T[i]$ the ith character of T. For integers i, j with $1 \leq i \leq j \leq |T|$, we denote by $T[i..j]$ the factor of T that starts at position i and ends at position j. For strings S, T, we denote by $lcp(S, T)$ the length of the longest common prefix (in short, lcp) of S and T. An integer p with $1 \leq p \leq |T|$ is called a period of T if $T[i] = T[i + p]$ holds for all i with $1 \leq i \leq |T| - p$. We call the smallest period of T the period of T. The border array Bord_T of a string T is an array of length $|T|$, where $\mathsf{Bord}_T[i]$ stores the length of the border of $T[1..i]$ for each $1 \leq i \leq |T|$ [15]. The border-group array BG_T of a string T is an array of length $|T|$ such that, for each $1 \leq i \leq |T|$, $\mathsf{BG}_T[i]$ stores the length of the shortest border of $T[1..i]$ whose smallest period equals that of $T[1..i]$ if such a border exists, and $\mathsf{BG}_T[i] = i$ otherwise [21]. It is known that the border array and the border-group array of a string T can be computed by $O(|T|)$ character comparisons [15,21].

The range maximum query (RMQ) over an integer array A of length N is, given a query range $[i, j] \subseteq [1, N]$, to output a position p such that $A[p]$ is a maximum value among sub-array $A[i..j]$. The range minimum query (RmQ) is defined analogously. The following result is known.

Lemma 1 (e.g., [3]). *There is a data structure of size $O(N)$ that can answer any RMQ (and RmQ) over an integer array of length N in $O(1)$ time. The data structure can be constructed in $O(N)$ time.*

In what follows, we fix an arbitrarily *non-unary* string T of length n for our purpose. This is because the longest unbordered factor of a unary string $\mathtt{a} \cdots \mathtt{a}$ is simply $\mathtt{a}$.

3   Tools for RLE Strings

This section provides some tools for RLE strings that are commonly used in Sect. 4.

3.1   Run-Length Encoding; RLE

The Run-Length Encoding (RLE) of a string T, denoted by $\mathrm{rle}(T)$, is a compressed representation of T that encodes every maximal character run $T[i..i + e - 1]$ in T by c^e if (1) $T[j] = c$ for all j with $i \leq j \leq i + e - 1$, (2) $i = 1$ or $T[i - 1] \neq c$, and (3) $i + e - 1 = n$ or $T[i + e] \neq c$. We simply call a maximal character run in T a run in T. Also, we call the number e of characters c in a run c^e the exponent of the run. The RLE size of string T, denoted by $r(T)$, is the number of runs in T. For each i with $1 \leq i \leq r(T)$, we denote by R_i the ith run of $\mathrm{rle}(T)$. Also, we denote by beg_i (resp., end_i) the beginning (resp., the ending) position of R_i, and by $\exp_i$ the exponent of R_i. A factor of T is said to be *RLE-bounded* if the factor starts at beg_i and ends at end_j for some i, j with $i \leq j$. In what follows, we use m to denote the RLE size of the given string T.

Example 1. The RLE of string $T = \mathsf{aaabbcccccabbbb}$ is $\mathsf{a}^3\mathsf{b}^2\mathsf{c}^5\mathsf{a}^1\mathsf{b}^4$. The exponents of the first two runs a^3 and b^2 are three and two, respectively. The factor $T[4..11] = \mathsf{bcccccca}$ of T is RLE-bounded since it starts at $\mathrm{beg}_2 = 4$ and ends at $\mathrm{end}_4 = 11$. The RLE size of T is 5.

The following lemma establishes a significant connection between unbordered strings and RLE strings.

Lemma 2. *Let u be a string of length at least two, and let $a = u[1]$ and $b = u[|u|]$. If u is unbordered, then both au and ub are also unbordered.*

Proof. For the sake of contradiction, assume that au is bordered. Let k be the length of the border of au. If $k = 1$, then the border of au is a, which implies $a = b$, contradicting the assumption that u is unbordered. If $k > 1$, let x be the border of au. Then, $x[2..k]$ is a border of u, a contradiction. Therefore, au must be unbordered. The proof for ub is symmetric.   □

From Lemma 2, any longest unbordered factor of a non-unary string T must be RLE-bounded. Furthermore, the number of occurrences of longest unbordered factors is at most $m - 1$ since any factor starting at the beginning position of the rightmost run is bordered. Also, the upper bound $m - 1$ is tight: For string $(\mathsf{a}^e\mathsf{b}^e)^{\frac{m}{2}}$ where e is a positive integer, all the occurrences of $\mathsf{a}^e\mathsf{b}^e$ and $\mathsf{b}^e\mathsf{a}^e$ are the longest unbordered factors. Similarly, the next observation holds:

Observation 1. *Let w be a non-empty string. If $w \cdot w[|w|]$ has a border of RLE size p, then w also has a border of RLE size p.*

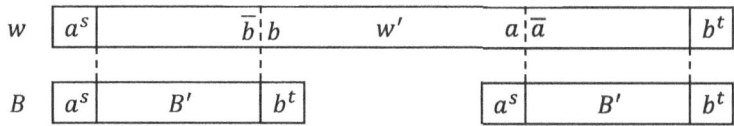

Fig. 1. Illustration for border $a^s B' b^t$ of w. String B' is an RLE-bounded border of w'.

3.2   RLE Shortest Border Array

We define the *RLE shortest border array* rSBord of T as follows: For each i with $1 \leq i \leq m$, rSBord$[i]$ stores the RLE size of the *shortest* border of the prefix $T[1..end_i]$ of T. For example, when $T = $ aaabbbbbaaaaaccaaaabbbaa, rSBord $= [1, 0, 1, 0, 1, 2, 1]$.

To design an efficient algorithm for computing rSBord, we give an observation. Let w be a non-unary string, and let a^s and b^t be the first and the last run of w, respectively. Further let w' be the factor of w such that $w = a^s w' b^t$. If w has a border B with $r(B) \geq 3$, then $B = a^s B' b^t$ holds for non-empty string B' that is a border of w'. Also, the occurrences of B' in w' as a suffix and as a prefix are RLE-bounded (see Fig. 1). Namely, each border B of w with $r(B) \geq 3$ can be obtained from some RLE-bounded border B' of w' by prepending a^s and appending b^t to B'.

From the above observation, we can compute rSBord as follows: for each $i \leq m$, check whether each RLE-bounded border of a string $R_2 \cdots R_{i-1}$ can be extended to the left by R_1 and to the right by R_i. A naïve implementation of this algorithm runs in $O(m^2)$ time because all RLE-bounded borders of all prefixes of $T' = R_2 \cdots R_{m-1}$ can be computed by considering rle(T') as a string of length $m - 2$ over the alphabet $\Sigma \times \mathbb{N}$ and constructing the border array of rle(T'). To speed up, we make use of the following well-known fact about the periodicity of borders:

Lemma 3 ([15]). *The set of borders of a string w can be partitioned into $O(\log |w|)$ groups according to their smallest periods.*

Within such a group of borders, the characters that follow prefix-occurrences of all borders except the longest one must be identical due to periodicity. Thus, at most two distinct characters can follow prefix-occurrences of borders within the group. The same holds for the number of distinct characters preceding suffix-occurrences of borders within a group. Finally, by utilizing the border array and the border-group array of rle(T'), we can compute rSBord in $O(m \log m)$ time:

Lemma 4. *Given* rle(T), *the RLE shortest border array* rSBord *of T can be computed in $O(m \log m)$ time.*

Proof. Let $T' = R_2 \cdots R_{m-1}$. We first construct the border array $\mathsf{Bord}_{\mathrm{rle}(T')}$ and the border-group array $\mathsf{BG}_{\mathrm{rle}(T')}$ of $\mathrm{rle}(T')$, considering $\mathrm{rle}(T')$ as a string of length $m-2$ over $\Sigma \times \mathbb{N}$. For each j with $1 \le j \le m-2$, we scan the borders of $\mathrm{rle}(T')[1..j] = \mathrm{rle}(T)[2..j+1]$ in decreasing order of their lengths, skipping some of them and processing the rest as follows: We check whether the current border B of $\mathrm{rle}(T')$ can be extended to the left by R_1 and to the right by R_{j+2}, by examining the runs immediately after the prefix-occurrence and before the suffix-occurrence of B in $\mathrm{rle}(T')[1..j]$. If the group to which B belongs has at least two borders, we perform the same procedure for the shortest border in the group. We then find the next group using the border-group array $\mathsf{BG}_{\mathrm{rle}(T')}$, update B to the longest border in that group, and repeat the above procedure. The time required to process each group is $O(1)$, and there are $O(\log m)$ groups for each prefix of $\mathrm{rle}(T')$ by Lemma 3. Therefore, the total running time is $O(m \log m)$. $\square$

3.3   Some Functions for RLE Strings

Given $\mathrm{rle}(T)$, we construct array ExpSum of size m such that $\mathsf{ExpSum}[j] = \sum_{k=1}^{j} \exp_k$ for each $1 \le j \le m$. Then, given a text position i with $1 \le i \le n$, we can compute the run to which the ith character $T[i]$ belongs in $O(\log m)$ time by performing binary search on ExpSum. Tamakoshi et al. [23] proposed an $O(m)$-space data structure based on RLE, called a *truncated RLE suffix array* (tRLESA). They showed that tRLESA enhanced with some additional information of size $O(m)$ supports several standard string queries, such as pattern matching. By applying tRLESA and related data structures in conjunction with RMQ and/or RmQ (Lemma 1), several additional queries can be efficiently supported, as detailed below.

For positive integers i, p, j, q satisfying $i \le m$, $p \le \exp_i$, $j \le m$, and $q \le \exp_j$, let $\mathrm{rlcp}(i, p, j, q)$ denote the length of the longest common prefix of $T[\mathrm{beg}_i + p - 1..n]$ and $T[\mathrm{beg}_j + q - 1..n]$.

Lemma 5. *After $O(m \log m)$-time and $O(m)$-space preprocessing for $\mathrm{rle}(T)$, the value $\mathrm{rlcp}(i, p, j, q)$ can be computed in $O(1)$ time for given integers i, p, j, and q.*

For positive integers x, y, h, ℓ with $h < x \le y \le m$, we define

$$\mathrm{ridx}_{x,y}(h, \ell) = \begin{cases} -1 & \text{if } lcp(T[\mathrm{end}_h..n], T[\mathrm{end}_z..\mathrm{end}_y]) > \ell \text{ for all } x \le z \le y, \\ \arg\max_{z:x \le z \le y}\{lcp(T[\mathrm{end}_h..n], T[\mathrm{end}_z..\mathrm{end}_y]) \le \ell\} & \text{otherwise.} \end{cases}$$

In words, $\mathrm{ridx}_{x,y}(h, \ell)$ is the index $x \le z \le y$ of a run such that $T[\mathrm{end}_z..\mathrm{end}_y]$ has the longest lcp with $T[\mathrm{end}_h..n]$ where the lcp length is at most ℓ, if such z exists.

Lemma 6. *After $O(m \log m)$-time and $O(m)$-space preprocessing for $\mathrm{rle}(T)$ and integers x, y with $1 \le x \le y \le m$, the value $\mathrm{ridx}_{x,y}(h, \ell)$ can be computed in $O(\log m)$ time for given integers h and ℓ with $h < x$.*

Algorithm 1. Algorithm for computing the longest unbordered factor

Input: String T of RLE size m.
Output: The set of longest unbordered factors of T.
```
 1: LUB ← LONGEST-SHORT-UB(T)          ▷ LUB: set of longest unbordered factors.
 2: ℓ* = max{|x| : x ∈ LUB}            ▷ ℓ*: length of the longest unbordered factor.
 3: Preprocess for RM-LONG-BORDERED
 4: for k ← 5 to ⌈√m⌉ do
 5:     Preprocess for CANDIDATEₖ
 6:     C ← {ε}
 7:     for i ← 1 to (k − 4)⌊√m⌋ do
 8:         C ← C ∪ {CANDIDATEₖ(i)}
 9:     end for
10:     U ← RM-LONG-BORDERED(k, C)     ▷ All strings in U are unbordered.
11:     ℓ ← max{|u| : u ∈ U}
12:     if ℓ < ℓ* then
13:         continue                    ▷ Do nothing and continue to the next stage.
14:     else if ℓ > ℓ* then
15:         LUB ← ∅                     ▷ Clear the current tentative solutions.
16:         ℓ* ← ℓ
17:     end if
18:     LUB ← LUB ∪ {u ∈ U : |u| = ℓ*}
19: end for
20: return LUB
```

4 Algorithm for Computing Longest Unbordered Factors

In this section, we prove our main theorem:

Theorem 1. *Given an RLE encoded string* $\mathrm{rle}(T)$ *of RLE size* m*, we can compute the set of longest unbordered factors of* T *in* $O(m\sqrt{m}\log^2 m)$ *time using* $O(m\log^2 m)$ *space.*

The high-level strategy of our algorithm, presented in Algorithm 1, is essentially the same as that of Gawrychowski et al. [10]. We divide the input string T into $\lceil\sqrt{m}\rceil$ blocks $J_1, J_2, \ldots, J_{\lceil\sqrt{m}\rceil}$, where each block J_k is RLE-bounded and has RLE size $\lfloor\sqrt{m}\rfloor$ for every $1 \le k < \lceil\sqrt{m}\rceil$. Throughout this section, we refer to a border of RLE size at most $\sqrt{m}$ as a *short* border, and otherwise as a *long* border. Let ρ_T denote the RLE size of a longest unbordered factor of T. Also, we refer to a factor starting at beg_i and ending within the kth block J_k as an (i,k)-*factor*. Note that the longest unbordered factor of T must be an (i,k)-factor for some i and k since it is RLE-bounded (see Lemma 2).

Let us look at Algorithm 1. The set LUB represents the current tentative solution and the variable ℓ^* represents the length of an element in LUB. First of all, we invoke the subroutine LONGEST-SHORT-UB, which outputs the longest unbordered factors of RLE size at most $4\sqrt{m}$, and tentatively update LUB and ℓ^* (lines 1–2). The main part of our algorithm consists of $O(\sqrt{m})$ stages, corresponding to the outer **for** loop (lines 4–19). In each stage, say the kth stage, we first compute a set C of *candidates* for longest unbordered factors that end within J_k by calling the subroutine CANDIDATE$_k$ $O(m)$ times (lines 7–9). Here, as we will show later in Lemma 7, the subroutine CANDIDATE$_k(i)$ returns one of the following three strings: (1) the longest unbordered (i,k)-factor, if such a factor exists; (2) the empty string, if all (i,k)-factors have short borders; or (3) an (i,k)-factor that has no short border but has a long border, otherwise. Then, the set C is guaranteed to contain a longest unbordered factor of T if (i) $\rho_T > 4\sqrt{m}$ and (ii) there is a longest unbordered factor ending within J_k. We then eliminate from C all factors that have a long border by calling the subroutine RM-LONG-BORDERED(k, C), which checks whether each string in C has a long border and removes it if so (line 10). If any candidates remain, we select the longest ones and update $\ell^\star$ and LUB if necessary (lines 11–18). After the outer **for** loop, we have the final answer LUB, thus output it.

The correctness of Algorithm 1 is straightforward from the properties of the three subroutines LONGEST-SHORT-UB, CANDIDATE$_k$, and RM-LONG-BORDERED. In what follows, we describe how to implement the subroutines efficiently.

4.1   Implementation of LONGEST-SHORT-UB

To implement LONGEST-SHORT-UB, we simply apply Lemma 4 for all RLE-bounded factors of RLE size $4\sqrt{m}$. By doing so, we can compute the longest unbordered factors of RLE size at most $4\sqrt{m}$ in a total of $O(m\sqrt{m}\log m)$ time.

4.2   Implementation of CANDIDATE$_k$

Throughout this subsection, we fix an integer $5 \le k \le \lceil \sqrt{m} \rceil$ arbitrarily. The definition of function CANDIDATE$_k(i)$ is as follows: CANDIDATE$_k(i)$ returns the longest (i,k)-factor that has no short border, if it exist; or the empty string, otherwise. This definition leads to the following properties:

Lemma 7. *(1) If there is an unbordered (i,k)-factor, then* CANDIDATE$_k(i)$ *returns the longest unbordered (i,k)-factor. (2) If all (i,k)-factors have short borders, then* CANDIDATE$_k(i)$ *returns the empty string. (3) Otherwise,* CANDIDATE$_k(i)$ *returns an (i,k)-factor that has no short border but has a long border.*

Let J_k.first and J_k.last be the indices of runs such that $T[\mathrm{beg}_{J_k.\mathrm{first}}..\mathrm{end}_{J_k.\mathrm{last}}] = J_k$. Let $D_k = J_{k-1}J_k$, $x = J_{k-1}$.first, $y = J_k$.first, and $z = J_k$.last. Namely, $T[\mathrm{beg}_x..\mathrm{end}_z] = D_k$ and $T[\mathrm{beg}_y..\mathrm{end}_z] = J_k$ hold. Let P_i be the longest prefix of $T[\mathrm{end}_i..\mathrm{end}_z]$ that occurs in D_k. If $P_i = \varepsilon$, then CANDIDATE$_k(i)$ returns $T[\mathrm{beg}_i..\mathrm{end}_z]$ since it has no short border. Otherwise, let $p = r(P_i)$. If $p = 1$, CANDIDATE$_k(i)$ can be easily computed by comparing the characters of the ith run and the zth run. Thus, we assume $p > 1$ in the following. Let $P_i = aub^{e_1}$ where $a = P_i[1]$, b^{e_1} is the last run of P_i, and $u \in \Sigma^\star$ is the rest. Further let Γ be the set of exponents of b following some occurrences of au in D_k. If $\min \Gamma > e_1$, the next character of aub^{e_1} in D_k is always b. Then, any factor starting at end_i

and ending at $\mathsf{end}_{j'}$ for some $y \leq j' \leq z$ can not have a short border of RLE size exactly p because if such a border exists, the border forms aub^e with $e \leq e_1$, contradicts that $\min \Gamma > e_1$. If $\min \Gamma \leq e_1$, we define $e_2 = \max\{\gamma \in \Gamma \mid \gamma \leq e_1\}$. Let end_t be the starting position of an occurrence of aub^{e_2} in D_k. We further define $F(t,j) = T[\mathsf{end}_t..\mathsf{end}_z]\$T[\mathsf{beg}_x..\mathsf{end}_{y+j-1}]$ for j with $1 \leq j \leq \sqrt{m}$, where $\$$ is a special character with $\$ \notin \Sigma$. The next lemma holds:

Lemma 8. *Assume* $\min \Gamma \leq e_1$ *holds. For each* $1 \leq j \leq \sqrt{m}$, $T[\mathsf{end}_i..\mathsf{end}_{y+j-1}]$ *has a short border of RLE size* p *if and only if* $F(t,j)$ *has a short border of RLE size* p.

Proof. Let $j' = y + j - 1$. ($\Rightarrow$) If $T[\mathsf{end}_i..\mathsf{end}_{j'}]$ has a short border of RLE size $p \leq \sqrt{m}$, the border forms $aub^{\mathsf{exp}_{j'}}$ and it holds that $\mathsf{exp}_{j'} \leq e_1$. Also, $\mathsf{exp}_{j'} \leq e_2$ holds since $e_2 < \mathsf{exp}_{j'} \leq e_1$ contradicts the definition of e_2. Therefore, $F(t,j)$ has border $aub^{\mathsf{exp}_{j'}}$ since $F(t,j)$ starts with aub^{e_2}. ($\Leftarrow$) If $F(t,j)$ has a short border of RLE size $p \leq \sqrt{m}$, the border forms $aub^{\mathsf{exp}_{j'}}$ and it holds that $\mathsf{exp}_{j'} \leq e_2$. Also, $\mathsf{exp}_{j'} \leq e_2 \leq e_1$ holds by the definition of e_2. Therefore, $T[\mathsf{end}_i..\mathsf{end}_{j'}]$ has border $aub^{\mathsf{exp}_{j'}}$ since $T[\mathsf{end}_i..\mathsf{end}_{j'}]$ starts with aub^{e_1}. □

Let $t^\star = t$ if $\min \Gamma \leq e_1$; otherwise, let $t^\star$ be the starting position of an occurrence of aub in D_k. Regardless of the value of $\min \Gamma$, for each $1 \leq j \leq \sqrt{m}$, the set of short borders of $T[\mathsf{end}_i..\mathsf{end}_{y+j-1}]$ of RLE size q is equivalent to that of $F(t^\star, j)$ for any $q < p$, since such borders are prefixes of their common prefix au. Also, the RLE size of any short border of $T[\mathsf{end}_i..\mathsf{end}_{y+j-1}]$ is upper bounded by p from the definition of P_i. To summarize, the set of short borders of $T[\mathsf{end}_i..\mathsf{end}_{y+j-1}]$ is equivalent to the set of short borders of $F(t^\star, j)$ of RLE size at most $p-1$ if $\min \Gamma > e_1$; otherwise, it is equivalent to the set of short borders of $F(t^\star, j)$ of RLE size at most p by Lemma 8.

Next, for $1 \leq j \leq \sqrt{m}$, let us consider the short borders of $T[\mathsf{beg}_i..\mathsf{end}_{j'}]$ where $j' = y+j-1$. A short border of $T[\mathsf{beg}_i..\mathsf{end}_{j'}]$ can be obtained by extending a short border of $T[\mathsf{end}_i..\mathsf{end}_{j'}]$ to the left by $\mathsf{exp}_i - 1$ characters. Consider all the short borders $B_1, B_2, \ldots, B_g$ of $T[\mathsf{end}_i..\mathsf{end}_{j'}]$, which satisfy that $r(B_s) \leq \min\{p, \sqrt{m}\}$ for all s. Note that all such borders are also borders of $F(t^\star, j)$ as discussed above. Let e_s be the exponent of the first run of the suffix-occurrence of B_s for each $1 \leq s \leq g$. Let $\mathcal{E}_{j'}^p$ be the set of such e_s for all $B_1, B_2, \ldots, B_g$. If $e_s < \mathsf{exp}_i$ for all $e_s \in \mathcal{E}_{j'}^p$, i.e., $\max \mathcal{E}_{j'}^p < \mathsf{exp}_i$, then $T[\mathsf{beg}_i..\mathsf{end}_{j'}]$ cannot have a short border. Conversely, if $\max \mathcal{E}_{j'}^p \geq \mathsf{exp}_i$, then $T[\mathsf{beg}_i..\mathsf{end}_{j'}]$ has a short border.

Based on the observations above, we design an algorithm for computing CANDIDATE$_k(i)$.

Preprocessing. We construct a data structure for $\mathsf{ridx}_{x,z}(\cdot, \cdot)$ by using Lemma 6. Next, let us *conceptually* consider a $\sqrt{m} \times \sqrt{m}$ table M_τ defined as follows: $M_\tau[r][j] = \infty$ for all r if the first and the last characters of $F(\tau, j)$ are the same; otherwise, $M_\tau[r][j]$ stores the maximum exponent among the first runs of the suffix-occurrences of those borders of $F(\tau, j)$ whose RLE size is at most r; if there is no such a border, then $M_\tau[r][j] = 0$ (see also the left part of Fig. 2).

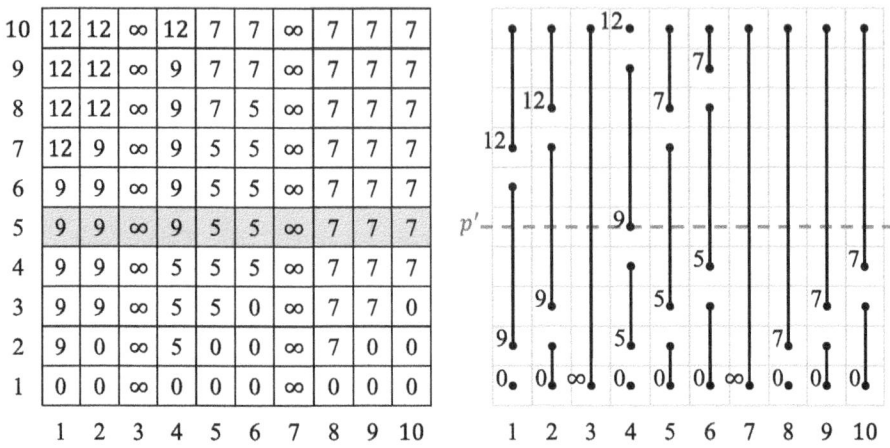

Fig. 2. Left: An example of table M_τ. Each column is non-decreasing from bottom to top. The ∞ symbols in the 3rd and 7th columns indicate that the first and last characters of $F(\tau,3)$ and $F(\tau,7)$ are the same, respectively. Right: Line segments corresponding to the runs in all columns of M_τ. If $p' = 5$ and $e = 7$, the largest $j^\star$ such that $M_\tau[p'][j^\star] < e$ is 6.

Note that we do not explicitly construct such $\sqrt{m} \times \sqrt{m}$ tables. The details of their implementation will be described later.

Query. Given a position i, we first compute $\alpha = \mathsf{ridx}_{x,z}(i,\infty)$, which satisfies that the lcp of $T[\mathsf{end}_i..n]$ and $T[\mathsf{end}_\alpha..\mathsf{end}_z]$ equals P_i. Also, we compute $e_1 = |P_i| - |au|$ and $p = r(P_i)$. If $\exp_{\alpha+p-1} = e_1$, then we set $t = \alpha$ since $e_2 = e_1$. Otherwise, we compute $\beta = \mathsf{ridx}_{x,z}(\alpha,|P_i|)$. Next, we compute $L = lcp(T[\mathsf{end}_i..n], T[\mathsf{end}_\beta..\mathsf{end}_z])$ by calling $\mathsf{rlcp}(i,\exp_i,\beta,\exp_\beta)$. If $|L| \leq |au|$, then $\min \Gamma > e_1$ holds and thus we set $t = \alpha$. Otherwise, we set $t = \beta$ since $e_2 = |L| - |au| \leq e_1$. Furthermore, we set $p' = \min\{p-1,\sqrt{m}\}$ if $\min \Gamma > e_1$; otherwise, set $p' = \min\{p,\sqrt{m}\}$. Next, we find the largest $j^\star$ such that $M_t[p'][j^\star] < \exp_i$. If there is no such $j^\star$, $\mathrm{CANDIDATE}_k(i)$ returns ε. Otherwise, it returns $T[\mathsf{beg}_i..\mathsf{end}_{y+j^\star-1}]$ since $T[\mathsf{beg}_i..\mathsf{end}_\iota]$ has a short border for all ι with $y + j^\star - 1 < \iota \leq z$, and hence, by Observation 1, $T[\mathsf{beg}_i..q]$ has a short border for all text positions q with $\mathsf{beg}_{y+j^\star} \leq q \leq \mathsf{end}_z$.

Implementing Table M_τ. The remaining task is to efficiently implement M_τ so that $j^\star$ can be found quickly. For each column of a table M_τ, say jth column, the values are non-decreasing by definition. Also, there are only $O(\log m)$ distinct values due to periodicity of borders of $T[\mathsf{end}_\tau+1..\mathsf{end}_z]\$T[\mathsf{beg}_x..\mathsf{end}_{y+j-2}]$. Namely, there are $O(\log m)$ runs of integers in the column. We define a (vertical) line segment that corresponds to each run, and assign the integer representing a run to each segment as its weight (see Fig. 2). Then we have $O(\sqrt{m}\log m)$ weighted line segments for M_τ in total. By using such weighted line segments, we can compute, for given p' and e, the largest $j^\star$ such that $M_\tau[p'][j^\star] < e$ as follows: find the rightmost line segment that intersects the horizontal line $r = p'$ and has weight less than e, then j^* is the j-coordinate of the line segment (again, see Fig. 2).

The set S_τ of such line segments can be computed in $O(\sqrt{m} \log m \log \log m)$ time by adapting the idea of the construction algorithm for the RLE shortest border array (Lemma 4) as follows: For each prefix of $F(\tau, z)$, we enumerate all $O(\log m)$ possible exponents of the first runs of the suffix-occurrences of borders, and for each exponent E, compute the minimum RLE size of a border whose first run has exponent E. By sorting these values by their RLE size in ascending order and then scanning their exponents, we can obtain the desired segments.

For each $x \le \tau \le z$, we construct a data structure of the *weighted lowest stabbing query (WLSQ)* on S_τ, where S_τ is rotated $90°C$ to the right, as defined below:

Definition 1 (Weighted Lowest Stabbing Query; WLSQ) . *A set S of weighted horizontal segments over $N \times N$ grid are given for preprocessing. The query is, given integers v, w_1, and w_2, to report the lowest segment s such that s is stabbed by vertical line $x = v$ and the weight of s is between w_1 and w_2.*

Given i, we can obtain j^* by answering WLSQ on S_t for $v = p'$, $w_1 = 0$, and $w_2 = \exp_i - 1$.

Very recently, Akram and Mieno proposed an algorithm for a problem called the *2D top-k stabbing query with weight constraint* (Definition 8 in [1]), which subsumes WLSQ as a special case. Although not stated explicitly, their data structure can be constructed in $O(|S| \log^2 |S|)$ time. We propose a simpler data structure specialized for WLSQ and show that it can be constructed slightly faster, in $O(|S| \log |S|)$ time:

Lemma 9. *A set S of segments is given as the input of WLSQ. We can build a data structure of size $O(|S| \log |S|)$ in $O(|S| \log |S|)$ time that can answer any WLSQ in $O(\log |S|)$ time.*

Thus, we can implement CANDIDATE$_k$ so that CANDIDATE$_k(i)$ can be computed in $O(\log m)$ time for each i after $O(\sum_{x \le \tau \le z} |S_\tau| \log |S_\tau|) = O(m \log^2 m)$ time and space preprocessing.

4.3   Implementation of RM-LONG-BORDERED

We define a new notion called the *RLE pseudo period* as follows: for a string w of RLE size r, the RLE pseudo period $pp(w)$ of w is the value $r - b$ where b is the RLE size of the border of w. Note that the RLE size p of the prefix (or suffix) of w whose length equals the period of w is not always equal to $pp(w)$, but it holds that $p - 1 \le pp(w) \le p$.

Example 2. Consider string $w = \text{abaababa}$ of RLE size $r = 7$. For this string, $pp(w) = 4$ holds since the border of w is aba of RLE size $b = 3$. The period of

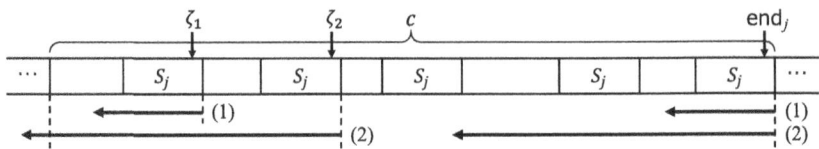

Fig. 3. String $c \in C_j$ is the candidate for an unbordered factor that ends at end_j we focus on. (1) The lcs of $T[1..\mathsf{end}_j]$ and the prefix $T[1..\zeta_1]$ of T that ends at the leftmost occurrence of S_j in the figure does not reach the starting position of c. We cannot yet determine whether c is bordered. (2) The lcs of $T[1..\mathsf{end}_j]$ and the prefix $T[1..\zeta_2]$ of T that ends at the second leftmost occurrence of S_j in this figure reaches the starting position of c, which reveals that c is bordered.

w is 5. On the one hand, the RLE size p_1 of the length-5 prefix abaab of w is 4, and then $pp(w) = p_1$ holds. On the other hand, the RLE size p_2 of the length-5 suffix ababa of w is 5, and then $pp(w) = p_2 - 1$ holds.

For each $j > \sqrt{m}$, let S_j be the longer string within (1) the shortest suffix of $\$T[1..\mathsf{end}_j]$ whose RLE pseudo period is greater than $\sqrt{m}/2 - 1$, and (2) $T[\mathsf{beg}_{j-\sqrt{m}+1}..\mathsf{end}_j]$. Note that S_j is well-defined since the RLE pseudo period of $\$T[1..\mathsf{end}_j]$ must be greater than $\sqrt{m}/2 - 1$ for $j > \sqrt{m}$.

Lemma 10. *If the shortest border of $T[\mathsf{beg}_i..\mathsf{end}_j]$ is a long border, then S_j is a suffix of each border of $T[\mathsf{beg}_i..\mathsf{end}_j]$.*

Proof. Let B be the shortest border of $T[\mathsf{beg}_i..\mathsf{end}_j]$. Since B is unbordered, the pseudo period of B equals the RLE size of B, which is greater than $\sqrt{m}$. Thus, $|S_j| \leq |B|$ holds, and thus S_j is a suffix of B. Therefore, S_j is a suffix of each border of $T[\mathsf{beg}_i..\mathsf{end}_j]$. □

Lemma 11. *The number of occurrences of S_j in T is in $O(\sqrt{m})$.*

Lemma 12. *Given $j > \sqrt{m}$, string S_j and its all occurrences can be computed in $O(\sqrt{m} \log m)$ time after $O(m \log m)$ time preprocessing.*

Preprocessing. We construct the data structure of Lemma 12.

Query Algorithm. Again, let us consider the kth stage and let $x = J_{k-1}$.first, $y = J_k$.first, and $z = J_k$.last. $D_k = T[\mathsf{beg}_x..\mathsf{end}_z]$ and $J_k = T[\mathsf{beg}_y..\mathsf{end}_z]$. For each j with $y \leq j \leq z$, we compute S_j and its occurrences, and sort them by using Lemma 12. Let L_j be the sorted list of occurrences of S_j. Given a set C of candidates for the longest unbordered factors, we first radix sort the elements of C using their starting points as the primary key and their ending points as the secondary key. Let C' be the sorted list of the elements of C. In the following, we scan elements in C' in order throughout the algorithm. We denote by C_j the list of elements in C' whose end positions are end_j. Then, C_j is a consecutive sub-list of C' from the condition of the radix sorting. For every j with $y \leq j \leq z$, we execute the following, scanning C_j and L_j from left to right: Let variables c and occ store elements in C_j and L_j we focus on, respectively. By using the data structure of Lemma 5 for the reversal of T, compute the longest common suffix (lcs) of $T[1..\text{end}_j]$ and the $T[1..\zeta]$ where ζ is the ending position of occ (see Fig. 3).

- If the lcs does not reach the starting position of c, then update occ to the next element in L_j. If such an element in L_j does not exists, then the current c is the longest string in C_j that has no long border by Lemma 10.
- If the lcs reaches the starting position of c, then c is bordered and hence update c to the next element in C_j that starts after ζ. If such an element in C_j does not exists, there is no unbordered factor in C_j. So we update c to the next element in C' and increment j accordingly.

The query algorithm can be performed in $O(|C|\log|C| + \sum_{y \leq j \leq z}|L_j|) = O(m \log m)$ time with $O(m)$ working space for each stage k.

Putting all pieces together, we complete the proof of Theorem 1.

Acknowledgments. This work was supported by JSPS KAKENHI Grant Number JP24K20734 (TM).

References

1. Akram, W., Mieno, T.: Sorted consecutive occurrence queries in substrings. In: 36th Annual Symposium on Combinatorial Pattern Matching, CPM 2025, June 17-19, 2025, Milan, Italy. LIPIcs, vol. 331, pp. 24:1–24:15. Schloss Dagstuhl - Leibniz-Zentrum für Informatik (2025). https://doi.org/10.4230/LIPICS.CPM.2025.24
2. Assous, R., Pouzet, M.: Une caracterisation des mots periodiques. Discret. Math. **25**(1), 1–5 (1979). https://doi.org/10.1016/0012-365X(79)90146-8
3. Bender, M.A., Farach-Colton, M.: The LCA problem revisited. In: LATIN 2000: Theoretical Informatics, 4th Latin American Symposium, Punta del Este, Uruguay, April 10-14, 2000, Proceedings. Lecture Notes in Computer Science, vol. 1776, pp. 88–94. Springer (2000). https://doi.org/10.1007/10719839_9
4. Cording, P.H., Gagie, T., Knudsen, M.B.T., Kociumaka, T.: Maximal unbordered factors of random strings. Theor. Comput. Sci. **852**, 78–83 (2021). https://doi.org/10.1016/J.TCS.2020.11.019
5. Crochemore, M., Mignosi, F., Restivo, A., Salemi, S.: Text compression using anti-dictionaries. In: Automata, Languages and Programming, 26th International Colloquium, ICALP'99, Prague, Czech Republic, July 11-15, 1999, Proceedings. Lecture Notes in Computer Science, vol. 1644, pp. 261–270. Springer (1999). https://doi.org/10.1007/3-540-48523-6_23
6. Duval, J.: Relationship between the period of a finite word and the length of its unbordered segments. Discret. Math. **40**(1), 31–44 (1982). https://doi.org/10.1016/0012-365X(82)90186-8
7. Duval, J., Harju, T., Nowotka, D.: Unbordered factors and Lyndon words. Discret. Math. **308**(11), 2261–2264 (2008). https://doi.org/10.1016/J.DISC.2006.09.054

8. Duval, J., Lecroq, T., Lefebvre, A.: Linear computation of unbordered conjugate on unordered alphabet. Theor. Comput. Sci. **522**, 77–84 (2014). https://doi.org/10.1016/J.TCS.2013.12.008

9. Ehrenfeucht, A., Silberger, D.M.: Periodicity and unbordered segments of words. Discret. Math. **26**(2), 101–109 (1979). https://doi.org/10.1016/0012-365X(79)90116-X

10. Gawrychowski, P., Kucherov, G., Sach, B., Starikovskaya, T.: Computing the longest unbordered substring. In: String Processing and Information Retrieval - 22nd International Symposium, SPIRE 2015, London, UK, September 1-4, 2015, Proceedings. Lecture Notes in Computer Science, vol. 9309, pp. 246–257. Springer (2015). https://doi.org/10.1007/978-3-319-23826-5_24

11. Harju, T., Nowotka, D.: Minimal Duval extensions. Int. J. Found. Comput. Sci. **15**(2), 349–354 (2004). https://doi.org/10.1142/S0129054104002467

12. Harju, T., Nowotka, D.: Periodicity and unbordered words: a proof of the extended Duval conjecture. J. ACM **54**(4), 20 (2007). https://doi.org/10.1145/1255443.1255448

13. Holub, S.: A proof of the extended Duval's conjecture. Theor. Comput. Sci. **339**(1), 61–67 (2005). https://doi.org/10.1016/J.TCS.2005.01.008

14. Holub, S., Nowotka, D.: The Ehrenfeucht-Silberger problem. J. Comb. Theory A **119**(3), 668–682 (2012). https://doi.org/10.1016/J.JCTA.2011.11.004

15. Knuth, D.E., Jr., J.H.M., Pratt, V.R.: Fast pattern matching in strings. SIAM J. Comput. **6**(2), 323–350 (1977). https://doi.org/10.1137/0206024

16. Kociumaka, T., Kundu, R., Mohamed, M., Pissis, S.P.: Longest unbordered factor in quasilinear time. In: 29th International Symposium on Algorithms and Computation, ISAAC 2018, December 16-19, 2018, Jiaoxi, Yilan, Taiwan. LIPIcs, vol. 123, pp. 70:1–70:13. Schloss Dagstuhl - Leibniz-Zentrum für Informatik (2018). https://doi.org/10.4230/LIPICS.ISAAC.2018.70

17. Kociumaka, T., Radoszewski, J., Rytter, W., Walen, T.: Internal pattern matching queries in a text and applications. SIAM J. Comput. **53**(5), 1524–1577 (2024). https://doi.org/10.1137/23M1567618

18. Loptev, A., Kucherov, G., Starikovskaya, T.: On maximal unbordered factors. In: Combinatorial Pattern Matching - 26th Annual Symposium, CPM 2015, Ischia Island, Italy, June 29 - July 1, 2015, Proceedings. Lecture Notes in Computer Science, vol. 9133, pp. 343–354. Springer (2015). https://doi.org/10.1007/978-3-319-19929-0_29

19. Margaritis, D., Skiena, S.: Reconstructing strings from substrings in rounds. In: 36th Annual Symposium on Foundations of Computer Science, Milwaukee, Wisconsin, USA, 23-25 October 1995, pp. 613–620. IEEE Computer Society (1995). https://doi.org/10.1109/SFCS.1995.492591

20. Mignosi, F., Zamboni, L.Q.: A note on a conjecture of Duval and Sturmian words. RAIRO Theor. Inform. Appl. **36**(1), 1–3 (2002). https://doi.org/10.1051/ITA:2002001

21. Mitani, K., Mieno, T., Seto, K., Horiyama, T.: Shortest cover after edit. In: 35th Annual Symposium on Combinatorial Pattern Matching, CPM 2024, June 25-27, 2024, Fukuoka, Japan. LIPIcs, vol. 296, pp. 24:1–24:15. Schloss Dagstuhl - Leibniz-Zentrum für Informatik (2024). https://doi.org/10.4230/LIPICS.CPM.2024.24

22. Sekizaki, S., Mieno, T.: Longest unbordered factors on run-length encoded strings. CoRR abs/2507.16285 (2025). https://doi.org/10.48550/ARXIV.2507.16285
23. Tamakoshi, Y., Goto, K., Inenaga, S., Bannai, H., Takeda, M.: An opportunistic text indexing structure based on run length encoding. In: Algorithms and Complexity - 9th International Conference, CIAC 2015, Paris, France, May 20-22, 2015. Proceedings. Lecture Notes in Computer Science, vol. 9079, pp. 390–402. Springer (2015). https://doi.org/10.1007/978-3-319-18173-8_29

Longest Common Subsequence in K-Length Substrings for Run-Length Encoded Strings

B. Riva Shalom[1]([✉])[iD], Eitan Kondratovsky[2][iD], and Ely Porat[3][iD]

[1] Shenkar College, Ramat Gan, Israel
rivash@shenkar.ac.il
[2] Open University, Ra'anana, Israel
eitan.k@openu.ac.il
[3] Bar Ilan University, Ramat Gan, Israel
porately@biu.ac.il

Abstract. The *Longest Common Subsequence (LCS)* problem is a classical problem in computer science. Numerous variants of the LCS problem have been proposed due to its significance. In this paper, we study a new variant of the LCS problem that combines two settings: the input strings are compressed using run-length encoding (RLE) and the solution must consist of matching substrings of length k, as in the LCSk problem. We propose two algorithms to address the challenges of the k-length substring constraint. The first requires $O(nM + mN)$ time where n and m are the lengths of the uncompressed input strings, N and M are the lengths of the compressed input strings. The second requires $O(N + M + \min\{n + |C| \log \log m, \ m + |C| \log \log n\})$ time, where $|C|$ denotes the number of required k-length matching substrings between the input strings.

Keywords: LCS · Run-length-encoded strings · Similarity

1 Introduction

The *Longest Common Subsequence (LCS)* problem is a classical problem in computer science. Given two strings over the same alphabet, the goal is to find the length of the longest subsequence that appears in both strings, where a subsequence is defined as a sequence derived from another by deleting some elements.

Numerous solutions have been proposed for the LCS problem; for surveys, see [12,25]. Over the years, many variants of the problem have also been introduced. These variants can be grouped into three categories:

(1) Variants where the input strings have specific structures, such as [1–4,6, 14,15,24].
(2) Variants that incorporate special distance measures or allow mismatches, such as [16,23,32,34].

G. Badkobeh et al. (Eds.): SPIRE 2025, LNCS 16073, pp. 248–264, 2026.
https://doi.org/10.1007/978-3-032-05228-5_20

(3) Variants imposing constraints on the resulting common subsequence, including [7–9, 11, 17, 18, 20, 22, 26, 33, 36].

In this paper, we introduce a novel variant of the problem, called $RLCSk$, which combines two well-studied extensions: the *LCS for RLE strings* and the *LCS with k-length substrings*. Both variants are defined hereafter.

LCS for Run-Length-Encoded Strings (RLCS). Let X be a string. $X[i..j]$ denotes the substring of X that begins at index i and ends at index j, $(i \leq j)$.

Definition 1. *[Run Length Encoding (RLE)] A substring $X[i..j]$, with $i \leq j$ is called a run if all symbols in the substring are identical, and the run is maximal, i.e., $X[i-1] \neq X[i]$ and $X[j+1] \neq X[j]$. The run-length of a run is the number of repeated symbols it contains, i.e., $j-i+1$. Given a string X, its RLE replaces each run with a pair of the repeated symbol and its run-length.*

For example, the RLE of the string $aadcccaaaa$ is $a^2d^1c^3a^4$. When the LCS problem is applied to run-length encoded (RLE) strings, it yields the $RLCS$ problem. The RLCS problem can achieve sub-quadratic performance in certain cases, depending on the number of runs in the compressed input.

Longest Common Subsequence in k-Length Substrings (LCSk). Benson et al. [11] observed that the classical LCS measure allows matching symbols to be arbitrarily spaced within the input strings. As a result, it may yield a misleading similarity score in biological contexts such as RNA analysis, where meaningful similarity depends on the alignment of fixed-length codons, each consisting of three nucleotides. For example, consider the RNA strands $A = (\text{GTG})^{n/3}$ and $B = (\text{TCC})^{n/3}$. Although LCS$(A, B) = n/3$, none of the matched characters appear in consecutive positions within the strings, rendering the similarity biologically irrelevant. To address this limitation, they proposed the LCSk problem.

Definition 2. *[The Longest Common Subsequence in k-Length Substrings (LCSk)]: Given two strings X and Y over an alphabet Σ, compute the maximum ℓ such that there exist ℓ common substrings of length k, denoted $X[i_f..i_f+k-1]$ and $Y[j_f..j_f+k-1]$ where for every $f \in [\ell-1]$, $i_f+k-1 < i_{f+1}$ and $j_f+k-1 < j_{f+1}$.*

The requirement that the common subsequence be composed of substrings of length k is referred to as the k-length substring constraint. (k-LS constraint).

For example, let $X = \mathbf{GG}C\mathbf{GT}G\mathbf{T}$, $Y = CC\mathbf{GGG T}$ and let $k = 2$. Then, $LCS2(X, Y) = 2$, and the output is formed by the bolded segments in both strings. Additional LCSk example can be seen in Fig. 1.

Longest Common Subsequence in k-Length Substrings for RLE Strings (RLCSk). The LCSk measure can provide more biologically meaningful insights than the standard LCS measure in several contexts: (1) Detection of conserved sequences, motifs, or binding regions in DNA and RNA. (2) Identification of conserved functional domains in proteins. (3) Simplification of sequence

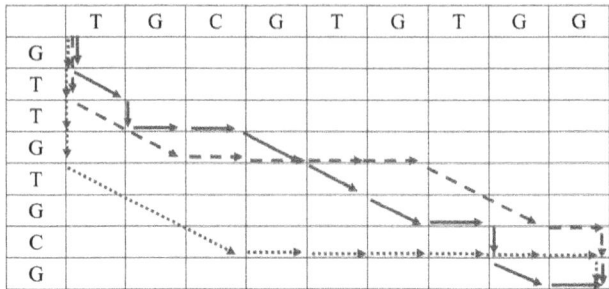

Fig. 1. An LCSk example. Diagonal arrows represent matches between the strings. Full arrows represent a possible LCS ($k = 1$) solution with no restriction on the length of matching substrings (the output common subsequence is of length 5). Dashed arrows represent a possible solution for $LCS2$ ($k = 2$) where every match must be a substring of length 2 (the output solution consists of two common substrings of length 2). Dotted arrows represent a possible solution for $LCS3$ ($k = 3$) where every match must be a substring of length 3 (the output solution is a single common subsequence of length 3)

alignment in large genomic databases by reducing the number of potential alignments or matches, improving computational efficiency and biological specificity.

Given the large size of biological sequences, it is desirable to work with compressed representations. To this end, we propose combining the LCSk framework with algorithms designed for compressed strings. As the first work in this research direction, we focus on the most natural compression scheme, RLE. Moreover, since some biological sequences are already stored in RLE format [29], the problem is not only of theoretical interest but also of practical relevance.

Definition 3. The LCS in k-Length Sub-Strings for RLE Strings *(RLCSk)*: *Given two RLE strings X and Y over an alphabet Σ, compute their LCSk.*

Traditional LCS techniques do not naturally accommodate the k-LS constraint, requiring specialized adaptations. The use of RLE further complicates matters, as efficiency must be maintained with respect to the compressed input. We suggest a hybrid approach that combines RLCS and LCSk to achieve subquadratic complexity in the average case.

Contributions of this paper:

- We define a novel and practical LCS variant, the RLCSk problem.
- We establish core structural properties of the LCSk problem, which serve as a foundation for designing algorithms to solve both its various problem variants and versions adapted to different compression schemes.
- We present two distinct algorithms for the RLCSk problem. The first is optimized for cases with many k-matches in the input. The second method is effective when the input strings exhibit few and short matching runs.

2 Notations and Previous Work

For an integers i we denote by $[i] = \{1..i\}$ the set of integers between 1 and i. A string X of length n over an alphabet Σ is a sequence of n symbols $X = X[1]X[2]\ldots X[n]$. The length of the string is denoted by $|X|$. For every $1 \leq i \leq j \leq n$, the *substring* $X[i..j] = X[i]X[i+1]\ldots X[j]$. If $i = 1$ we say that $X[i..j]$ is a prefix, and if $j = n$, it is a suffix.

The RLE string X is denoted by $X_1 X_2 \ldots X_N$, where X_i represents the ith run. Let $n_i = |X_i|$ denote the length of X_i. Then, the length of string X is $n = |X| = \sum_{i=1}^{N} n_i$. $\sigma(X_i)$ denotes the repetitive symbol of X_i. The RLE string of Y is $Y_1 Y_2 .. Y_M$, where Y_j is the j th run of Y and $m_j = |Y_j|$, then $m = |Y| = \sum_{j=1}^{M} m_i$. Let $X_i .. X_h$ denote the substring of the RLE string X consisting of the consecutive runs $X_i, X_{i+1}, .., X_h$.

For each index in an input string, we define its *location* based on its position within the runs of the string. The location of the i'th symbol in run X_i is denoted by $X_i(i')$. For every run X_i, we store its length as $len(X_i) = n_i$. For Example, let $X = A^4 G^3$. Then $len(X_1) = 4$ and $len(X_2) = 3$ and $X[6] = X_2(2)$.
Similarly, for the j'th symbol of run Y_j, the location is denoted by $Y_j(j')$.

A k-match is a common substring of length k, i.e., $X[i..i+k-1] = Y[j..j+k-1]$. A block (X_i, Y_j) represents an alignment between run X_i and run Y_j. When considering an entry $[X_i(i'), Y_j(j')]$ in the dynamic programming (DP) table, it defines a *relevant block*, which is the block (X_i, Y_j) restricted to the first i' rows and the first j' columns of the block. If $\sigma(X_i) = \sigma(Y_j)$, the block is called a *match-block*, otherwise it is a *mismatch-block*.

Let $\mathscr{L}[i,j]$ denote the LCS value of the prefixes $X[1..i]$ and $Y[1..j]$, $R\mathscr{L}[i,j]$ denote the RLCS value, and $R\mathscr{L}k[i,j]$ denote the RLCSk value for the same prefixes. Although $\mathscr{L}[i,j] = R\mathscr{L}[i,j]$, the underlying algorithmic problem differs depending on the string encoding. Therefore, we use different representations for the functions.

LCS k Previous Work. Several algorithms have been proposed to solve the LCSk problem. The time complexities of these algorithms are expressed in terms of the following parameters: n and m, the lengths of the input strings, ℓ, the length of the resulting LCSk , and R, the number of k-matches between the two strings. The complexities of the main proposed algorithms are summarized as follows: $O(nm)$ time by [9]. $O(n + R \log \ell)$, $O(n^2/k + n(k \log n)^{2/3})$, and $O(n^2/\log n)$ time by [17]. $O(n^2)$ time by [36]. $O(n + \ell(m - k\ell) + R)$ and $O(n + \ell(m - k\ell) \log n)$ time by [20].

Several variants of the LCSk problem have been proposed. Benson et al. [10] extended LCSk to the LCS with substrings of length at least k problem, where the restriction to matching substrings of exactly length k is relaxed. Instead, matched substrings are required to be of length at least k, preserving the proximity constraint. This problem is denoted by $LCS_{\geq k}$ or LCSk+. Further studies on $LCS_{\geq k}$ appear in [19,27].

Paveti et al. [28] introduced the LCSk++ measure, which further extends LCSk by allowing the common substrings of length k to overlap. Further work and applications of this variant are presented in [30,31,35]

RLCS Previous Work. The problem of finding LCS (not LCSk) for RLE strings was considered in several papers, among which are [4,6,14]. Some of them are based on the work of Bunke and Csirik [14], which introduces a block-based decomposition of the DP-table, where each block represents the alignment of two runs from the input strings. In this representation, dark blocks denote match-blocks, while light blocks indicate mismatch-blocks. An example of block division is provided in Fig. 2.

A common approach for RLCS [5] is to compute the $R\mathcal{L}$ only of entries from the bottom row and the rightmost column of each block. A recursive rule for such computing of $R\mathcal{L}$ is presented in Lemma 1.

Lemma 1. *[5] Let $[a, b] = [X_i(i'), Y_j(j')]$ be a computed entry in the DP-table. Then, $R\mathcal{L}[a, b]$ can be computed using the following rule:*

$$R\mathcal{L}[a, b] = \begin{cases} \max \left\{ \begin{array}{l} R\mathcal{L}[a, b - j'] \\ R\mathcal{L}[a - i', b] \end{array} \right\} & \text{if } \sigma(X_i) \neq \sigma(Y_j) \\ R\mathcal{L}[a - s, b - s] + s, \text{ where } s = \min\{i', j'\} & \text{if } \sigma(X_i) = \sigma(Y_j) \end{cases}$$

3   The k-LS Constraint in RLE Strings

Even though the LCSk and RLCS problems each have multiple solutions, combining these solutions is not straightforward due to the k-LS constraint. We present several observations and lemmas that illustrate this difficulty.

Observation 1. *When considering a match-block (X_i, Y_j), the LCS between the runs X_i and Y_j within the block is $\min\{|X_i|, |Y_j|\} = \min\{n_i, m_j\}$.*

Observation 2. *Observation 1 does not hold in the context of $R\mathcal{L}k$ computation.*

For example, refer to Fig. 2(b), where $\min\{|X_1|, |Y_1|\} = 4$. However, since $k = 3$, only a single 3-match is added to the constructed LCSk.

Observation 3. *Let $[a, b] = [X_i(i'), Y_j(j')]$ be an entry in the DP-table, , $R\mathcal{L}[a, b]$ is the maximum among a set of values described in Lemma 1. If multiple entries attain the maximum, any one of them may be chosen arbitrarily.*

Observation 4. *Observation 3 does not hold in the context of $R\mathcal{L}k$ computation.*

Proof. Observation 3 follows from the fact that, in the RLCS setting, all possible matches encountered thus far are fully utilized in every previous entry. However, in the RLCSk problem, if the length of a match within a match-block is not divisible by k, some matching symbols may remain unused.

As a result, two entries may contain different numbers of matched symbols yet yield the same $R\mathcal{L}k$. Such entries may differ in their potential to extend unused matching symbols, via diagonally adjacent matches, into a valid future k-match. Therefore, the decision of which entry to extend when computing $R\mathcal{L}k$ becomes critical.   □

For example, consider the DP-table shown in Fig. 3. To compute the value of entry $[10, 5]$, we examine entries $[10, 4]$ and $[9, 5]$. Both have an $R\mathcal{L}k$ of 1. However, entry $[10, 4]$ has no unused matches, as the k-match it represents includes the symbol $X[10]$. In contrast, entry $[9, 5]$ retains one unused matching symbol, as it corresponds to a k-match ending at $[3, 4]$, leaving $X[10]$ available for extension.

Lemma 2. *Let $[a, b] = [X_i(i'), Y_j(j')]$ be a computed entry in the DP-table. In addition to the source entries specified in Lemma 1, when computing $R\mathcal{L}k[a, b]$, the entry $[a - i', b - j']$, denoted by $diag^*(a, b)$, must also be considered as a potential source.*

Proof. When computing $R\mathscr{L}[a, b]$, where $[a, b] = [X_i(i'), Y_j(j')]$ with $i' \neq j'$, aligning the match within the block to its upper-left position does not affect the value of $R\mathscr{L}[a, b]$. Nevertheless, in the $R\mathscr{L}k$ setting, aligning the match within the current block to its upper-left entry by following the $diag^*$ source entry, can lead to a higher $R\mathscr{L}k$. This is because the match may be concatenated with unused matching symbols from the previous diagonal block, potentially forming an additional k-match.   □

For example, in Fig. 2(a), matching $X[12]$ with $Y[9]$ yields the same value for $R\mathscr{L}[13, 9]$ as matching $X[13]$ with $Y[9]$. However, this equivalence does not hold in the $R\mathscr{L}k$ case, as illustrated in Fig. 2(b).

Lemma 3. *When computing the RLCSk using a DP algorithm, values propagated from entries to the left or above a given entry must be considered for both match-blocks and mismatch-blocks (unlike in Lemma 1).*

Proof. The k-LS constraint implies that a k-match may terminate within an inner row or column of a block. Therefore, to ensure that each entry contains the optimal $R\mathscr{L}k$ allowing a new k-match to follow immediately after the previous one ends, we must account for such propagation within match-blocks as well.   □

For example, consider entry $[6, 15]$ of Fig. 3. The score of the bottom-rightmost entry of block $(2, 5)$ is $R\mathscr{L}k[5, 11] = 1$, and no additional k-match occurs in the first row of block $(3, 6)$. However, entry $[6, 11]$ holds a score of 2, indicating that a better score can be propagated from a block to the left of the current block.

Observations 2, 4 and Lemmas 2, 3 imply that computing $R\mathscr{L}k$ using existing RLCS-based approaches requires substantial modifications in order to satisfy the k-LS constraint. An example of an RLCS algorithm that does not yield an optimal $R\mathscr{L}k$

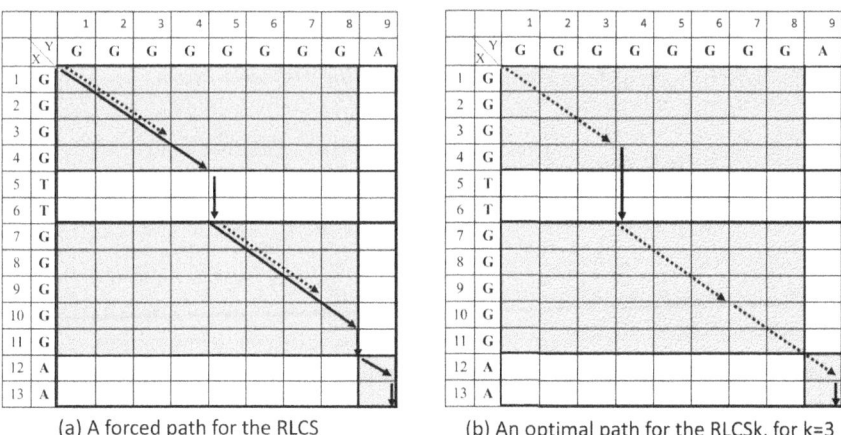

(a) A forced path for the RLCS   (b) An optimal path for the RLCSk, for k=3

Fig. 2. An example of a DP-tables divided into match-blocks (gray) and mismatch-blocks (white). Diagonal arrows represent symbol matches, while dotted arrows indicate *3-matches*. Figure (a) illustrates a forced path that is optimal for the $R\mathscr{L}$ problem; however, it includes only two *3-matches*. Figure (b) shows an optimal solution for the $R\mathscr{L}k$ problem with $k = 3$, which contains three *3-matches* (Color figure online)

solution is the one proposed by Apostolico et al. [6]. They defined *forced paths* in the block-based DP-table. Such a path begins at the upper-left corner of a match-block and exits through one of its sides. For each match-block, the algorithm evaluates the forced path originating from its upper-left corner. Figure 2 contrasts the optimal solution for $R\mathscr{L}$ with the optimal solution for $R\mathscr{L}k$, highlighting the limitation of their approach.

The primary challenge in computing $R\mathscr{L}k$ lies in efficiently utilizing partial regions of the DP-table, while maximizing the optimal $\mathscr{L}k$ value by carefully accounting for how each decision influences subsequent entries in the table.

4   RLCSk First Solution

In this section, we propose an efficient dynamic programming (DP) algorithm for computing $R\mathscr{L}k$, particularly suited for instances with a large number of match-blocks. We follow the commonly used framework of [14], in which the DP-table is partitioned into match and mismatch-blocks and only the bottom row and the leftmost column of each block are computed. As a result, we cannot adopt previously proposed LCSk algorithms, such as [9], since they require computing internal entries within blocks. We begin by introducing several definitions. As stated in Observation 4, the choice of the entry from which the value of $R\mathscr{L}(a,b)$ is propagated, plays a crucial role in determining the optimality of the solution. We refer to such preceding entry as a *source entry* of the current $R\mathscr{L}k$ computation. According to Lemmas 1 and 2, the source entry for computing $R\mathscr{L}k$ can be one of the previously computed entries: directly above, to the left, diagonally above-left of the current entry, or the bottom-rightmost entry of the block diagonally preceding the current block.

Since only partial entries are computed within each block, the position of certain source entries depends on whether the current entry lies in the block's bottom row or rightmost column. Definition 4 introduces functions that determine the indices of the potential source entries for a given entry $[a,b]$.

Definition 4. *[left,up, diag, diag* * functions]*
Let $[a,b] = [X_i(i'), Y_j(j')]$ be a computed entry in the DP-table, then,

$$left(a,b) = \begin{cases} (a, b-1) & i' = n_i \quad \text{(bottom row)} \\ (a, b-m_j)) & i' \neq n_i \quad \text{(hence } j' = m_j) \end{cases}$$

$$up(a,b) = \begin{cases} (a-1, b) & j' = m_j \quad \text{(rightmost column)} \\ (a-n_i, b) & j' \neq m_j \quad \text{(hence } i' = n_i) \end{cases}$$

$$diag(a,b) = (a - min\{i', j'\}, b - min\{i', j'\})$$
$$diag^*(a,b) = (a - i', b - j')$$

free value. According to Observation 4, ensuring the optimality of the $R\mathscr{L}k$ computation requires tracking matching substrings that were excluded from the current entry's $R\mathscr{L}k$ due to the k-LS constraint. To address this, we introduce the notion of a *free* value, which is associated with each computed entry in the DP-table.

Definition 5. *[free]* *Let $[a,b]$ be a computed entry in the DP-table, then the value $f(a,b)$ is the length of the longest common substring that serves as a suffix of $X[1..a]$ and $Y[1..b]$, and that has not yet been included in $R\mathscr{L}k[a-q, b-q']$, for some $q \in [a]$ and $q' \in [b]$, hence, $0 \le f(a,b) < k$. For computed mismatch-block entries, the free value is set to zero.*

For example, consider Fig. 3 with $k = 3$. We observe that $R\mathcal{L}3[4,5] = 1$, since the longest common subsequences between $X[1..4]$ and $Y[1..5]$ is A^4, which contains a single 3-match. The remaining matching symbol A, which is not included in the current $R\mathcal{L}3$ due to the k-LS constraint, contributes to the *free* value, thus, $f(4,5) = 1$.

The assignment of the *free* value to a table entry depends on the number of matched symbols within the current match-block, in case a k-match is formed in that block. Otherwise, the *free* value additionally depends on the number of matched symbols carried over from the preceding diagonal match-block that were not included in any k-match contributing to the current $\mathscr{RL}k$.

To track the number of such symbols, we introduce two additional offset variables: $hoff(a,b)$ and $voff(a,b)$, stored for each computed entry in the DP-table. They represent the horizontal and vertical distances, respectively, between the indices of the current entry $[a,b]$ and those of the entry that concluded the most recent k-match included in $R\mathcal{L}k[a,b]$ calculation. If there is a k-match in the current block ending at entry $[g,h]$, the offset values are set as follows: $hoff(a,b) = b - h$ and $voff(a,b) = a - g$. Otherwise, the offset values are determined based on the offsets values of the source entry of $R\mathcal{L}k[a,b]$, with the addition of the distance between the current entry and the source entry.

Figure 3 shows the DP table with *(hoff, voff)* values annotated for all computed entries. The formal computation of offsets, along with the proof of the *free* value computation based on the offsets, presented in Corollary 1, is conceptually intuitive but involves numerous details. For this reason, these proofs are deferred to the complete version of the paper, due to space constraints.

Having the offset values, we can compute the *free* value of an entry, as described in Corollary 1.

Corollary 1. *Let $[a,b] = [X_i(i'), Y_j(j')]$ be a computed entry in the DP-table that corresponds to a match-block, with associated offset values $hoff(a,b)$ and $voff(a,b)$ and let $w = \min\{i', voff(a,b), j', hoff(a,b)\}$ then, $f(a,b)$ can be computed as follows:*

$$f(a,b) = \begin{cases} (w + f(a - \min\{i', j'\}, b - \min(\{i', j'\})) \bmod k & \text{if } voff(a,b) \geq i' \,\& \, hoff(a,b) \geq j' \\ w \bmod k & else \end{cases}$$

Maximizing Options. In the context of maximizing $R\mathscr{L}k$ from source entries, recall that two DP-table entries with the same $R\mathscr{L}k$ may not necessarily enable the same set of future matches, as discussed in the proof of Observation 4. To address this, we define a specialized relation max^k, where the primary criterion is a higher $\mathscr{L}k$. As a secondary criterion, we consider the potential to enable future k-matches by utilizing matching symbols within the block that were not previously consumed, captured by a higher *free* value. If both entries share the same $\mathscr{L}k$ and *free* value, preference is given to the entry with the larger offset in both strings. If all compared values are equal, ties can be broken arbitrarily.

Definition 6. *Let $[a,b]$ and $[a',b']$ be two computed entries in the DP-table with $R\mathscr{L}k$, hoff, voff and free values. We write $R\mathscr{L}k[a,b] >^k R\mathscr{L}k[a',b']$ if one of the following conditions holds:*

1. $R\mathscr{L}k[a,b] > R\mathscr{L}k[a',b']$
2. $R\mathscr{L}k[a,b] = R\mathscr{L}k[a',b']$ and $f(a,b) > f(a',b')$
3. $R\mathscr{L}k[a,b] = R\mathscr{L}k[a',b']$ and $f(a,b) = f(a',b')$ and $\min\{hoff(a,b), voff(a,b)\} > \min\{hoff(a',b'), voff(a',b')\}$

The $>^k$ relation extends the standard 'greater than' ($>$) comparison to a 3-tuple $(R\mathscr{L}k, free, \min\{hoff, voff\})$. Thus, $>^k$ is transitive, allowing us to define max^k as a maximization function that operates according to this relation.

4.1   The DP Algorithm

Lemma 4 preents the dynamic programming formulation of the suggested algorithm, applied to entries along the bottom row and rightmost column of each block. The entries are processed in increasing order of indices, following the initialization of the zero row and zero column with values of 0.

Lemma 4. *Let* $[a, b] = [X_i(i'), Y_j(j')]$ *be a computed entry in the DP-table. Then,* $R\mathscr{L}k[a, b]$ *can be computed using the following rule:*

$$R\mathscr{L}k[a, b] = max^k \begin{cases} R\mathscr{L}k[left(a, b)] \\ R\mathscr{L}k[up(a, b)] \\ R\mathscr{L}k[diag(a, b)] + ID(X_i, Y_j) \cdot \lfloor \min\{i', j'\}/k \rfloor \\ R\mathscr{L}k[diag^*(a, b)] + ID(X_i, Y_j) \cdot s \\ \qquad where\ s = \lfloor (\min\{i', j'\} + f(diag^*(a, b)))/k \rfloor \end{cases}$$

where $ID(X_i, Y_j)$ *returns 1 if* $\sigma(X_i) = \sigma(Y_j)$, *otherwise returns 0.*

Proof. The source entries are selected according to Lemmas 1, 2, and 3. The k-LS constraint imposes that the length of the added common substring, in the case of a match-block, must be divided by k. For a diagonal source, this length is $\min\{i', j'\}$. In the case of a $diag^*$ source, the $\min\{i', j'\}$ matching symbols extend diagonally aligned *free* values, as described in the proof of Corollary 1.   □

Algorithm 1 details the computation of a match-block entry according to Lemma 4. Figure 3 illustrates an $R\mathscr{L}k$ DP-table filled according to Lemma 4.

Algorithm 1: $R\mathscr{L}k(a, b)$ Computation for Match-Blocks

1: Let $option1 \leftarrow left$, $option2 \leftarrow up$, $option3 \leftarrow diag$ and $option4 \leftarrow diag^*$
2:
3: **For** $z = 1$ to 4 **do:**
4:     $R\mathscr{L}k^z[a, b] \leftarrow R\mathscr{L}k$ according to source entry $[optionz(a, b)]$
5:     $(hoff^z(a, b), voff^z(a, b))$ are assigned
6:     $f^z(a, b)$ is assigned according to Corollary 1
7: $z \leftarrow argmax^k_{z \in [4]}\{R\mathscr{L}k(a, b)\}$
8: $RLk[a, b] \leftarrow RLk^z[a, b]$
9: $(hoff(a, b), voff(a, b)) \leftarrow (hoff^z(a, b), voff^z(a, b))$
10: $f(a, b) \leftarrow f^z(a, b)$
11: **if** $\sigma(X_i) \neq \sigma(Y_j)$ **then** $f(a, b) \leftarrow 0$

Theorem 1. *The RLCSk problem can be solved in $O(nM + mN)$ time and $O(\min\{m + Mn_{\max}, n + Nm_{\max}\})$ space, where N and M denote the number of blocks in X and Y, respectively, $n_{\max}$ and $m_{\max}$ represent the maximum run lengths in X and Y, and n and m are the lengths of the corresponding uncompressed strings.*

			Y_1	Y_2				Y_3	Y_4		Y_5			Y_6				Y_7		Y_8
		Y	1	2	3	4	5	6	7	8	9	10	11	12	13	14	15	16	17	18
	X		T	A	A	A	A	C	T	T	A	A	A	T	T	T	T	A	A	C
X_1	1	A	0, f=0 (1,1)				0, f=1 (1,5)	0, f=0 (1,6)		0, f=0 (1,8)			0, f=1 (1,11)				0, f=0 (1,11)		0, f=0 (1,17)	0, f=0 (1,18)
	2	A	0, f=0 (2,1)				0, f=2 (2,5)	0, f=0 (2,6)		0, f=0 (2,8)			0, f=2 (2,11)				0, f=0 (2,11)		0, f=0 (2,17)	0, f=0 (2,18)
	3	A	0, f=0 (3,1)			*	1, f=0 (0,1)	1, f=0 (0,2)		1, f=0 (0,4)			*1, f=0 (0,7)				1, f=0 (0,11)		1, f=0 (0,13)	1, f=0 (0,14)
	4	A	0, f=0 (4,1)	0, f=1 (4,2)	0, f=2 (0,0)	1, f=0 (1,0)	1, f=1 (1,1)	1, f=0 (1,2)	1, f=0 (1,3)	1, f=0 (1,4)	1, f=1 (1,5)	1, f=1 (1,6)	1, f=1 (1,7)	1, f=0 (1,8)	1, f=0 (1,9)	1, f=0 (1,10)	1, f=0 (1,11)	1, f=1 (1,12)	1, f=1 (1,13)	1, f=0 (1,14)
X_2	5	C	0, f=0 (5,1)	0, f=0 (5,2)	0, f=0 (5,3)	1, f=0 (2,0)	1, f=0 (2,1)	1, f=2 (2,2)	1, f=0 (2,3)	1, f=0 (2,4)	1, f=0 (2,5)	1, f=0 (2,6)	1, f=0 (2,7)	1, f=0 (2,8)	1, f=0 (2,9)	1, f=0 (2,10)	1, f=0 (2,11)	1, f=0 (2,12)	1, f=0 (2,13)	*1, f=2 (2,13)
X_3	6	T	0, f=1 (6,1)				1, f=0 (3,1)	1, f=0 (3,2)	*	2, f=1 (0,1)			2 f=0 (0,4)				2 f=0 (0,8)		2, f=0 (0,10)	2, f=0 (0,11)
	7	T	0, f=1 (7,1)				1, f=0 (4,1)	1, f=0 (4,2)		2, f=1 (1,1)			2, f=0 (1,4)				2, f=1 (1,8)		2, f=0 (1,10)	2, f=0 (1,11)
	8	T	0, f=1 (8,1)	0, f=0 (8,2)	0, f=0 (8,3)	1, f=0 (5,0)	1, f=0 (5,1)	1, f=0 (5,2)	2, f=0 (2,0)	2, f=1 (2,1)	2, f=0 (2,2)	2, f=0 (2,3)	2, f=0 (2,4)	2, f=1 (2,5)	2, f=2 (2,6)	*2, f=2 (2,7)	2, f=2 (2,8)	2, f=0 (2,9)	2, f=0 (2,10)	2, f=0 (2,11)
X_4	9	A	0, f=0 (9,1)				1, f=1 (6,1)	1, f=0 (6,2)		2, f=0 (3,1)			2, f=1 (3,4)				2, f=1 (3,9)	*	3, f=0 (0,1)	3, f=0 (0,2)
	10	A	0, f=0 (10,1)	0, f=1 (10,2)	*1, f=0 (0,0)	1, f=0 (0,1)	1, f=1 (7,1)	1, f=0 (7,2)	2, f=0 (4,0)	2, f=0 (4,1)	2, f=1 (4,2)	*3, f=0 (0,0)	3, f=0 (0,1)	3, f=0 (0,2)	3, f=0 (0,3)	3, f=0 (0,4)	3, f=0 (0,5)	3, f=0 (0,6)	3, f=1 (1,1)	3, f=0 (1,3)

Fig. 3. An example of a DP-table for $R\mathscr{L}k$ where $k = 3$. Each entry displays the computed $R\mathscr{L}k$, the *free* value f, and the offsets *hoff* and *voff*, shown in brackets. Asterisks (*) indicate the end of a 3-match. The colors represent the length of $R\mathscr{L}k$.

Proof. To identify, for every entry $[a, b]$, its location-based representation $(X_i(i'), Y_j(j'))$, we maintain the following variables: $numrun_X$ and $numrun_Y$, which indicate the current run number in strings X and Y, respectively, and $indexrun_X$ and $indexrun_Y$, which refer to the index of the current entry within the current run. Since the DP algorithm processes entries in increasing order of indices, these variables can be updated in constant time by consulting $len(numrun_X)$ and $len(numrun_Y)$.

The dynamic programming algorithm based on Lemma 4 computes $R\mathscr{L}k$ only for entries located in the bottom row or rightmost column of each block. Accordingly, for each X_i, the bottom row of all (X_i, Y_j) blocks, where $j \in [M]$, is computed, resulting in $O(m)$ entries per row and a total of $O(mN)$ entries. A similar argument holds for the rightmost columns, giving an additional $O(nM)$ entries. Each entry is computed in $O(1)$ time, involving the maximization over four values and constant-time updates to the offsets and the *free* value. Therefore, the overall time complexity of the algorithm is $O(nM + mN)$.

Regarding space, the DP table only references entries from blocks to the left, above, or diagonally above-left of the current block—so only two rows of blocks are needed for the computation. For each row of blocks, all bottom rows are stored, resulting in $O(m)$ entries. Additionally, the rightmost columns of all blocks in the row are computed, summing up to (Mn_{max}) entries. An alternative approach is filling the DP-table column by column. This method requires storing only two columns of blocks at a time, resulting in a space complexity of $O(n + Nm_{max})$ based on similar arguments. □

Note, that while some RLCS solutions compute only the bottom-rightmost entry in mismatch-blocks, this simplification is not valid for $R\mathscr{L}k$, as a k-match may end at an internal position within the block.

5   RLCS*k* Second Solution Via Range Maximum Queries

Iliopoulos and Rahman [22] proposed solving the LCS problem by identifying all *matches*, i.e., positions where $X[a] = Y[b]$, and computing $\mathscr{L}[a, b]$ only for those entries in the DP-table that correspond to matches. They used the following rule to compute the LCS: $\mathscr{L}[a, b] = 1 + \max_{a', b'}\{\mathscr{L}[a', b'] \mid X[a'] = Y[b']$ and $a' < a$ and $b' < b\}$. The maximization is performed efficiently using a range maximum query (RMQ) data structure constructed over the matches.

Adapting this framework to the $R\mathscr{L}k$ problem requires fundamental modifications. The first involves considering k-matches instead of exact matches. It is essential to extract these k-matches, referred to as *candidates*, directly from the RLE-compressed input strings, without decompressing them.

Observation 5. *If a k-match starting at* $[a, b]$ *is part of the optimal solution, and* $[a + k, b + k]$ *marks the start of another k-match, then the sequence of 2 consecutive k-matches starting at* $[a, b]$ *can also be included in the optimal solution.*

Observation 5 implies that, under the k-LS constraint, candidates should be of maximal length, as defined in Definition 7.

Definition 7. *[Candidates and Candidate Set]*
Let $[a, b] = [X_i(i'), Y_j(j')]$ *be a DP-table entry of a match-block. A candidate* $c_{[a,b]}$ *represents a maximal sequence of consecutive k-matches that **begins** at entry* $[a, b]$. *For each candidate, we store the number of k-matches it represents in a field called len. There are two types of candidates:*
(1) block candidate - representing k-matches within a single match-block.
(2) chain candidate - representing k-matches that spans multiple match-blocks, origi-nating from (X_i, Y_j) *and extending to* (X_{i+1}, Y_{j+1}). *If the obtained match is shorter than k, the sequence can be extended across additional match-blocks—each diagonally aligned—until at least one complete k-match is formed.*
The candidate set C_a *is the set of all candidates starting in row a:* $C_a = \{(c_{[a,b]}) \mid b \in [m]$ *where* $c_{[a,b]}$ *is a candidate*$\}$. C_a *is stored in increasing order of column indices* (b).

For example, consider the strings shown in Fig. 3 with $k = 3$. $C_1 = \{c_{[1,2]}, c_{[1,3]}, c_{[1,9]}\}$, where all candidates are block candidates, where $C_8 = \{c_{[8,1]}, c_{[8,8]}, c_{[8,15]}\}$ con-sists entirely of chain candidates, each is a chain of 2 match-blocks. All these candidates have $len = 1$ as each consists of a single 3-match.

Lemma 5. *Given a Match-block* (X_i, Y_j), $O(k)$ *chain candidates are required for the computation of an optimal RLCSk solution.*

Proof. Due to the k-LS constraint, k-matches may terminate within the interior of a match block, including chain matches that end in the current block. Therefore, to ensure all possible opportunities for subsequent k-matches are considered, it is necessary to examine not only the maximal chain candidate, but also its suffixes. However, only additional $k - 1$ potential starting positions need to be considered. Each corresponds to a k-match beginning l positions diagonally before the bottom-right corner of the block (X_i, Y_j), for $l \in [\min\{n_i, m_j, k - 1\}]$. If a chain candidate starts with k matching symbols within the block, these initial symbols are already accounted for by another candidate: either by a block candidate, when the preceding k-match source lies to the left or above, or by another chain candidate whose k-matches terminate within the current block (X_i, Y_j), due to the maximality of candidates length.   □

Lemma 6. *The time required to extract block and chain candidates starting in a match-block (X_i, Y_j) is $O(1)$ per a block-candidates and $O(k)$ for the chain-candidates.*

Proof. For each match-block (X_i, Y_j), and for each entry $[a, b] = [X_i(i'), Y_j(j')]$, let $match = \min\{n_i - i', m_j - j'\} + 1$. If $match \geq k$, then a candidate $c_{[a,b]}$ is inserted into C_a with length $len = \lfloor match/k \rfloor$.

Regarding chain candidates, if the block (X_{i+1}, Y_{j+1}) is also a match-block, then $match = \min\{n_i, m_j\} + \min\{n_{i+1}, m_{j+1}\}$. A chain candidate is formed if $match \geq k$, with length $len = \lfloor match/k \rfloor$. If $match < k$, the chain can be extended with an additional block (X_{i+2}, Y_{j+2}) provided that: (1) the block (X_{i+2}, Y_{j+2}) is a match-block, and (2) the previous block is square, i.e., $n_{i+1} = m_{j+1}$. In this case, the match length is updated as: $match += \min\{n_{i+2}, m_{j+2}\}$.

This process continues iteratively until one of the following conditions is met:

1. $match \geq k$,
2. The next diagonally aligned block is a mismatch-block, or
3. A rectangular (non-square) match-block is encountered.

Thus, extracting the maximal chain candidate requires $O(k)$ time.

The additional $k - 1$ chain candidates can be efficiently obtained by updating the match length as $match = match - \min\{n_i, m_j\} + \min\{n_i, m_j, k - 1\}$, and then decrementing the match length by one for each subsequent chain candidate, as we consider candidates that start progressively closer to the bottom-right corner of block (X_i, Y_j). If the updated match length becomes smaller than k, the chain extension proceeds according to the conditions mentioned above. However, across all $k - 1$ additional chain candidates, at most $O(k)$ extension operations are required. This is because requiring k extensions would imply that the block sizes are $O(1)$. Therefore, the total time needed to extract all chain candidates beginning within a single match-block is $O(k)$. $\square$

BoundedHeap Data Structure. The RMQ data structure used by [21] is BoundedHeap proposed by [13]. The *BoundedHeap H*, supports the following queries, each in $O(\log \log u)$ time, where the positions are drawn from $[u]$. It requires $O(u)$ space.

(a) *Insert(H,Pos,Value,Data)*: Insert into the H the position *Pos* with value *Value* and associated information *Data*.
(b) *IncreaseValue(H,Pos,Value,Data)*: If H does not already contain the position *Pos*, perform *Insert(H, Pos, Value, Data)*. Otherwise, set this position value to $\max\{Value, previous\ Value\ in\ the\ position\}$. Also, update Data accordingly.
(c) *BoundedMax(H,Pos)*: Return the item in H with the maximum value among those whose position is less than *Pos*. If no such item exists, return 0.

For LCS computation, [22] maintained only two *BoundedHeaps*, H_{a-1} and H_a, at any given time. The heap H_a is initialized by setting $H_a \leftarrow H_{a-1}$. Then, every match from C_a is inserted into H_a, implying that by the end of iteration a, H_a stores the best $\mathcal{L}$ value per column index of a match up to the ath row.

However, a candidate $c_{[a,b]}$ with $c_{[a,b]}.len = q$ represents a k-match that extends up to row $a + q - 1$. Therefore, its $R\mathcal{L}k$ must be stored in heap H_{a+q-1} by applying: $Insert(H_{a+q-1}, b+q-1, R\mathcal{L}k[a+q-1, b+q-1], q)$. Yet, H_{a+q-1} remains uninitialized at this point, since heap initialization is deferred until the preceding heap has been fully constructed. To address this, we maintain linked lists $wait_{a'}$ for all $a < a' \leq n$, which store the ending locations of such k-matches along with their associated $R\mathcal{L}k$.

At the beginning of the iteration for row a', we first traverse the list $wait_{a'}$, applying an *Insert* operation for each of its elements into the heap $H_{a'}$. Only after all deferred insertions have been processed do we proceed to iterate over the candidate set $C_{a'}$ and perform the corresponding *BoundedMax* queries.

Reduced Number of Candidates.

Exploiting the structural properties of match-blocks we reduce the number of relevant candidates, thereby improving the time complexity of the $R\mathscr{L}k$ computation.

Lemma 7. *Let (X_i, Y_j) be a match-block. Then, the number of block candidates start-ing within this block that are required for computing $R\mathscr{L}k$ is $O(n_i - k + m_j - k)$.*

Proof. Consider a match-block (X_i, Y_j). The longest possible match within this block begins at its top-left corner. To handle cases where a previous k-match has already consumed part of the Y_jth run, ending in the middle of the block from above, we must treat all entries in the first row as candidates, as they correspond to suffixes of the Y_jth run. A similar argument applies when a previous k-match consumes symbols from the X_ith run, in such cases, all entries in the first column must also be considered as candidates, where applicable.

Let $[a, b] = [X_i(1), Y_j(1)]$. An inner entry of the block, such as $[a + z', b + z'']$, can initiate a k-match of equal or shorter length compared to those starting at $[a, b + z'']$ or $[a + z', b]$, since its distance to the bottom-right corner of the block is strictly smaller. Consequently, any such k-match can be replaced by one starting at either $[a, b + z'']$ or $[a + z', b]$, and thus, inner entries are not required for optimal computation.

Additionally, no k-match can start within the last $k-1$ rows or columns of the block. Consequently, the total number of required block candidates is $n_i - k + 1 + m_j - k + 1$. An exception arises for match-blocks that have no match-block directly above them. In such cases, only the first-column candidates are relevant. Similarly, for match-blocks with no match-block to their left, only the first-row candidates are considered. □

As mentioned, additional modification of the LCS solution, is that in the RLCSk problem the score obtained due to a candidate (saving the beginning of a k-match), is saved in the entry where the k-match ends. hence, for a candidate c, we seek the max-imum $R\mathscr{L}k$ value obtained from a k-match that ends in a row and column preceding those of c, thereby avoiding overlaps between the k-matches. Lemma 8 presents the adaptation of the recursive rule from [22] to the RLCSk problem.

Lemma 8. *Let $[a, b] = [X_i(i'), Y_j(j')]$ be a candidate entry with $c_{[a,b]}.len = q$ then, $R\mathscr{L}k[a + q - 1, b + q - 1] = q + \max_{e,f}\{R\mathscr{L}k[e + q' - 1, f + q' - 1] | where\ c_{[e,f]} \in C_e, c_{[e,f]}.len = q', e + q' \leq a\ and\ f + q' \leq b\}.$*

Proof. The first candidate to be considered represents the initial k-match, therefore, its length serves as the optimal score. Using induction over the candidate indices, let $[a, b] = [X_i(i'), Y_j(j')]$ be a candidate entry with $c_{[a,b]}.len = q$. Suppose, for the sake of contradiction, that $R\mathscr{L}k[a, b]$ is not optimal. According to Lemma 7, no additional candidate's k-match can end at the entry $[a + q - 1, b + q - 1]$. Since the maximum $R\mathscr{L}k$ of entries preceding $[a, b]$ is added to q, this implies that not using the k-match beginning at $[a, b]$ would yield a better score. Therefore, it must be better to use a candidate $c_{[e,f]}$ representing a k-match starting at entry $[e, f]$, where $c_{[e,f]}.len = q'$ and $e \leq a$ or $f \leq b$, but such that $e + q' - 1 > a$ and $f + q' > b$. Otherwise, the k-match starting at $[a, b]$ could extend the one starting at $[e, f]$, contradicting the assumption of optimality.

As a consequence, the last k-match of $c_{[e,f]}$ and $c_{[a,b]}$ must overlap. Therefore, the earlier match uses fewer than k distinct symbols not already present in the latter, so, if the former yields a higher $R\mathscr{L}k[a + q - 1, b + q - 1]$ t would contradict the maximality of the previous $R\mathscr{L}k$ computations.

This leads to a contradiction of the initial assumption.   □

Lemmas 5, 7 and 8 yield the following Theorem.

Theorem 2. *The RLCSk problem can be solved in $O(N + M + \min\{n + |C| \log \log m, \ m + |C| \log \log n\})$ time and $O(|C| + \min\{n, m\})$ space where $|C| \leq \sum_{MB_{(X_i, Y_j)}} (n_i - k + m_j)$ and n_i, m_j are the run-lengths of the match-block (MB) (X_i, Y_j).*

Proof. Match-blocks can be identified in $O(N + M)$ time, by jointly sorting the runs of the compressed input strings. For each candidate set C_a, processed in increasing order of a, we iterate over its candidates in increasing order of their column indices. For each candidate $c_{[a,b]}$ encountered, the *BoundedMax* and *Insert* operations are performed, with the corresponding $R\mathscr{L}k$ value stored in a linked list between these operations, in constant time. Each *BoundedHeap* operation takes $O(\log \log m)$ time, where positions are drawn from the indices of Y. Additionally, constant-time maintenance is required for each C_a. The same arguments apply when maintaining candidate sets per column instead of per row, resulting in a time complexity of $O(m + \log \log n)$.

The *BoundedHeap* requires $O(m)$ space for row-oriented candidate sets, or $O(n)$ for column-oriented ones. Additionally, up to $O(|C|)$ items may be stored in the *wait* lists.   □

Note that in the worst-case scenario, such as when $|\Sigma| = 2$, up to half of the blocks may be match-blocks. If the block dimensions exceed k, the number of block candidates can reach $O(Mn + Nm)$. In such cases, the dynamic programming solution described in Sect. 4 is more efficient. However, when the number of match-blocks is relatively small or when the block sizes are smaller than k, the number of candidates is significantly less than $O(Mn + Nm)$, and in the latter case, it is even bounded by bk, where b is the number of match-blocks. In these cases, the algorithm described here is more efficient than the one previously proposed.

In any case, the number of candidates extracted is substantially smaller than the total number of k-matches, which are explicitly enumerated in LCSk algorithms such as [17].

6   Conclusion and Open Problems

In this paper, we addressed a novel problem: finding the LCSk for RLE strings. We have identified fundamental structural properties of the LCSk problem, which facilitate the development of algorithms suited to different compression schemes. We proposed two distinct algorithms to solve the RLCSk problem. The first algorithm is efficient when the input strings contain many matching blocks and the run lengths exceed k. The second algorithm performs better in other scenarios.

Future work includes establishing a lower bound for the RLCSk problem, developing more efficient algorithms suited to different compression schemes and exploring new LCS variants.

Disclosure of Interests. The authors have no competing interests to declare that are relevant to the content of this article.

References

1. Ahsan, S.B., Moosa, T.M., Rahman, M.S., Shahriyar, S.: Computing a longest common subsequence of two strings when one of them is run length encoded. INFOCOMP J. Comput. Sci. **10**(3), 48–55 (2011)
2. Amir, A., Gotthilf, Z., Shalom, B.R.: Weighted LCS. J. Discrete Algorithms **8**(3), 273–281 (2010)
3. Amir, A., Hartman, T., Kapah, O., Shalom, B.R., Tsur, D.: Generalized LCS. Theoret. Comput. Sci. **409**(3), 438–449 (2008)
4. Ann, H.Y., Yang, C.B., Tseng, C.T., Hor, C.Y.: A fast and simple algorithm for computing the longest common subsequence of run-length encoded strings. Inf. Process. Lett. **108**(6), 360–364 (2008)
5. Ann, H.Y., Yang, C.B., Tseng, C.T., Hor, C.Y.: Fast algorithms for computing the constrained LCS of run-length encoded strings. Theoret. Comput. Sci. **432**, 1–9 (2012)
6. Apostolico, A., Landau, G.M., Skiena, S.: Matching for run-length encoded strings. J. Complex. **15**(1), 4–16 (1999). https://doi.org/10.1006/JCOM.1998.0493
7. Bannai, H., I, T., Kociumaka, T., Köppl, D., Puglisi, S.J.: Computing longest Lyndon subsequences and longest common Lyndon subsequences. Algorithmica **86**(3), 735–756 (2024)
8. Becerra, D., Soto, W., Nino, L., Pinzón, Y.: An algorithm for constrained LCS. In: ACS/IEEE International Conference on Computer Systems and Applications-AICCSA 2010, pp. 1–7. IEEE (2010)
9. Benson, G., Levy, A., Maimoni, S., Noifeld, D., Shalom, B.R.: LCSk: a refined similarity measure. Theor. Comput. Sci. **638**, 11–26 (2016). https://doi.org/10.1016/J.TCS.2015.11.026
10. Benson, G., Levy, A., Maimoni, S., Noifeld, D., Shalom, B.R.: LCSk: a refined similarity measure. Theor. Comput. Sci. **638**, 11–26 (2016)
11. Benson, G., Levy, A., Shalom, B.R.: Longest common subsequence in k length substrings. In: Brisaboa, N., Pedreira, O., Zezula, P. (eds.) SISAP 2013. LNCS, vol. 8199, pp. 257–265. Springer, Heidelberg (2013). https://doi.org/10.1007/978-3-642-41062-8_26
12. Bergroth, L., Hakonen, H., Raita, T.: A survey of longest common subsequence algorithms. In: de la Fuente, P. (ed.) Seventh International Symposium on String Processing and Information Retrieval, SPIRE 2000, A Coruña, Spain, September 27-29, 2000, pp. 39–48. IEEE Computer Society (2000). https://doi.org/10.1109/SPIRE.2000.878178
13. Brodal, G.S., Kaligosi, K., Katriel, I., Kutz, M.: Faster algorithms for computing longest common increasing subsequences. In: Annual Symposium on Combinatorial Pattern Matching, pp. 330–341. Springer (2006)
14. Bunke, H., Csirik, J.: An improved algorithm for computing the edit distance of run-length coded strings. Inf. Process. Lett. **54**(2), 93–96 (1995)
15. Cygan, M., Kubica, M., Radoszewski, J., Rytter, W., Waleń, T.: Polynomial-time approximation algorithms for weighted LCS problem. Discret. Appl. Math. **204**, 38–48 (2016)

16. Das, D., Saha, B.: Approximating LCS and alignment distance over multiple sequences. arXiv preprint arXiv:2110.12402 (2021)
17. Deorowicz, S., Grabowski, S.: Efficient algorithms for the longest common subsequence in k-length substrings. Inf. Process. Lett. **114**(11), 634–638 (2014)
18. Gotthilf, Z., Hermelin, D., Lewenstein, M.: Constrained LCS: hardness and approximation. In: Ferragina, P., Landau, G.M. (eds.) CPM 2008. LNCS, vol. 5029, pp. 255–262. Springer, Heidelberg (2008). https://doi.org/10.1007/978-3-540-69068-9_24
19. Huang, G.F., Yang, C.B., Tseng, K.T., Huang, K.S.: Diagonal algorithms for the longest common subsequence problems with t-length and at least t-length substrings. In: Proceedings of the 37th Workshop on Combinatorial Mathematics and Computation Theory, pp. 119–127
20. Huang, G.F., Yang, C.B., Tseng, K.T., Huang, K.S.: Diagonal algorithms for the longest common subsequence problems with t-length and at least t-length substrings. In: Proceedings of the 37th Workshop on Combinatorial Mathematics and Computation Theory, pp. 119–127 (2020)
21. Iliopoulos, C.S., Rahman, M.S.: Algorithms for computing variants of the longest common subsequence problem. Theoret. Comput. Sci. **395**(2–3), 255–267 (2008)
22. Iliopoulos, C.S., Rahman, M.S.: New efficient algorithms for the LCS and constrained LCS problems. Inf. Process. Lett. **106**(1), 13–18 (2008)
23. Kociumaka, T., Radoszewski, J., Starikovskaya, T.: Longest common substring with approximately k mismatches. Algorithmica **81**(6), 2633–2652 (2019)
24. Mozes, S., Tsur, D., Weimann, O., Ziv-Ukelson, M.: Fast algorithms for computing tree LCS. Theoret. Comput. Sci. **410**(43), 4303–4314 (2009). https://doi.org/10.1016/j.tcs.2009.07.011
25. Nath, D., Kurmi, J., Rawat, V.: A survey on longest common subsequence. Int. J. Res. Appl. Sci. Eng. Technol. (IJRASET) **6**, 4552–4556 (2018)
26. Pavetic, F., Katanic, I., Matula, G., Zuzic, G., Sikic, M.: Fast and simple algorithms for computing both LCS_k and LCS_{k+}. In: Holub, J., Zdárek, J. (eds.) Prague Stringology Conference 2018, Prague, Czech Republic, 2018, pp. 50–62. Czech Technical University in Prague, Faculty of Information Technology, Department of Theoretical Computer Science (2018)
27. Pavetic, F., Katanic, I., Matula, G., Zuzic, G., Sikic, M.: Fast and simple algorithms for computing both lcs_k and lcs_{k+}. In: Holub, J., Zdárek, J. (eds.) Prague Stringology Conference 2018, Prague, Czech Republic, 2018, pp. 50–62. Czech Technical University in Prague, Faculty of Information Technology, Department of Theoretical Computer Science (2018). http://www.stringology.org/event/2018/p06.html
28. Pavetić, F., Žužić, G., Šikić, M.: *lcsk++*: Practical similarity metric for long strings. arXiv preprint arXiv:1407.2407 (2014)
29. Priyanka, Goel, S.: A compression algorithm for DNA that uses ascii values. In: 2014 IEEE International Advance Computing Conference (IACC), pp. 739–743 (2014). https://doi.org/10.1109/IAdCC.2014.6779416
30. Prodanov, T., Bansal, V.: Sensitive alignment using paralogous sequence variants improves long-read mapping and variant calling in segmental duplications. Nucleic Acids Res. **48**(19), e114–e114 (2020)
31. Prodanov, T., Plender, E.G., Seebohm, G., Meuth, S.G., Eichler, E.E., Marschall, T.: Locityper: targeted genotyping of complex polymorphic genes. bioRxiv, pp. 2024–05 (2025)
32. Rubinstein, A., Seddighin, S., Song, Z., Sun, X.: Approximation algorithms for LCS and LIS with truly improved running times. SIAM J. Comput. (0), FOCS19–276 (2023)

33. Soto, D., Soto, W.: The constrained longest common subsequence: theory and experiments. In: 2019 IEEE 39th Central America and Panama Convention (CON-CAPAN XXXIX), pp. 1–6. IEEE (2019)
34. Tsujimoto, T., Shibata, H., Mieno, T., Nakashima, Y., Inenaga, S.: Computing longest common subsequence under cartesian-tree matching model. In: International Workshop on Combinatorial Algorithms, pp. 369–381. Springer (2024)
35. Weerakoon, M., Saunders, C.T., Heaton, H.: LCSKPOA: enabling banded semi-global partial order alignments via efficient and accurate backbone generation through extended LCSK++. bioRxiv, pp. 2024–07 (2024)
36. Zhu, D., Wang, L., Wang, T., Wang, X.: A space efficient algorithm for the longest common subsequence in k-length substrings. Theoret. Comput. Sci. **687**, 79–92 (2017)

Practical Algorithms for Hierarchical Overlap Graphs

Saumya Talera, Parth Bansal, Shabnam Khan, and Shahbaz Khan(✉)⬤

Department of Computer Science and Engineering, Indian Institute of Technology
Roorkee, Roorkee, India
{saumya_t,parth_b,shabnam_k,shabhaz.khan}@cs.iitr.ac.in

Abstract. One of the most prominent problems studied in bioinformatics is genome assembly, where, given a set of overlapping substrings of a source string, the aim is to compute the source string. Most classical approaches to genome assembly use assembly graphs built using this set of substrings to compute the source string efficiently. Prominent such graphs present a tradeoff between scalability and avoiding information loss. The space efficient (hence scalable) de Bruijn graphs come at the price of losing crucial overlap information. On the other hand, complete overlap information is maintained by overlap graphs at the expense of quadratic space. Hierarchical overlap graphs (HOG) were introduced by Cazaux and Rivals [IPL20] to overcome these limitations, i.e., avoiding information loss despite using linear space. However, their algorithm required superlinear space and time. After a series of suboptimal improvements, two optimal algorithms were simultaneously presented by Khan [CPM2021] and Park et al. [CPM2021].

We empirically analyze all the algorithms for computing HOG, where the optimal algorithm [CPM2021] outperforms the previous algorithms as expected. However, it is still based on relatively complex arguments for its formal proof and uses relatively complex data structures for its implementation. We present an *intuitive*, *optimal* algorithm requiring linear space and time, which uses only *elementary arrays*. The superior performance of the optimal algorithm [CPM2021] over previous algorithms comes at the expense of extra memory. Our algorithm empirically proves to be even better for both time and memory over all the algorithms, highlighting its significance in both theory and practice.

We also explore the applications of the HOG to solve the variants of suffix-prefix queries on a set of strings, studied by Loukides et al. [CPM2023]. They presented state-of-the-art algorithms requiring complex black-box data structures, making them seemingly impractical. Our algorithms, despite failing to match their theoretical bounds, answer queries in 0.002-100 ms for datasets having around a billion characters, improving 18-1300× over KMP for complex queries. Our result also unknowingly answered an open question regarding the construction of HOG by Kikuchi and Inenaga [SPIRE2024].

Full paper available at [27].

G. Badkobeh et al. (Eds.): SPIRE 2025, LNCS 16073, pp. 265–280, 2026.
https://doi.org/10.1007/978-3-032-05228-5_21

Keywords: Hierarchical Overlap Graphs · String algorithms ·
Genome assembly · All Pairs Suffix Prefix · KMP · Empirical Study

1   Introduction

Genome assembly is one of the most fundamental problems in Bioinformatics. For a source string (*genome*), practical limitations allow sequencing of only a collection of its substrings (*reads*) instead of the whole string. Thus, genome assembly aims to reconstruct the source genome using the sequenced reads. This is possible only because the sequencing ensures coverage of the entire string so that we find sufficient overlap among the substrings. This overlap information plays a fundamental role in solving the genome assembly problem. In most practical approaches [2–4, 20, 25, 26, 31], this overlap information is efficiently processed by representing the substrings in the form of *assembly graphs*, such as *de Bruijn graphs* [24] and *overlap graphs* (or string graphs [19]). The de Bruijn graphs present a trade-off of achieving scalability (space efficiency) at the expense of a loss in overlap information. This is possible by considering all possible substrings (*k-mers*) of the reads having length k, and storing the limited overlap information amongst them, thereby ensuring linear space. On the other hand, overlap graphs store the maximum overlap between every pair of reads *explicitly*, thereby requiring quadratic size, making them impractical for large data sets.

Hierarchical Overlap Graphs (HOG) were formally introduced by Cazaux and Rivals [7] as an alternative to overcome the limitations of the existing assembly graphs, maintaining the complete overlap information *implicitly* using optimal space (linear). Structurally, it is similar to AC Trie [1], where HOG has nodes corresponding to the original substrings and the maximum overlap between every pair of these substrings only (instead of all prefixes as in AC Trie). Notably, the linear size of HOG (despite storing complete overlap information) highlights the redundancy of data stored in overlap graphs. Several applications of HOG have been formerly studied in [5,6]. For a given set P of k strings with total length n, they [7] compute the HOG in $O(n + k^2)$ time using superlinear space. Later, Park et al. [23] presented an algorithm requiring $O(n \log k)$ time and linear space. However, they assumed the character set of the strings to be of constant size. Finally, Khan [11] and Park et al. [22] independently presented optimal algorithms for computing the HOG in $O(n)$ time. While the former algorithm only used standard data structures (as lists, stacks), the latter reduced the problem to that of computing borders from the classical KMP algorithm [14].

Related Work. All the above algorithms, in theory, remove the non-essential nodes from an AC Trie (ACT) [1]. In practice, they can use an intermediate *Extended HOG* (EHOG) to reduce the memory requirement for these algorithms. Thus, the *fundamental* problem is marking those nodes of an ACT (or EHOG), which represent the set of maximum overlaps between every ordered pair of strings in P which is a special case of the All Pairs Suffix Prefix (APSP) problem (not to be confused with the classical graph problem of computing All Pairs Shortest Paths). APSP aims to calculate the maximum overlap between every pair of strings among a set of strings. Gusfield et al. [9] optimally solved this classical problem in $O(n + k^2)$ time and $O(n)$ space using generalized suffix tree [30]. Later, other optimal algorithms were presented using generalized suffix array [18] ([21,28]) and Aho Corasick automaton [1] ([16]). Moreover, in practice, several algorithms having suboptimal bounds, such as Readjoiner [8], SOF [10], and a recent algorithm by Lin and Park [15], have far superior performance for APSP, demonstrating the stark difference between theory and practice. Thus, any empirical analysis cannot trivially overlook the suboptimal algorithms. Since the APSP problem *explicitly* reports the maximum overlap for each of the k^2 ordered pairs, the above bound is optimal. However, when only the set of *maximum overlaps* is required (as in the case of HOG), the $O(k^2)$ factor proves suboptimal.

Loukides et al. [17] have studied several practically relevant variants of the APSP problem and presented state-of-the-art algorithms for the same. However, most of these algorithms involve complex black-box data structures, making them seemingly impractical. A variant of the APSP problem was solved by Ukkonen [29], which reports all pairwise overlaps (not just maximum) in the decreasing order of lengths. Notably, this problem can be solved by EHOG by reporting the internal nodes bottom-up. Recently, Kikuchi and Inenaga [12] studied a dynamic data structure for APSP queries. Notably, their static algorithm for computing APSP is similar to our algorithm (using *CompTrie* instead of *favoured* nodes) with incomparable analysis. However, our algorithm was developed independently as our preliminary version [27] predates theirs [13].

Our Results. Our results can be succinctly outlined as follows.

1. **Novel optimal algorithm for HOG.** We present arguably a very *intuitive* algorithm, which is also *optimal*. Our algorithm runs in linear time and space. Despite the simplicity of the existing optimal algorithms, they still use relatively complex arguments for their formal proof and relatively complex data structures for implementation. However, our algorithm uses only *elementary arrays* for implementation with a very *intuitive* proof.
2. **Empirical evaluation of practical HOG algorithms.** We analyzed all the seemingly practical algorithms on random and real datasets. On real datasets, the previous optimal algorithm improves the other algorithms by 1.3-8× using 1.5-6× more memory. However, the proposed optimal algorithm improves the best time by $1.3 - 1.7\times$ and memory by $\approx 1.5\times$. The relative performances improve with the size of the dataset. On random datasets having 10^5 strings with a total length of 10^7, the optimal algorithm improves the previous state of the art by over $2 - 4\times$ times, while our algorithm improves the optimal by 2.5–3×. Further, we find that the performance mainly depends on an intermediate data structure EHOG, used by all the algorithms. Our results prove that the proposed algorithm performs superior to the previous algorithms on both random and real datasets, highlighting its significance in both theory and practice.
3. **Applications of HOG.** We also explore the applications of HOG to solve various variants of the APSP problem. Despite not theoretically matching

the state of the art [17], which requires complex black-box data structures, our algorithms require 0.002-100 ms for queries on a data set of a billion characters, improving 18-1300$\times$ over classical KMP [14] for complex queries.

Outline of the Paper. We first describe the basic notations and definitions used throughout the paper in Sect. 2. In Sect. 3, we briefly describe the previous results, highlighting the aspects affecting practical performance. This is followed by describing our proposed algorithm in Sect. 4. Then, in Sect. 5, we present the experimental setup and evaluation of the practical algorithms on real and random datasets, justifying the observations using further evaluation. The details of experimental evaluation are available in the full paper. Section 6 discussed some applications of HOG on variants of APSP queries. We finally present the conclusions and scope of future work in Sect. 7.

2   Preliminaries

We are given a set $P = \{p_1, p_2, ...p_k\}$ of k non-empty strings over a finite-set alphabet. The length of a string p_i is denoted by $|p_i|$. Let the total length of all strings $p_i \in P$ be $||\mathcal{P}|| = \sum_{i=1}^{k} |p_i| = n$ ($\geq k$ as non-empty strings). An empty string is denoted by ϵ. A substring of a string p starting from the first character of p is a *prefix* of p, while a substring ending at the last character of p is known as a *suffix* of p. A prefix or suffix of p is called *proper* if it is not the same as p. For an ordered pair (p, q), an *overlap* is a string that is both a proper suffix of p and a proper prefix of q, and $ov(p, q)$ denotes the maximum such overlap. We denote $Ov(P)$ as the set of $ov(p_i, p_j)$ for all $p_i, p_j \in P$.

For a given P, the ACT $\mathcal{A}$, EHOG $\mathcal{E}$, and HOG $\mathcal{H}$ are defined using different node sets as follows. The nodes of $\mathcal{A}$ are all possible prefixes of the strings in P (hence including P and ϵ). The nodes of $\mathcal{E}$ are $P \cup \{\epsilon\}$ and all possible (not necessarily maximum) overlaps of strings in P. The nodes of $\mathcal{H}$ are further restricted to the maximum overlaps $Ov(P) \cup P \cup \{\epsilon\}$. Hence for a given P, it holds $V(\mathcal{H}) \subseteq V(\mathcal{E}) \subseteq V(\mathcal{A})$. For all the above, we add the following types of edges: (a) *Tree edges* (y, x) for all $x \in V$ with label z, where y is the longest proper prefix of x in V and $x = yz$, and (b) *Suffix links* (x, y) represented by dotted red edges, for all $x \in V$ where y is the longest proper suffix of x in V (see Fig. 1). Also, for each of these structures the number of nodes is $O(n)$.

We abuse the notation to interchangeably refer to a node as a string (corresponding to it) and vice-versa. The name *tree edges* allows us to abuse the notation and treat each of $\mathcal{A}, \mathcal{E}$, and $\mathcal{H}$ as a tree with additional suffix links. This makes the outgoing tree edge neighbors of a node its children, which extends to the notion of descendants and ancestors. This also makes the *leaves* of the tree a *subset* of P, the empty string ϵ as the root, and the remaining strings as *internal nodes*. Note that some string $p_i \in P$ can also be an internal node if it is a prefix of some other string $p_j \in P$, which is why leaves of the tree are not necessarily the same as P. Further, starting from every node $p_i \in P$ and following the suffix links until we reach the root gives us the *suffix path* of p_i. We further have the following property.

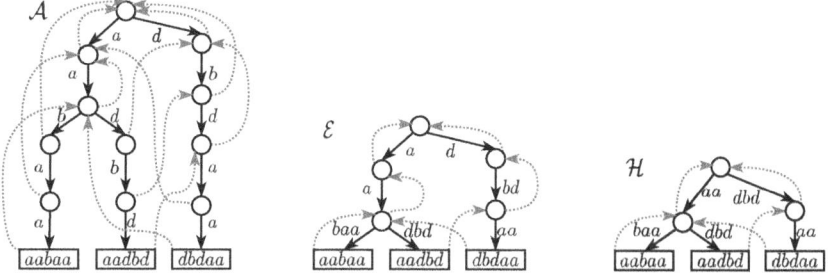

Fig. 1. The structures $\mathcal{A}$, $\mathcal{E}$ and $\mathcal{H}$ for $P = \{aabaa, aadbd, dbdaa\}$ (example from [11])

Lemma 1. (HOG Property [9]). *An internal node v in $\mathcal{A}$ of P, is $ov(p_i, p_j)$ for two strings $p_i, p_j \in P$ iff v is an overlap of (p_i, p_j) and no descendant of v is an overlap of (p_i, p_j).*

3   Previous Work

We now describe the relevant previous results related to the computation of HOG. For a given set of strings P, the ACT [1] can be computed in linear time and space. The classical algorithm builds a Trie and adds the suffix links efficiently using dynamic programming. EHOG [7] can be computed by simply traversing the suffix path of every string in P in linear time, and removing the untraversed nodes. The corresponding labels of tree edges and suffix links with its labels are also updated simultaneously.

Given an ACT (or EHOG) of P, its HOG can be constructed by removing the nodes not satisfying Lemma 1, and updating edge labels and suffix links as in EHOG. Various approaches to compute $Ov(P)$ are

1. **Cazaux and Rivals [7] requiring $O(n+k^2)$ time.** This algorithm computes a list $R_l(u)$ containing all leaves having u as its suffix by traversing over all suffix paths. A bit vector maintains for each internal node x, the strings in P having overlaps in the descendants of x. The bit vector can be updated in a bottom-up fashion, and compared with $R_l(u)$ to decide whether u is added to $\mathcal{H}$. Along with suboptimal time, it also uses super-linear space, seemingly limiting its scalability.
2. **Park et al. [23] requiring $O(n \log k)$ time.** This algorithm sorts the strings in P lexicographically, which makes a contiguous set of leaves the descendant of each internal node. This allows them to use the practical *segment tree* data structure, allowing interval update in $O(\log k)$ time. For each node in the suffix path of $p_i \in P$, they simply count the number of descendant leaves (eligible prefixes) that are not covered by a larger prefix of p_i. This allows them to process each node of every suffix path (total $O(n)$) in a single segment tree query and update requiring linear space.

3. **Khan [11] requiring $O(n)$ time.** This algorithm also computes $R_l(u)$ as in [7] (referred to as $\mathcal{L}_u$) and performs a traversal of $\mathcal{A}$ which maintains the lowest internal node having each $p_i \in P$ as a suffix on top of a stack S_i. On reaching a leaf $p_j \in P$, each such internal node, the top of S_i, is the maximum overlap $ov(p_i, p_j)$. They only use stacks and lists for their implementation, giving a simple algorithm with a relatively complex proof.

4. **Park et al. [22] requiring $O(n)$ time.** The algorithm computes the overlaps by reducing the problem to that of computing borders or failure links in the classical KMP algorithm [14]. This allows each internal node to uniquely identify the ancestors whose presence in $\mathcal{H}$ is affected by them (recall Lemma 1) in linear time.

Note that, except for the last algorithm [22], all the algorithms can operate directly on $\mathcal{E}$ (instead of $\mathcal{A}$ or P), resulting in significant performance enhancement when $|\mathcal{E}| \ll |\mathcal{A}|$.

4 Proposed Algorithm

As described earlier, all the algorithms essentially mark the nodes $Ov(P)$ of $\mathcal{A}$ (or $\mathcal{E}$), which will be added to $\mathcal{H}$, followed by a single traversal [11] to generate $\mathcal{H}$. These algorithms take $\mathcal{A}$ (or $\mathcal{E}$) as input and produce $in\mathcal{H}$ array as an output. For any node v, $in\mathcal{H}[v]$ is a boolean such that if its value is true, then v should be in $\mathcal{H}$. We thus focus here on optimally computing this $in\mathcal{H}$ array. We describe our algorithm incrementally in three stages. *Firstly,* we describe how to mark all overlaps for pairs (p_i, p_j) for a fixed p_i and all $p_j \in P$. *Secondly,* perform the task in $O(|p_i| + k)$ using additional data structures resulting in $O(n + k^2)$ algorithm for computing $Ov(P)$. *Finally,* use a simple amortized analysis to prove that our algorithm requires optimal $O(n)$ time. We use the HOG Property (Lemma 1) to identify the nodes in $Ov(P)$. For the sake of simplicity, we assume no string in P is a prefix of another, making the nodes in P the leaves of $\mathcal{A}$. We shall later address how to handle this case separately.

4.1 Marking All Overlaps $ov(p_i, *)$ for a String p_i

Recall that the suffix path of p_i visits all the overlaps of p_i in decreasing order of depth (moving closer to the root). Further, an internal node x on this path can be an overlap of (p_i, p_j) only if p_j is a descendant of x (as x is a prefix of its descendants). Hence, for all the internal nodes which are the overlaps for (p_i, p_j), the node visited first (farthest from the root, hence maximum length) would be $ov(p_i, p_j)$. Given the above properties, we propose the following simple process by colouring the leaves from *white* to *black* as follows.

Starting with all leaves white, for every node x on the suffix path of p_i, we blacken all the white leaves in subtree rooted at x, marking x if any leaf was blackened while processing x.

Consider, for example, in Fig. 2, while processing p_1, the suffix path is shown in red. We first mark v_9 as it blackens p_4, then mark v_8 as it blackens p_2, and then mark v_6 as it blackens p_3. However, we do not mark v_3 as all the leaves in its subtree are already blackened. Thereafter, we mark v_2 since it blackens p_1, and finally leave v_0 unmarked. The correctness of the above algorithm is evident as x's are visited bottom up along the suffix path, and only the leaves p_j in the subtree of x can have x as an overlap of (p_i, p_j).

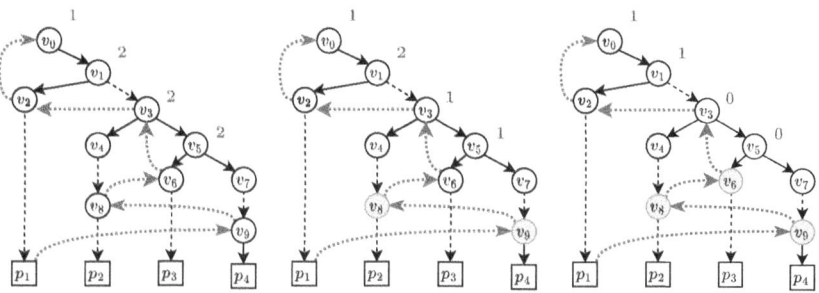

Fig. 2. Example showing *count* of v_0, v_1, v_3, v_5 while processing the suffix path of p_1, (a) before p_1 (left), (b) after v_8 (mid), and (c) after v_6 (right). When v_9 and v_8 are processed and marked, the count of favoured proper ancestors v_5 and v_3 is reduced to 1, respectively. When v_6 is processed and marked, the count of v_5 and v_3 is reduced to zero, and that of v_1 is reduced to 1. Example: p_1:*abaabaaabbaabaab*, p_2:*aabbaabaabab*, p_3:*aabaabab*, and p_4:*aabaaabbaabaabab*, with v_9:*aabaaabbaabaab*, v_8:*aabbaabaab*, v_7:*aabaaa*, v_6:*aabaab*, v_5:*aabaa*, v_4:*aabb*, v_3:*aab*, v_2:*ab*, v_1:*a*, v_0:*ϵ*.

4.2 Marking $ov(p_i, *)$ for a String p_i in $O(|p_i| + k)$ Time

Clearly, visiting all descendants repeatedly is inefficient. The main idea behind the optimization is to lazily *blacken* the entire subtree of each x in *constant* time. This is possible since future updates and queries about *blackened* nodes (due to some suffix x' of x) will only occur at nodes closer to the root than x, avoiding direct queries to x's descendants.

However, to limit the number of updates to *constant*, we face another issue due to paths having nodes with a single child (see Fig. 2). We circumvent it by maintaining such an entire path of nodes (having a single child) using its *favoured* descendant, where *favoured* refers to the root, the leaves, and the nodes having multiple children. Note a *favoured* node is its own favoured descendant. The entire path of nodes having the same favoured descendant is processed together as the number of descendant white leaves for the nodes in the path is the same. For example (see Fig. 2), each x in path from v_4 to p_2 has $favDesc[x] = p_2$, $favPAnc[x] = v_3$. By definition, the number of favoured nodes is $O(k)$. We thus maintain the following data to use the above approach efficiently.

- *count*$[x]$: For a leaf x, it denotes whether x is white (1) or black (0).
  For internal node x, the number of child subtrees of x having white leaves.

- $favPAnc[x]$: The closest *proper* ancestor of x that is favoured, else root.
- $favDesc[x]$: The closest descendant of x (possibly x) that is favoured.
- V_m : List of nodes whose $count[\cdot]$ is modified while visiting a suffix path.

Initialization. Initially, all the leaves are *white*, we initialize $V_m = \emptyset$ and $count[x]$ as the number of children of x in the tree $\mathcal{A}$. Clearly, using a single traversal over the tree $\mathcal{A}$, we can initialize $count[\cdot]$, $favPAnc[\cdot]$ and $favDesc[\cdot]$ in $O(n)$ time. Additionally, $in\mathcal{H}[\cdot]$ is initialized to *false* for all the nodes.

Processing Suffix Path of p_i. Now, starting from the longest suffix of p_i in $\mathcal{A}$, we process each node x on the suffix path in order. The process of *blackening* all the leaves of the subtree of x can be performed *lazily* because the future nodes on the suffix path of p_i cannot be the descendants of x. Assuming the data structure is correctly maintained, the x is marked in $in\mathcal{H}$ if $count[x] \neq 0$, implying that some leaves can be *blackened* while processing x (recall the simple process in Sect. 4.1). To *blacken* all the leaves in the subtree of x, we set the count of its representative (nodes having a single child in a path are equivalent) $favDesc[x]$ to *zero*. Recall that the suffix path of p_i is processed in the decreasing order of depth, hence we don't need to update the count of any other descendant of x. Further, the entire data structure can be updated by updating the count of the ancestors of x whose *count* is affected. This includes each ancestor y of x having a child subtree containing white leaves only among the descendants of x. We thus reduce the count of y ($favPAnc[x]$) and continue reducing *count* of $favPAnc[y]$ onwards only if y does not have any other child subtrees with white leaves, i.e., $count[y]$ became zero. In case *count* of y (or some y' later) remains non-zero after the update, no further updates are required in the ancestors of x.

Consider Fig. 2, while processing p_1 we first visit v_9 having $favDesc[v_9] = p_4$ with non-zero count. We thus mark v_9 and set $count[p_4] = 0$. We also reduce the count of $favPAnc[v_9] = v_5$, which becomes 1. Similarly, we process v_8, marking v_8 and setting $count[p_2] = 0$ and reducing count of $favPAnc[v_8] = v_3$ to 1. While processing v_6, we again mark v_6 setting $count[p_3] = 0$ and reduce count of $favPAnc[v_6] = v_5$. Since $count[v_5]$ becomes zero, we continue reducing the count of $favPAnc[v_5] = v_3$, which again becomes zero. This continues till $favPAnc[v_3] = v_1$ whose count becomes 1. Next, we process v_3, which is its own favoured descendant having count zero, hence it is not marked. Then we process v_2, marking v_2 and setting $count[p_1] = 0$. Thereafter, we reduce the count of $favPAnc[v_2] = v_1$, which becomes zero, and hence we continue to reduce the count of v_0 to zero. Finally, we reach the root v_0, and processing for p_1 ends.

Analysis. While processing v on the suffix path of $p_i \in P$, in addition to *constant* time operations, count is updated for one favoured descendant and multiple favoured ancestors. However, among the favoured ancestors for multiple ancestors, the count is reduced to zero, and exactly one remains non-zero. Since the count for any favoured node can become zero exactly once while processing a $p_i \in P$, the time required to update all the counts is $O(k)$. Note that it also bounds $|V_m| = O(k)$ and hence restoring the counts also requires $O(k)$ time after processing p_i is completed. Thus, the total time for processing p_i is $O(|p_i| + k)$ for traversing the suffix path in $O(|p_i|)$ and updating counts in $O(k)$.

Algorithm 1: $Compute\mathcal{H}(\mathcal{A})$

Initialize $in\mathcal{H}, count, favDesc, favPAnc;$ **Process**(v):

$in\mathcal{H}[root] \leftarrow true;$

foreach *node* x *in* P **do**

    $in\mathcal{H}[x] \leftarrow true;$

    $v \leftarrow$ Suffix link of x;

    $V_m \leftarrow \emptyset;$

    while $v \neq root$ **do**

       Process(v);

       $v \leftarrow$ Suffix link of v;

    foreach *node* $v \in V_m$ **do**

       Restore $count[v]$;

return $in\mathcal{H}$;

Process(v):

if $count[favDesc[v]] \neq 0$ **then**

    $in\mathcal{H}[v] \leftarrow true;$

    $u \leftarrow favDesc[v];$

    $count[u] \leftarrow 0;$

    Add u to V_m;

    $u \leftarrow favPAnc[v];$

    while $u \neq root$ **do**

       $count[u] \leftarrow count[u] - 1;$

       Add u to V_m;

       if $count[u] > 0$ **then** *break*

       $u \leftarrow favPAnc[u];$

Thus, overall all strings $p_i \in P$ are processed after $O(n)$ for initialization in $O(\sum_{i=1}^{k} |p_i| + k) = O(n + k^2)$ time.

4.3 Improved Analysis for a String p_i Requiring $O(|p_i|)$ Time

For each x on the suffix path of $p_i \in P$, we update the *count* of one favoured descendant and multiple favoured proper ancestors. Among these for multiple such ancestors y, the *count* became zero, and at most one ancestor y^*, the *count* remained non-zero. Since each favoured ancestor y (except root) initially has $count[y] \geq 2$ (by definition), we can use an amortized analysis argument to limit the cost to a constant for each node on the suffix path. In particular, we assign cost 2 to x for updating $count[y^*]$, 1 for updating count while processing x, and 1 for future when $count[y^*]$ becomes zero. Now, each ancestor y whose count became zero while processing x was once the unique node whose count remained non-zero after being reduced while processing some x'. Thus, the extra charge while processing x' now pays for updating the count of y while processing x. This limits the time required to update counts to $O(|p_i|)$ for each $p_i \in P$. Note that it also bounds $|V_m| = O(|p_i|)$ and hence restoring the counts also requires $O(|p_i|)$ time. Thus, for processing all the strings, $O(n)$ is required for initialization and $O(\sum_{i=1}^{k} |p_i|) = O(n)$ for processing the strings.

Recall Fig. 2, while processing v_6 (x) on the suffix path of p_1 we update the count of the ancestors v_5 and v_3 (y) which becomes zero, and v_1 (y^*) which remains non-zero. This extra cost of updating counts of v_5 and v_3 while processing v_6 can be associated with the processing of v_9 and v_8 (x'), respectively, when their count was updated but remained non-zero.

Remark: The case of a string $p_i \in P$ being a prefix of another in P can be handled as follows. Clearly, despite not being a leaf of the tree $\mathcal{A}$, the simple algorithm may blacken it. Hence, on including p_i, the effective *count* of the parent of p_i differs from that of its favoured descendant. Thus, each such $p_i \in P$ can be accounted for in the algorithm simply by increasing the count of its parent in $\mathcal{A}$ and treating it as *favoured*.

5   Experimental Evaluation

We now evaluate the algorithms for computing HOG on real datasets and randomly generated datasets (see full paper [27] for details).

5.1   Algorithms

We evaluated the following algorithms (see Sect. 3).

1. CazauxR: The simple algorithm by Cazaux and Rivals [7] requiring $O(n+k^2)$.
2. ParkCPR: The improved algorithm by Park et al. [23] requiring $O(n \log k)$.
3. Khan: The optimal algorithm by Khan [11] requiring $O(n)$.
4. ParkPCPR: The optimal algorithm by Park et al. [22] requiring $O(n)$.
5. New: Our proposed algorithm requiring $O(n)$.

   The code is available on GitHub[1].

5.2   Performance Measures and Environment

Since all the algorithms compute HOG, which is unique for a given set of strings P, the only relevant parameters evaluated are the *time* and *memory*. All the algorithms were implemented in C++ with g++ compiler (v9.4.0) and use $-O3$ optimization flag. The performances are evaluated on an AMD Ryzen 9 7950X3D processor using 128 GB RAM running Linux Ubuntu 20.04.6 LTS.

5.3   Evaluation on Real and Randomly Generated Datasets

Our evaluation can be summarized as follows:

Observation 1. *On real EST and bacterial genome datasets evaluated, we have:*

1. *ParkCPR improves CazauxR by 20-180×, Khan improves ParkCPR by 1.3-8× using 1.3-5× more memory, and New improves Khan in time by 1.3-1.7× and in memory by 1.4-5×.*
2. *With an increase in the size of the dataset, the relative performances of all algorithms improve.*
3. *Though theoretically optimal ParkPCPR in unable to exploit computation using $\mathcal{E}$ making its performance strongly dependent on ratio of $|\mathcal{A}|$ and $|\mathcal{E}|$.*

Observation 2. *On randomly generated datasets evaluated, we have:*

1. *Khan improves ParkCPR $2 - 4×$, while New improves Khan by $2.4 - 3×$.*
2. *ParkPCPR shows no consistent trend and is strongly dependent on input data, particularly the relative sizes of $\mathcal{A}$ and $\mathcal{E}$.*
3. *The performance of ParkCPR, Khan and New are strongly dependent on $|\mathcal{E}|$, while that of ParkPCPR is strongly dependent on $|\mathcal{A}|$, instead of other parameters evaluated.*

---

[1] https://github.com/shahbazk/HoG.

6   Applications to String Problems

We consider variants of the All Pairs Suffix Prefix problem (APSP) which were recently studied in [17], giving state-of-the-art theoretical algorithms. However, they require complex black-box algorithms/data structures for their implementation, making them impractical for use in practice. We thus formulate simple solutions (see Table 1) for the following problems using HOG $\mathcal{H}$.

$OneToOne(i,j)$: report the string $ov(p_i, p_j)$
$OneToAll(i)$: report $ov(p_i, p_j)$ for every $p_j \in P$.
$Top(i,c)$: report all $j \in [1, k]$ corresponding to any c highest values of $|ov(p_i, p_j)|$.
$Count(i,l)$: report the count of all $j \in [1, k]$ where $|ov(p_i, p_j)| \geq l$ for $l \geq 0$.
$Report(i,l)$: report all $j \in [1, k]$ where $ov(p_i, p_j) \geq l$ for $l \geq 0$.

Table 1. Comparison of the classical and state-of-the-art algorithms with proposed

Problem	KMP [14]	State of the art [17]	Using HOG
$OneToOne(i,j)$	$O(|p_i| + |p_j|)$	$O(\log \log k)$	$O(\min\{k^2, |p_i|\})$
$OneToAll(i)$	$O(k|p_i| + n)$	$O(k)$	$O(k + \min\{k^2, |p_i|\})$
$Top(i,c)$	$O(k|p_i| + n)$	$O(\log^2 n / \log \log n + output)$	$O(k + \min\{k^2, |p_i|\})$
$Count(i,l)$	$O(k|p_i| + n)$	$O(\log n / \log \log n)$	$O(k + \min\{k^2, |p_i|\})$
$Report(i,l)$	$O(k|p_i| + n)$	$O(\log n / \log \log n + output)$	$O(k + \min\{k^2, |p_i|\})$

Given a dictionary P of k strings $\{p_1, p_2, \cdots, p_k\}$ of total length n. We assume the strings in P are lexicographically sorted; else, we sort them in $O(||P||)$ using ACT $\mathcal{A}$. We compute $\mathcal{H}$ and compute the following for each node $v \in \mathcal{H}$

$SubTreeMinMax[v]$: Minimum and maximum index of strings in P having the prefix v, i.e., in the subtree of v in $\mathcal{H}$.

Since the strings in P are lexicographically sorted, each internal node in $x \in \mathcal{H}$ has a range of indices $[min, max]$ such that $\{p_{min}, \cdots, p_{max}\}$ has x as its prefix. Note that these strings are the leaves of the subtree rooted at x in $\mathcal{A}$ or $\mathcal{H}$. Also, note that the root (ϵ or empty string) has this range $[1, n]$, having all the leaves. We can preprocess $\mathcal{H}$ to compute and store these ranges for all the internal nodes in linear time using linear space.

6.1   Algorithms

Our solutions can be categorized as (see Algorithm 2 for pseudocodes)

Paired String query: $OneToOne(i,j)$: report the string $ov(p_i, p_j)$
To compute the query, we start with p_i and iterate over each node v in its suffix path, evaluating whether $j \in [min, max]$ of $SubTreeMinMax[v]$. We simply report the first node for which the above property holds true. Since the range

for root is $[1, k]$, the algorithm will always terminate with a solution. Also, we process each suffix v of p_i in decreasing order of length; the first v that is also a prefix of p_j is $ov(p_i, p_j)$.

Now, the number of nodes in the suffix path of a string is bounded by the length of the string p_i. Using precomputed data structure $SubTreeMinMax[v]$, each evaluation for v takes $O(1)$ time, resulting in an overall complexity of $O(|p_i|)$. Also, note that the internal nodes in $\mathcal{H}$ are only the maximum overlaps between k strings, limiting the internal nodes (and hence suffix path) to $O(k^2)$. This results in a complexity of $O(\min\{k^2, |p_i|\})$. Also, on average the length $|p_i|$ is $O(n/k)$, so average cost of the query over all strings is $O(\min\{k^2, \frac{n}{k}\})$.

Note: The above data structure can also be computed for $\mathcal{A}$ (or $\mathcal{E}$) and the corresponding algorithm will give the complexity $O(|p_i|)$. However, in practice, the difference in the number of nodes of $\mathcal{A}, \mathcal{E}, \mathcal{H}$ significantly impacts the performance as compared to the theoretical limit of $O(k^2)$ for $\mathcal{H}$.

Multi String Queries:
$OneToAll(i)$: report $ov(p_i, p_j)$ for every $p_j \in P$
$Top(i, c)$: report all $j \in [1, k]$ corresponding to any c highest values of $|ov(p_i, p_j)|$.
$Count(i, l)$: report the count of all $j \in [1, k]$ where $|ov(p_i, p_j)| \geq l$ for $l \geq 0$.
$Report(i, l)$: report all $j \in [1, k]$ where $ov(p_i, p_j) \geq l$ for an integer $l \geq 0$.

We first describe $OnetoAll(i)$ query where we need to find $ov(p_i, p_j)$, for every $p_j \in P$ for given p_i. We traverse each node v in the suffix path of p_i, marking v as the answer $ans[j]$ for each string p_j in the subtree of v (hence having prefix v), if not marked previously. Clearly, since we process nodes on a suffix path in decreasing order of length of v, we store the maximum overlaps in $ans[\cdot]$ with p_i (recall the simple process of blackening leaves described in Sect. 4.1).

To perform this task efficiently, we store the next potentially unmarked leaf in $next[j]$ if j is marked; else, we have $next[j] = j$. Hence, while processing v on the suffix path, in case $next[j] = j$ for some leaf, we store $ans[j] = v$ and move to $next[j]$. Further, note that while processing v, none of the leaves in its subtree will be left unmarked and hence can be skipped while processing further nodes on the suffix path after v. This is achieved by updating $next[min] = max + 1$ where $[min, max]$ is $SubTreeMinMax[v]$. This is sufficient to skip the above range in the future because any node x on the suffix path trying to mark a leaf in the above range would surely be an ancestor of v, hence will first attempt to mark min in this range, thereby skipping processing up to $max + 1$.

Again, the number of nodes in the suffix path is bounded by $O(|p_i|)$, and we use total $O(k + |p_i|)$ time to mark and skip using $next[\cdot]$. For each v in the suffix path, it visits $next[i]$ of either the minimum index, or of a leaf that will never be processed again (skipped in the future). Thus, $next[\cdot]$ is processed once for each leaf and once for each node in the suffix path, requiring $O(k + |p_i|)$ time. Again, the length of the suffix path is also bounded by the number of internal nodes in $\mathcal{H}$, i.e., $O(k^2)$ resulting in the complexity of $O(k + \min\{k^2, |p_i|\})$. And the average cost of the query across all strings is $O(k + \min\{k^2, \frac{n}{k}\})$.

The queries $Top(i, c), Count(i, l)$, and $Report(i, l)$ terminate the above process early when the number of reported strings exceeds c or the length of string

v on the suffix path becomes smaller than l. Thus, the above queries have the same time complexity in the worst case.

6.2   Classical KMP Algorithm

Note that each overlap query $ov(p_i, p_j)$ can be naively computed by comparing every suffix of p_i with p_j, requiring $O(|p_i| \times |p_j|)$ time. However, using the classical KMP algorithm [14] with pattern p_j and text p_i it can be answered in $O(|p_i| + |p_j|)$. For multiple string queries, we repeat for all $p_j \in P$, requiring total $O(\sum_{j=1}^{k} |p_i| + |p_j|) = O(k|p_i| + n)$ time. Note that the average cost of the query over all possible strings p_i is $O(k \times \frac{n}{k} + n) = O(n)$.

6.3   Comparisons with the State-of-the-Art

Current algorithms with HOG are easier to implement than [17], though clearly, it has a lot of scope for improvement. Also, we are able to answer multiple types of queries with only HOG, while in [17] they have used different data structures for different types of queries, which makes them harder to implement and will practically take more time for construction. Table 1 summarizes the complexities of the proposed algorithms for the studied problems.

Algorithm 2: APSP Queries using HOG.

Paired String Query:
$OneToOne(i, j)$:

$v \leftarrow p_i$;

while *true* **do**
  $(min, max) \leftarrow SubTreeMinMax(v)$;
  if $j \geq min$ **and** $j \leq max$ **then**
  | **return** v
  else $v \leftarrow$ Suffix link of v

Multi String Query:

$OneToAll(i)$ / $Report(i, l)$ /
$Count(i, l)$ / $Top(i, c)$:

foreach $j \in [1, k]$ **do**
  $ans_o[j] \leftarrow \epsilon$;
  $next[j] \leftarrow j$;
$ans_r \leftarrow \emptyset \mid ans_c \leftarrow 0 \mid ans_t \leftarrow \emptyset$;
$v \leftarrow p_i$;

while $v \neq \epsilon$ *and* $|v| \geq l \mid$ *and* $|v| \geq l \mid$
and $|ans_t| \leq c$ **do**
  $(min, max) \leftarrow SubTreeMinMax(v)$;
  $j \leftarrow min$;

  while $j \leq max$ **do**
    while $next[j] \neq j$ *and*
    $next[j] \leq max$ **do**
    | $j \leftarrow next[j]$

    if $next[j] == j$ **then**
      $ans_o[j] \leftarrow v \mid$
      $ans_r \leftarrow ans_r \cup \{j\} \mid$
      $ans_c \leftarrow ans_c + 1 \mid$
      $ans_t \leftarrow ans_t \cup \{j\}$;

    $j \leftarrow j + 1$;

  $next[min] \leftarrow max + 1$;
  $v \leftarrow$ suffix link of v;

return ans;

6.4   Experimental Evaluation

Despite failing to match the state-of-the-art algorithms [17] theoretically by a large margin, our algorithms performed well in practice (see full paper [27]). Our algorithm required $0.002 - 0.010\ ms$ to answer paired string queries, and $0.5 - 100\ ms$ to answer multi string queries, for a data set having around a billion characters. For comparison, the classical KMP algorithm answers paired string queries $2 - 20\times$ faster, where *OneToOne* queries are tailor-made for the KMP algorithm. However, the remaining multi string queries are $18 - 1300\times$ slower. Among the multi string queries, *OneToAll* has the least improvement $(18 - 106\times)$, and *Count* has the most improvement $(199 - 1300\times)$, as KMP is unable to exploit any structural information and computes all overlaps. Since the state-of-the-art algorithms [17] require complex black box algorithms and data structures, their evaluation is beyond the scope of this paper.

7   Conclusion

We proposed a new algorithm for computing HOG and performed an empirical evaluation of all the practical algorithms. While the previous optimal algorithm [11] improves the other algorithms in time, it comes at the expense of increased memory usage. Our algorithm improves both time and memory with respect to all previous algorithms. Moreover, our algorithm is also arguably more intuitive and easier to implement using only elementary arrays.

To highlight the significance of our results, we also considered applications of the HOG on variants of the APSP problem and demonstrated its practicality by showing that it requires acceptable time despite the large size of the dataset.

In the future, faster algorithms based on HOG for the studied variants of APSP might prove useful. Possibly, the structure of HOG can result in superior algorithms, both in theory and practice. For example, Kikuchi and Inenaga [12] independently developed an algorithm for the static APSP problem, which uses a similar algorithm as our proposed algorithm, with an incomparable analysis. Our result unknowingly answered an open question by [12] regarding the construction of an HOG using similar techniques. It remains an open problem whether HOG can be maintained efficiently under the insertion and/or deletion of strings.

References

1. Aho, A.V., Corasick, M.J.: Efficient string matching: an aid to bibliographic search. Commun. ACM **18**(6), 333–340 (1975). https://doi.org/10.1145/360825.360855
2. Alanko, J.N., Puglisi, S.J., Vuohtoniemi, J.: Small searchable κ-spectra via subset rank queries on the spectral burrows-wheeler transform. In: Berry, J.W., Shmoys, D.B., Cowen, L., Naumann, U. (eds.) SIAM Conference on Applied and Computational Discrete Algorithms, ACDA 2023, Seattle, WA, USA, May 31 - June 2, 2023, pp. 225–236. SIAM (2023)
3. Antipov, D., Korobeynikov, A.I., McLean, J.S., Pevzner, P.A.: hybridspades: an algorithm for hybrid assembly of short and long reads. Bioinform. **32**(7), 1009–1015 (2016). https://doi.org/10.1093/BIOINFORMATICS/BTV688

4. Bankevich, A., et al.: Spades: a new genome assembly algorithm and its applications to single-cell sequencing. J. Comput. Biol. **19**(5), 455–477 (2012). https://doi.org/10.1089/CMB.2012.0021
5. Cánovas, R., Cazaux, B., Rivals, E.: The compressed overlap index. CoRR abs/1707.05613 (2017)
6. Cazaux, B., Cánovas, R., Rivals, E.: Shortest DNA cyclic cover in compressed space. In: Bilgin, A., Marcellin, M.W., Serra-Sagristà, J., Storer, J.A. (eds.) 2016 Data Compression Conference, DCC 2016, Snowbird, UT, USA, March 30 - April 1, 2016, pp. 536–545. IEEE (2016). https://doi.org/10.1109/DCC.2016.79
7. Cazaux, B., Rivals, E.: Hierarchical overlap graph. Inf. Process. Lett. **155** (2020). https://doi.org/10.1016/J.IPL.2019.105862
8. Gonnella, G., Kurtz, S.: ReadJoiner: a fast and memory efficient string graph-based sequence assembler. BMC Bioinform. **13**, 82 (2012). https://doi.org/10.1186/1471-2105-13-82
9. Gusfield, D., Landau, G.M., Schieber, B.: An efficient algorithm for the all pairs suffix-prefix problem. Inf. Process. Lett. **41**(4), 181–185 (1992). https://doi.org/10.1016/0020-0190(92)90176-V
10. Haj Rachid, M., Malluhi, Q.: A practical and scalable tool to find overlaps between sequences. Biomed. Res. Int. **2015**, 905261 (2015). https://doi.org/10.1155/2015/905261
11. Khan, S.: Optimal construction of hierarchical overlap graphs. In: Gawrychowski, P., Starikovskaya, T. (eds.) 32nd Annual Symposium on Combinatorial Pattern Matching, CPM 2021, 5-7 July 2021, Wrocław, Poland. LIPIcs, vol. 191, pp. 17:1–17:11. Schloss Dagstuhl - Leibniz-Zentrum für Informatik (2021). https://doi.org/10.4230/LIPICS.CPM.2021.17
12. Kikuchi, M., Inenaga, S.: All-pairs suffix-prefix on dynamic set of strings. In: Lipták, Z., de Moura, E.S., Figueroa, K., Baeza-Yates, R. (eds.) String Processing and Information Retrieval - 31st International Symposium, SPIRE 2024, Puerto Vallarta, Mexico, September 23-25, 2024, Proceedings. LNCS, vol. 14899, pp. 192–203. Springer, Cham (2024). https://doi.org/10.1007/978-3-031-72200-4_15
13. Kikuchi, M., Inenaga, S.: All-pairs suffix-prefix on dynamic set of strings. CoRR abs/2407.17814 (2024)
14. Knuth, D.E., Jr., J.H.M., Pratt, V.R.: Fast pattern matching in strings. SIAM J. Comput. **6**(2), 323–350 (1977). https://doi.org/10.1137/0206024
15. Lim, J., Park, K.: A fast algorithm for the all-pairs suffix-prefix problem. Theor. Comput. Sci. **698**, 14–24 (2017). https://doi.org/10.1016/J.TCS.2017.07.013
16. Loukides, G., Pissis, S.P.: All-pairs suffix/prefix in optimal time using Aho-Corasick space. Inf. Process. Lett. **178**, 106275 (2022). https://doi.org/10.1016/J.IPL.2022.106275
17. Loukides, G., Pissis, S.P., Thankachan, S.V., Zuba, W.: Suffix-prefix queries on a dictionary. In: Bulteau, L., Lipták, Z. (eds.) 34th Annual Symposium on Combinatorial Pattern Matching, CPM 2023, 26-28 June 2023, Marne-la-Vallée, France. LIPIcs, vol. 259, pp. 21:1–21:20. Schloss Dagstuhl - Leibniz-Zentrum für Informatik (2023). https://doi.org/10.4230/LIPICS.CPM.2023.21
18. Manber, U., Myers, E.W.: Suffix arrays: a new method for on-line string searches. SIAM J. Comput. **22**(5), 935–948 (1993). https://doi.org/10.1137/0222058
19. Myers, E.W.: The fragment assembly string graph. In: ECCB/JBI 2005 Proceedings, Fourth European Conference on Computational Biology/Sixth Meeting of the Spanish Bioinformatics Network (Jornadas de BioInformática), Palacio de Congresos, Madrid, Spain, September 28 - October 1, 2005, p. 85 (2005). https://doi.org/10.1093/BIOINFORMATICS/BTI1114

20. Nurk, S., Meleshko, D., Korobeynikov, A.I., Pevzner, P.A.: MetaSpades: a new versatile de novo metagenomics assembler. In: Singh, M. (ed.) Research in Computational Molecular Biology - 20th Annual Conference, RECOMB 2016, Santa Monica, CA, USA, 17-21 April 2016, Proceedings. Lecture Notes in Computer Science, vol. 9649, p. 258. Springer (2016), https://link.springer.com/content/pdf/bbm%3A978-3-319-31957-5%2F1.pdf

21. Ohlebusch, E., Gog, S.: Efficient algorithms for the all-pairs suffix-prefix problem and the all-pairs substring-prefix problem. Inf. Process. Lett. **110**(3), 123–128 (2010). https://doi.org/10.1016/J.IPL.2009.10.015

22. Park, S., Park, S.G., Cazaux, B., Park, K., Rivals, E.: A linear time algorithm for constructing hierarchical overlap graphs. In: Gawrychowski, P., Starikovskaya, T. (eds.) 32nd Annual Symposium on Combinatorial Pattern Matching, CPM 2021, July 5-7, 2021, Wrocław, Poland. LIPIcs, vol. 191, pp. 22:1–22:9. Schloss Dagstuhl - Leibniz-Zentrum für Informatik (2021). https://doi.org/10.4230/LIPICS.CPM.2021.22

23. Park, S.G., Cazaux, B., Park, K., Rivals, E.: Efficient construction of hierarchical overlap graphs. In: Boucher, C., Thankachan, S.V. (eds.) String Processing and Information Retrieval - 27th International Symposium, SPIRE 2020, Orlando, FL, USA, 13-15 October 2020, Proceedings. LNCS, vol. 12303, pp. 277–290. Springer, Cham (2020). https://doi.org/10.1007/978-3-030-59212-7_20

24. Pevzner, P.A.: l-Tuple DNA sequencing: computer analysis. J. Biomol. Struct. Dyn. **7**(1), 63–73 (1989)

25. Pevzner, P.A., Tang, H., Waterman, M.S.: An Eulerian path approach to DNA fragment assembly. Proc. Natl. Acad. Sci. U.S.A. **98**(17), 9748–9753 (2001)

26. Simpson, J.T., Durbin, R.: Efficient construction of an assembly string graph using the FM-index. Bioinform. **26**(12), 367–373 (2010). https://doi.org/10.1093/BIOINFORMATICS/BTQ217

27. Talera, S., Bansal, P., Khan, S., Khan, S.: Practical algorithms for hierarchical overlap graphs. CoRR abs/2402.13920 (2024)

28. Tustumi, W.H.A., Gog, S., Telles, G.P., Louza, F.A.: An improved algorithm for the all-pairs suffix-prefix problem. J. Discrete Algorithms **37**, 34–43 (2016). https://doi.org/10.1016/J.JDA.2016.04.002

29. Ukkonen, E.: A linear-time algorithm for finding approximate shortest common superstrings. Algorithmica **5**(3), 313–323 (1990). https://doi.org/10.1007/BF01840391

30. Weiner, P.: Linear pattern matching algorithms. In: 14th Annual Symposium on Switching and Automata Theory, Iowa City, Iowa, USA, 15-17 October 1973, pp. 1–11. IEEE Computer Society (1973). https://doi.org/10.1109/SWAT.1973.13

31. Zerbino, D.R., Birney, E.: Velvet: algorithms for de novo short read assembly using de Bruijn graphs. Genome research **18**(5), 821–829 (2008). https://doi.org/10.1101/gr.074492.107

Counting Distinct (Non-)crossing Substrings

Haruki Umezaki[1], Hiroki Shibata[2] , Dominik Köppl[3] , Yuto Nakashima[4] ,
Shunsuke Inenaga[4($\boxtimes$)] , and Hideo Bannai[5]

[1] Department of Information Science and Technology, Kyushu University,
Fukuoka, Japan
umezaki.haruki.314@s.kyushu-u.ac.jp
[2] Joint Graduate School of Mathematics for Innovation, Kyushu University,
Fukuoka, Japan
shibata.hiroki.753@s.kyushu-u.ac.jp
[3] Department of Computer Science and Engineering, University of Yamanashi,
Kofu, Japan
dkppl@yamanashi.ac.jp
[4] Department of Informatics, Kyushu University, Fukuoka, Japan
{nakashima.yuto.003,inenaga.shunsuke.380}@m.kyushu-u.ac.jp
[5] M&D Data Science Center, Institute of Integrated Research, Institute of Science
Tokyo, Tokyo, Japan
hdbn.dsc@tmd.ac.jp

Abstract. Let w be a string of length n. The problem of counting factors crossing a position - Problem 64 from the textbook "125 Problems in Text Algorithms" [Crochemore, Leqroc, and Rytter, 2021], asks to count the number $\mathcal{C}(w,k)$ (resp. $\mathcal{N}(w,k)$) of distinct substrings in w that have occurrences containing (resp. not containing) a position k in w. The solutions provided in their textbook compute $\mathcal{C}(w,k)$ and $\mathcal{N}(w,k)$ in $O(n)$ time *for a single position* k in w, and thus a direct application would require $O(n^2)$ time for *all positions* $k = 1, \ldots, n$ in w. Their solution is designed for constant-size alphabets. In this paper, we present new algorithms which compute $\mathcal{C}(w,k)$ in $O(n)$ total time for general ordered alphabets, and $\mathcal{N}(w,k)$ in $O(n)$ total time for linearly sortable alphabets, for all positions $k = 1, \ldots, n$ in w.

Keywords: string algorithms · distinct substrings · runs · LPF arrays

1 Introduction

Let w be a string of length n. The problem of counting factors crossing a position - Problem 64 from the textbook "125 Problems in Text Algorithms" [3], asks to count the number $\mathcal{C}(w,k)$ (resp. $\mathcal{N}(w,k)$) of distinct substrings in w that have occurrences containing (resp. not containing) a position k in w. According to the textbook [3], the notions of $\mathcal{C}(w,k)$ and $\mathcal{N}(w,k)$ are inspired by the notion of *string attractors* [8], which form a set $\mathcal{P} = \{p_1, \ldots, p_\gamma\}$ of γ positions such that

G. Badkobeh et al. (Eds.): SPIRE 2025, LNCS 16073, pp. 281–290, 2026.
https://doi.org/10.1007/978-3-032-05228-5_22

any substring of w has an occurrence containing a position $p_i \in \mathcal{P}$. Besides this origin, how efficiently one can compute $\mathcal{C}(w, k)$ and $\mathcal{N}(w, k)$ for a given string w, is an intriguing stringology question.

The solutions provided in the textbook [3] compute $\mathcal{C}(w, k)$ and $\mathcal{N}(w, k)$ in $O(n)$ time *for a single position k* in w for constant-size alphabets. Thus, a direct application of their solutions to the *all-position variant* of the problems, which ask to compute $\mathcal{C}(w, k)$ and $\mathcal{N}(w, k)$ for *all positions* $k = 1, \ldots, n$ in w, requires $O(n^2)$ total time.

In this paper, we present new algorithms which compute for all positions $k = 1, \ldots, n$, $\mathcal{C}(w, k)$ in $O(n)$ total time and space for general ordered alphabets, and $\mathcal{N}(w, k)$ in $O(n)$ total time and space for linearly sortable alphabets. Our solution for computing $\mathcal{C}(w, k)$ for $k = 1, \ldots, n$ exploits the combinatorial property of the problem and utilizes the *runs* (a.k.a. *maximal repetitions*) [9] occurring in w, which is completely different from the original solution from the textbook [3].

2   Preliminaries

2.1   Strings

Let Σ be an ordered alphabet. An element of Σ^* is called a *string*. The length of a string $w \in \Sigma^*$ is denoted by $|w|$. The *empty string* ε is the string of length 0. Let $\Sigma^+ = \Sigma^* \setminus \{\varepsilon\}$. For string $w = xyz$, x, y, and z are called a *prefix*, *substring*, and *suffix* of w, respectively. Let $\mathsf{Substr}(w)$ and $\mathsf{Suffix}(w)$ denote the sets of substrings and suffixes of w, respectively. For a string w of length n, $w[i]$ denotes the ith character of w and $w[i..j] = w[i] \cdots w[j]$ denotes the substring of w that begins at position i and ends at position j for $1 \le i \le j \le n$. For convenience, let $w[i..j] = \varepsilon$ for $i > j$.

For two non-empty strings s and w, let $\mathsf{occ}(s, w) = \{i \mid w[i..i + |s| - 1] = s\}$ denote the set of occurrences of s in w, where we identify an occurrence of s with its starting position. For each position $1 \le k \le |w|$ in w, let

$$\mathsf{cocc}_k(s, w) = \{i \in \mathsf{occ}(s, w) \mid i \le k \le i + |s| - 1\}$$
$$\mathsf{ncocc}_k(s, w) = \{i \in \mathsf{occ}(s, w) \mid i + |s| - 1 < k \text{ or } k < i\}$$

denote the sets of occurrences of string s that cross (resp. do not cross) the position k in w. Let

$$\mathsf{C}(w, k) = \{s \in \Sigma^+ \mid \mathsf{cocc}_k(s, w) \ne \emptyset\}$$
$$\mathsf{N}(w, k) = \{s \in \Sigma^+ \mid \mathsf{ncocc}_k(s, w) \ne \emptyset\}$$
$$= \mathsf{Substr}(w[1..k - 1]) \cup \mathsf{Substr}(w[k + 1..|w|])$$

denote the sets of substrings s of string w that have crossing (resp. non-crossing) occurrence(s) for the position k in w.

Problem 1 (Counting distinct substrings with (non-)crossing occurrences). Given a string w of length n, compute $\mathcal{C}(w, k) = |\mathsf{C}(w, k)|$ and $\mathcal{N}(w, k) = |\mathsf{N}(w, k)|$ for all positions $k = 1, \ldots, n$ in w.

2.2 Repetitions and Runs

For a string s, an integer p $(1 \leq p \leq |s|)$ is a period of s if $s[i] = s[i + p]$ for all $1 \leq i \leq |s| - p$. The *exponent* of s is the rational $|s|/p$, where p is the smallest period of s. A string $s \in \Sigma^+$ is said to be *periodic* if the exponent of s is at least 2, or equivalently, s's smallest period is at most $|s|/2$. A maximal periodic substring $s = w[i..j]$ of w, i.e., the smallest period p of s does not extend to the left of position i nor to the right of position j, namely, $i = 1$ or $w[i - 1] \neq w[i + p - 1]$ and $j = |w|$ or $w[j + 1] \neq w[j - p + 1]$, is called a *maximal repetition*, or *run*, in w. We identify a run $w[i..j]$ with the smallest period p by a tuple $\langle i, j, p \rangle$. Let $\mathsf{Runs}(w) = \{\langle i, j, p \rangle \mid w[i..j] \text{is a run in} w\}$ denote the set of runs in w.

Theorem 1. ([1]) $|\mathsf{Runs}(w)| < n$ holds for any string w of length n.

Theorem 2. ([5]) $\mathsf{Runs}(w)$ can be computed in $O(n)$ time for any string $w[1..n]$ over an ordered alphabet.

2.3 Suffix Trees

The *suffix tree* [10] of a string w, denoted $\mathsf{STree}(w)$, is a path-compressed trie representing $\mathsf{Suffix}(w)$ such that (1) each internal node has at least two children, (2) each edge is labeled by a non-empty substring of w, and (3) the labels of out-going edges of the same node begin with distinct characters. Each leaf of $\mathsf{STree}(w)$ is associated with the occurrence of its corresponding suffix of w.

For a node v of $\mathsf{STree}(w)$, let $\mathsf{str}(v)$ denote the string label of the path from the root to v. Each node v stores its string depth $|\mathsf{str}(v)|$. The *locus* of a substring $s \in \mathsf{Substr}(w)$ in $\mathsf{STree}(w)$ is the position where s is spelled out from the root. The number of nodes in $\mathsf{STree}(w)$ is at most $2n - 1$, where $n = |w|$. We can represent $\mathsf{STree}(w)$ in $O(n)$ space by representing each edge label s with a pair (i, j) of positions in w such that $w[i..j] = s$.

Suppose that string w terminates with an end-marker \$ that does not occur anywhere else in w. Then, since $|\mathsf{occ}(y, w)| = 1$ holds for every suffix y of w, $\mathsf{STree}(w)$ has exactly $|w|$ leaves.

Theorem 3. ([6]) $\mathsf{STree}(w)$ can be built in $O(n)$ time for any string $w[1..n]$ over a linearly-sortable alphabet.

3   Computing $\mathcal{C}(w, k)$ for All Positions k in a String w

In this section, we show how to compute $\mathcal{C}(w, k)$ in $O(n)$ total time for all positions k in a given string w of length n over an ordered alphabet.

In our algorithm for computing $\mathcal{C}(w, k)$, we first compute the size of the multiset of substrings that cross position k in w, and then subtract the number $\mathcal{D}(w, k)$ of duplicates. Let $\mathsf{U}(w, k)$ be the multiset of substrings crossing k in a given string w. Since $|\mathsf{U}(w, k)|$ is equal to the number of intervals including k in w, $|\mathsf{U}(w, k)| = k(|w| - k + 1)$ holds: $[i, j]$ includes k iff $i \in [1, k]$ and $j \in [k, |w|]$.

Let us consider how to compute $\mathcal{D}(w, k)$. The following observation and lemma are a key.

Observation 1. *For any substring x and position k in string w, if $\mathsf{cocc}_k(x, w) \geq 2$, then x is a substring of a run of w with smallest period $p < |x|$.*

We use the following well-known result:

Lemma 1. *(Weak periodicity lemma [7]) If p and q are periods of a string w, then $\gcd(p, q)$ is also a period of w.*

Lemma 2. *For a run $r = \langle i, j, p \rangle$ of a string w, the distance d between any two consecutive occurrences of a substring x in r with $|x| \geq p$ must be p.*

Proof. Due to the periodicity of r, any two consecutive occurrences are at distance $d \leq p$. If $d < p$, it follows from the weak periodicity lemma that d and p are periods of a substring of length $d + |x| > d + p$, implying that $p' = \gcd(d, p) < p$ is a period of r, which contradicts the minimality of p. □

Let $\mathsf{Runs}(w, k) = \{\langle i, j, p \rangle \in \mathsf{Runs}(w) \mid i \leq k \leq j\}$ denote the set of runs in w that cross position k. For a run $\langle i, j, p \rangle \in \mathsf{Runs}(w, k)$, let

$$S(\langle i, j, p \rangle, k) = \{x \in \Sigma^+ \mid x = w[g..h], i \leq g \leq k \leq h \leq j, |x| = h - g + 1 > p\}$$

denote the set of substrings x of length at least $p + 1$ that occur in the run $\langle i, j, p \rangle$ and cross k. Let $\mathsf{dup}(\langle i, j, p \rangle, k) = \sum_{x \in S(\langle i, j, p \rangle, k)} (|\mathsf{cocc}_k(x, w)| - 1)$ be the number of duplicates contained in the run $\langle i, j, p \rangle$. From Observation 1 and Lemma 2, it follows that

$$\mathsf{dup}(\langle i, j, p \rangle, k) = \begin{cases} 0 & \text{if } i \leq k \leq i + p - 1, \ (1) \\ (k - i - p + 1)(j - p + 1 - k) & \text{if } i + p \leq k \leq j - p, \ (2) \\ 0 & \text{if } j - p + 1 \leq k \leq j. \ (3) \end{cases}$$

See Fig. 1. In Case (1), there is only one crossing occurrence for each substring x of length at least $p + 1$ in the run $\langle i, j, p \rangle$, and thus $\mathsf{dup}(\langle i, j, p \rangle, k) = 0$. Case (3) is symmetric. In Case (2), for each substring x of length at least $p + 1$ in the run $\langle i, j, p \rangle$, we count all occurrences crossing k except for the rightmost one. Notice that any substring x that starts in the dark gray region of length $k - i - p + 1$ and ends in the light gray region of length $j - p + 1 - k$ crosses k and has exactly one occurrence that starts in the region of length p between the two gray regions and crosses k. Therefore, a total of $(k - i - p + 1)(j - p + 1 - k)$ duplicate occurrences are counted in Case (2).

Lemma 3. $\mathcal{D}(w, k) = \sum_{\langle i, j, p \rangle \in \mathsf{Runs}(w, k)} \mathsf{dup}(\langle i, j, p \rangle, k)$.

Proof. It is clear that $\mathcal{D}(w, k) \leq \sum_{\langle i, j, p \rangle \in \mathsf{Runs}(w, k)} \mathsf{dup}(\langle i, j, p \rangle, k)$. We prove the lemma by showing that the same duplicates are counted in different runs. Assume for a contradiction that the same substring x is counted in $\mathsf{dup}(\langle i, j, p \rangle, k)$ and in $\mathsf{dup}(\langle i', j', p' \rangle, k)$ by two distinct runs $\langle i, j, p \rangle, \langle i', j', p' \rangle \in \mathsf{Runs}(w, k)$. Without loss of generality suppose that $p \leq p'$.

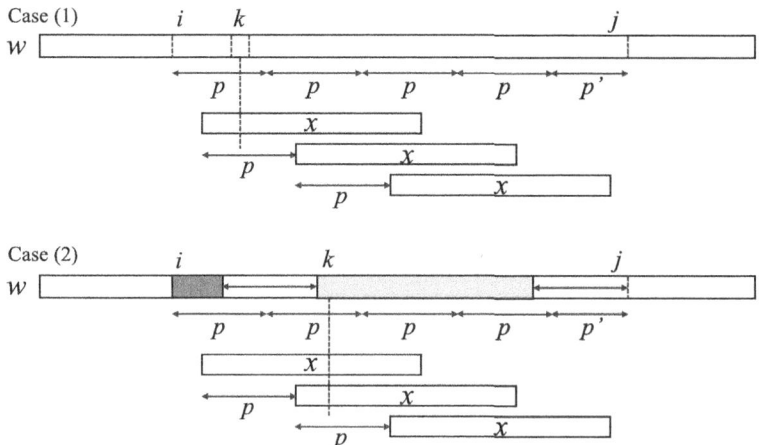

Fig. 1. Illustration for $\mathsf{dup}(\langle i,j,p\rangle,k)$ for Cases (1) and (2).

If x has only a single occurrence that crosses k within the run $\langle i,j,p\rangle$, then it is not counted in $\mathsf{dup}(\langle i,j,p\rangle,k)$ because of the definition $\mathsf{dup}(\langle i,j,p\rangle,k) = \sum_{x\in S(\langle i,j,p\rangle,k)}(|\mathsf{cocc}_k(x,w)|-1)$. The other case with $\langle i',j',p'\rangle$ is analogous.

For the case where x has at least two occurrences that cross k in each of the runs $\langle i,j,p\rangle$ and $\langle i',j',p'\rangle$, Lemma 2 implies that $p=p'$. However, since the runs overlap by at least p positions, the periodicity of one run extends into the other, contradicting their maximality.   □

After $O(n)$-time preprocessing for computing $\mathsf{Runs}(w)$ with Theorem 2, Observation 1 and Lemma 3 immediately lead us to an $O(n)$-time solution to compute $\mathcal{C}(w,k)$ for a *fixed* k.

Our strategy to compute $\mathcal{C}(w,k)$ for all $k=1,\dots,n$ is first to compute $\mathcal{C}(w,1)=|\mathsf{U}(w,1)|-\mathcal{D}(w,1)$ for $k=1$ in $O(n)$ time, and compute $\mathcal{C}(w,k)=|\mathsf{U}(w,k)|-\mathcal{D}(w,k)$ in amortized $O(1)$ time for increasing $k=2,\dots,n$. Since $|\mathsf{U}(w,k)|$ is computable in $O(1)$ time by a simple arithmetic for every k, in what follows we focus on how to compute $\mathcal{D}(w,k)$.

The next lemma exploits a useful structure of $\mathsf{dup}(\langle i,j,p\rangle,k)$ for the consecutive positions $k=i+p,\dots,j-p$.

Lemma 4. *For each run $\langle i,j,p\rangle$, consider the sequence*

$$num_{\langle i,j,p\rangle} = \mathsf{dup}(\langle i,j,p\rangle,i+p),\dots,\mathsf{dup}(\langle i,j,p\rangle,j-p)$$

of $j-i-2p+1$ integers. Then, $num_{\langle i,j,p\rangle}$ is an integer sequence whose difference sequence is an arithmetic progression that starts with $\mathsf{dup}(\langle i,j,p\rangle,i+p)$ and has common difference -2.

Proof. Let $b=\mathsf{dup}(\langle i,j,p\rangle,i+p)=j-i-2p+1$ from Case (2). Let $num_{\langle i,j,p\rangle}[a]$ denote the ath term in the sequence. Then $num_{\langle i,j,p\rangle}[a]=a(b+1-a)$, and hence

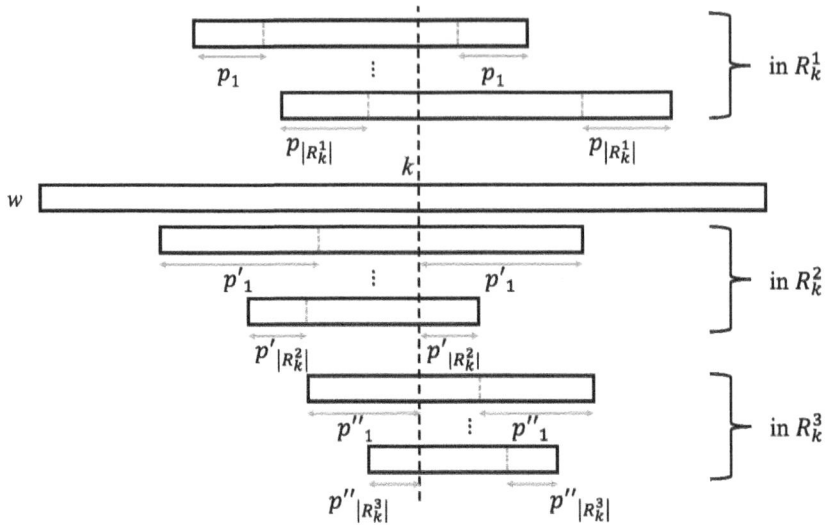

Fig. 2. Illustration for R_k^1, R_k^2, and R_k^3

$num_{\langle i,j,p\rangle}[a+1] - num_{\langle i,j,p\rangle}[a] = (a+1)(b-a) - a(b+1-a) = ab - a^2 + b - a - ab - a + a^2 = b - 2a$. This is the general term of the arithmetic progression that starts with b and has common difference -2. Therefore, $num_{\langle i,j,p\rangle}$ is an integer sequence whose difference sequence is an arithmetic progression that starts with $b = \mathsf{dup}(\langle i,j,p\rangle, i+p)$ and has common difference -2.   □

For each position k, let $R_k = \{\langle i,j,p\rangle \in \mathsf{Runs}(w,k) \mid i+p \le k \le j-p\}$ be the set of runs $\langle i,j,p\rangle$ such that $\mathsf{dup}(\langle i,j,p\rangle, k) > 0$. We divide runs $\langle i,j,p\rangle \in R_k \cup R_{k+1}$ into the following three disjoint subsets (see also Fig. 2):

$$R_k^1 = R_k \cap R_{k+1},$$
$$R_k^2 = R_k \setminus R_{k+1},$$
$$R_k^3 = R_{k+1} \setminus R_k.$$

Recall that for each $\langle i,j,p\rangle \in R_k$, we have $\mathsf{dup}(\langle i,j,p\rangle, k) = (k-i-p+1)(j-p+1-k)$. By Lemma 4, $num_{\langle i,j,p\rangle}$ is an integer sequence whose difference sequence is an arithmetic progression that starts with $\mathsf{dup}(\langle i,j,p\rangle, i+p)$ and has a common difference -2.

By Lemma 3, $\mathcal{D}(w, k) = \sum_{\langle i,j,p \rangle \in \mathsf{Runs}(w,k)} \mathsf{dup}(\langle i, j, p \rangle, k)$ holds. For computing $\mathcal{D}(w, k)$ for increasing k, we maintain the following invariants:

$$m_k = |R_k|,$$

$$f_k = \sum_{\langle i,j,p \rangle \in R_k} \mathsf{dup}(\langle i, j, p \rangle, i + p) = \sum_{\langle i,j,p \rangle \in R_k} num_{\langle i,j,p \rangle}[1],$$

$$d_k = \sum_{\langle i,j,p \rangle \in R_k} (k - (i + p)),$$

$$e_k = \sum_{\langle i,j,p \rangle \in R_k^2} \mathsf{dup}(\langle i, j, p \rangle, j - p) = \sum_{\langle i,j,p \rangle \in R_k^2} num_{\langle i,j,p \rangle}[1].$$

f_k is the sum of the first terms of $num_{\langle i,j,p \rangle}$, for $\langle i, j, p \rangle \in R_k$. d_k is the sum of the distances between k and $i + p$, for $\langle i, j, p \rangle \in R_k$. e_k is the sum of the last terms of $num_{\langle i,j,p \rangle}$, for $\langle i, j, p \rangle \in R_k^2$.

By Lemma 4, for a run $\langle i, j, p \rangle$, $\mathsf{dup}(\langle i, j, p \rangle, k + 1)$ can be maintained with the following recurrence:

$$\mathsf{dup}(\langle i, j, p \rangle, k + 1) = \mathsf{dup}(\langle i, j, p \rangle, k) + num_{\langle i,j,p \rangle}[1] - 2(k - (i + p)).$$

This leads to the following recurrence for $\mathcal{D}(w, k + 1)$:

$$\mathcal{D}(w, k + 1) = \sum_{\langle i,j,p \rangle \in \mathsf{Runs}(w,k+1)} \mathsf{dup}(\langle i, j, p \rangle, k + 1)$$

$$= \sum_{\langle i,j,p \rangle \in R_k^1} \mathsf{dup}(\langle i, j, p \rangle, k + 1) + \sum_{\langle i,j,p \rangle \in R_k^3} \mathsf{dup}(\langle i, j, p \rangle, k + 1)$$

$$= \sum_{\langle i,j,p \rangle \in R_k^1} (\mathsf{dup}(\langle i, j, p \rangle, k) + num_{\langle i,j,p \rangle}[1] - 2(k + 1 - (i + p)))$$

$$+ \sum_{\langle i,j,p \rangle \in R_k^3} num_{\langle i,j,p \rangle}[1]$$

$$= f_{k+1} + \sum_{\langle i,j,p \rangle \in R_k^1} (\mathsf{dup}(\langle i, j, p \rangle, k) - 2(k + 1 - (i + p)))$$

$$= f_{k+1} + \sum_{\langle i,j,p \rangle \in R_k} \mathsf{dup}(\langle i, j, p \rangle, k) - \sum_{\langle i,j,p \rangle \in R_k^2} \mathsf{dup}(\langle i, j, p \rangle, k)$$

$$- 2 \sum_{\langle i,j,p \rangle \in R_k^1} (k + 1 - (i + p))$$

$$= f_{k+1} + \mathcal{D}(w, k) - e_k - 2d_{k+1}.$$

Therefore, $\mathcal{D}(w, k + 1)$ can be computed with this recurrence relation $\mathcal{D}(w, k+1) = \mathcal{D}(w, k) + f_{k+1} - e_k - 2d_{k+1}$. We show how to compute m_k, f_k, d_k, e_k. First, $m_1 = 0, f_1 = 0, d_1 = 0, e_1 = 0$ because $R_1 = \emptyset$ and $R_1^2 = \emptyset$. Then, $m_{k+1}, f_{k+1}, d_{k+1}$ can be computed from m_k, f_k, d_k as follows: m_{k+1} can be computed from m_k by adding the number of runs $\langle i, j, p \rangle$ such that $i + p = k + 1$.

Algorithm 1: Compute $\mathcal{C}(w, k)$ for all positions

Input: a string $w[1..n]$ over an ordered alphabet
Output: $\mathcal{C}(w, k)$ for all $k = 1, 2, ..., n$

1 Compute the sorted list L of the runs $\langle i, j, p \rangle \in \mathsf{Runs}(w)$ in increasing order of $i + p$;

2 Compute the sorted list R of the runs $\langle i, j, p \rangle \in \mathsf{Runs}(w)$ in increasing order of $j - p$;

3 **for** *each* $\langle l_1, r_1, p_1 \rangle, \dots, \langle l_{|L|}, r_{|L|}, p_{|L|} \rangle \in$ L **do**

4 | $\mathsf{L}_l[q] \leftarrow l_q$, $\mathsf{L}_r[q] \leftarrow r_q$;

5 **end**

6 **for** *each* $\langle l_1, r_1, p_1 \rangle, \dots, \langle l_{|R|}, r_{|R|}, p_{|R|} \rangle \in$ R **do**

7 | $\mathsf{R}_l[q] \leftarrow l_q$, $\mathsf{R}_r[q] \leftarrow r_q$;

8 **end**

9 $y \leftarrow 1, z \leftarrow 1, m \leftarrow 0, d \leftarrow 0, f \leftarrow 0, \mathcal{D}(w, 0) \leftarrow 0$;

10 **for** *all* $k = 1, \dots, n$ **do**

11 | $e \leftarrow 0$;

12 | $d \leftarrow d + m$;

13 | **while** $\mathsf{L}_l[y] = k$ **do**

14 | | $f \leftarrow f + \mathsf{L}_r[y] - \mathsf{L}_l[y] + 1$;

15 | | $m \leftarrow m + 1$;

16 | | $y \leftarrow y + 1$;

17 | **end**

18 | **while** $\mathsf{R}_r[z] = k$ **do**

19 | | $f \leftarrow f - (\mathsf{R}_r[z] - \mathsf{R}_l[z] + 1)$;

20 | | $m \leftarrow m - 1$;

21 | | $z \leftarrow z + 1$;

22 | | $e \leftarrow e - (\mathsf{R}_r[z] - \mathsf{R}_l[z] + 1)$;

23 | **end**

24 | $d \leftarrow d - e$;

25 | $\mathcal{D}(w, k) \leftarrow \mathcal{D}(w, k - 1) + f - 2d - e$;

26 | $\mathcal{C}(w, k) \leftarrow k(n - k + 1) - \mathcal{D}(w, k)$;

27 **end**

A pseudo-code of the proposed algorithm is shown in Algorithm 1. Below, we describe our algorithm.

For each run $r = \langle i, j, p \rangle$, we call the interval $[i + p, j - p]$ the run interval for r. To find runs by the starting and ending positions of their run intervals, we create two sorted lists L and R of pairs composed of positions and runs. The list L (resp. R) is sorted by the positions, which are the starting positions (resp. the ending positions) of the run intervals of the respective runs. L and R help us to access a run in amortized constant time, when we process the string positions $k = 1, \dots, n$ in increasing order. The sorted lists L and R can be computed in linear time with an integer sorting algorithm.

f_{k+1} can be computed from f_k by adding $num_{i,j,p}[1]$ for runs $\langle i, j, p \rangle$ such that $i + p = k + 1$ and subtracting $num_{i,j,p}[1]$ for runs $\langle i, j, p \rangle$ such that $j - p = k$.

We have $d_k = \sum_{\langle i,j,p \rangle \in R_k} (k - (i + p))$ and $d_{k+1} = \sum_{\langle i,j,p \rangle \in R_{k+1}} (k + 1 - (i + p))$, and therefore the sum increases by $|R_k| = m_k$ and decreases by $\sum_{\langle i,j,p \rangle \in R_k^2} (k +$

$1 - (i+p)) = \sum_{\langle i,j,p \rangle \in R_k^2} (j - p + 1 - (i+p)) = \sum_{\langle i,j,p \rangle \in R_k^2} num_{i,j,p}[1] = e_k$. This is why d_{k+1} can be computed by recurrence relation $d_{k+1} = d_k + m_k - e_k$.

Finally, e_k can be directly computed by summing the last term of $num_{\langle i,j,p \rangle}$ for runs $\langle i,j,p \rangle$ such that $j - p = k$.

4   Computing $\mathcal{N}(w, k)$ for All Positions k in w

We show the following result.

Theorem 4. *Given a string $w[1..n]$ of length n over a linearly-sortable alphabet, we can sequentially output $\mathcal{N}(w, 1), \ldots, \mathcal{N}(w, n)$ such that the first value needs $O(n)$ time, but all subsequent values need constant-time delay.*

Let $A_x = \mathsf{Substr}(w[1..x])$ and $B_x = \mathsf{Substr}(w[x..n])$. Then, $\mathcal{N}(w, x) = |A_{x-1} \cup B_{x+1}|$. The idea is to compute, for increasing values of x, the two differences $|A_x \cup B_{x+1}| - |A_{x-1} \cup B_{x+1}|$ and $|A_x \cup B_{x+2}| - |A_x \cup B_{x+1}|$ so that $\mathcal{N}(w, x+1) = |A_x \cup B_{x+2}|$ can be computed from $\mathcal{N}(w, x)$ by adding these differences. If we can find the two differences in constant time for each x, then we can solve the addressed problem using the $O(n)$ textbook algorithm for $\mathcal{N}(w, 1)$.

We make use of the following two data structures that can be built in $O(n)$ time. The *longest previous non-overlapping factor table* (LPnF) of w is an integer array $\mathsf{LPnF}_w[1..n]$ whose i-th integer is the length of the longest prefix of $w[i..n]$ that has an occurrence in $w[1..i-1]$. The *longest next factor table* (LNF) of w is an integer array $\mathsf{LNF}_w[1..n]$ whose i-th integer is the length of the longest prefix of $w[i..n]$ that has an occurrence in $w[i+1..n]$.

Lemma 5. *([2,4]) We can build LPnF_w in $O(n)$ time.*

Lemma 6. *We can build LNF_w in $O(n)$ time.*

Proof. First, we build the suffix tree STree over w by Theorem 3. Next, we select sequentially the leaves of STree in ascending order with respect to their suffix numbers. For each such leaf λ with suffix number i, we move to its parent, write its string depth into $\mathsf{LNF}[i]$, delete λ, and continue the iteration. We keep the invariant that an internal node always has two children. In case that we deleted the penultimate leaf of a node, we merge this node with its remaining child. □

We first claim that having the arrays $\mathsf{LNF}_w, \mathsf{LPnF}_w$ for w at hand, $|A_x \cup B_{x+2}| - |A_x \cup B_{x+1}|$ can be computed in $O(1)$ time. Since $A_x \cup B_{x+2} \subseteq A_x \cup B_{x+1}$, we only need to count how many elements are removed, which must be prefixes of $w[x+1..n]$. The removed prefixes are the prefixes of $w[x+1..n]$ that do not occur in A_x and do not occur in B_{x+2}. From the definitions, $\alpha = \mathsf{LPnF}_w[i+1]$ is the length of the longest prefix of $w[x+1..n]$ that has an occurrence in $w[1..x]$ thus included in A_x, and $\beta = \mathsf{LNF}_w[i+1]$ is the length of the longest prefix of $w[x+1..n]$ that has an occurrence in $w[x+2..n]$ thus included in B_{x+2}. Therefore, the prefixes of $w[x+1..n]$ are removed if and only if they are longer than $\max(\alpha, \beta)$, and their number is $n - x - \max(\alpha, \beta)$.

The case for $|A_x \cup B_{x+1}| - |A_{x-1} \cup B_{x+1}|$ is symmetric and can be computed using the arrays LNF_{w^R} and LPnF_{w^R} for the reverse string w^R in a similar fashion.

Acknowledgments. This work was supported by JSPS KAKENHI Grant Numbers JP23K24808, JP23K18466 (SI), JP23H04378, JP25K21150 (DK), JP24K02899 (HB).

References

1. Bannai, H., I, T., Inenaga, S., Nakashima, Y., Takeda, M., Tsuruta, K.: The "runs" theorem. SIAM J. Comput. **46**(5), 1501–1514 (2017)
2. Crochemore, M., Ilie, L.: Computing longest previous factor in linear time and applications. Inf. Process. Lett. **106**(2), 75–80 (2008). https://doi.org/10.1016/j.ipl.2007.10.006
3. Crochemore, M., Lecroq, T., Rytter, W.: 125 Problems in Text Algorithms. Cambridge University Press, Cambridge (2021)
4. Crochemore, M., Tischler, G.: Computing longest previous non-overlapping factors. Inf. Process. Lett. **111**(6), 291–295 (2011). https://doi.org/10.1016/j.ipl.2010.12.005
5. Ellert, J., Fischer, J.: Linear time runs over general ordered alphabets. In: ICALP 2021. LIPIcs, vol. 198, pp. 63:1–63:16 (2021)
6. Farach-Colton, M., Ferragina, P., Muthukrishnan, S.: On the sorting-complexity of suffix tree construction. J. ACM **47**(6), 987–1011 (2000)
7. Fine, N.J., Wilf, H.S.: Uniqueness theorems for periodic functions. Proc. Am. Math. Soc. **16**(1), 109–114 (1965)
8. Kempa, D., Prezza, N.: At the roots of dictionary compression: string attractors. In: STOC 2018, pp. 827–840. ACM (2018)
9. Kolpakov, R.M., Kucherov, G.: Finding maximal repetitions in a word in linear time. In: FOCS 1999, pp. 596–604 (1999)
10. Weiner, P.: Linear pattern matching algorithms. In: 14th Annual Symposium on Switching and Automata Theory, pp. 1–11 (1973)

Faster Algorithm for Bounded Damerau–Levenshtein Distance

Ryosuke Yamano[1,2]($\boxtimes$)(iD) and Tetsuo Shibuya[2](iD)

[1] Department of Computer Science, Graduate School of Information Science and
Technology, The University of Tokyo, Bunkyō, Japan
ryoyamano15@g.ecc.u-tokyo.ac.jp
[2] Division of Medical Data Informatics, Human Genome Center, Institute of Medical
Science, The University of Tokyo, Minato, Japan
tshibuya@hgc.jp

Abstract. The Damerau–Levenshtein distance between two strings is
the minimum number of insertions, deletions, substitutions, and adjacent transpositions required to transform one string into the other. Unlike
the standard Levenshtein distance, it accounts for the common typing
error of adjacent character swaps. When edits are restricted so that no
substring is edited more than once, existing algorithms for Levenshtein
distance can be extended with relatively minor changes to support transpositions. However, in the unrestricted setting (i.e., edits may overlap or
interact arbitrarily), the problem becomes significantly more complex,
and existing techniques no longer apply directly. In this work, we show
that even in the unrestricted setting, the Damerau–Levenshtein distance
can be computed efficiently. We present two algorithms that extend the
classic $O(n + k^2)$ edit-distance frameworks of Myers (Algorithmica '86)
and Landau and Vishkin (JCSS '88), adapting them to accommodate
unrestricted transpositions. The first algorithm runs in $O(\sigma n + k^2)$ time,
where σ is the alphabet size. The second achieves $O(n + k^2 \log n)$ time
for integer alphabets. Here, n is the length of the input strings and
k is the distance threshold. Experimental results show that our algorithms achieve substantial speedups over k-independent methods when
k is small.

Keywords: Edit distance · Damerau–Levenshtein distance ·
Approximate string matching

1 Introduction

One of the standard methods for quantifying the similarity between two strings
A and B is to determine the minimum number of edit operations needed to
transform A into B. Depending on which operations are permitted, this leads to
various string distance measures. When the allowed operations include inserting
a character into A, deleting a character from A, substituting a character in A
with a different one, and transposing two adjacent characters in A, the resulting

© The Author(s), under exclusive license to Springer Nature Switzerland AG 2026
G. Badkobeh et al. (Eds.): SPIRE 2025, LNCS 16073, pp. 291–303, 2026.
https://doi.org/10.1007/978-3-032-05228-5_23

metric is known as the *Damerau–Levenshtein (DL) distance* [7]. Several well-known distance measures can be obtained by restricting the set of permitted operations in this general framework. For instance, the *Levenshtein distance* [13] allows only insertion, deletion, and substitution. The *Hamming distance* [21] considers substitutions alone and requires that the two strings have the same length. When only insertions and deletions are allowed, the resulting metric corresponds to the *Longest Common Subsequence distance* [1,18].

The DL distance has proven to be a versatile tool in various computational domains. Its applications range from classical tasks such as spelling error correction [2,5,14] and packet trace analysis [6], to more recent challenges in bioinformatics, including gene function prediction [15], clustering of RNA-seq reads [3], and code design for DNA-based storage [9].

Lowrance and Wagner [23] extended the DL distance by allowing each edit operation to have a distinct cost. Under the assumption that $2T \geq I + D$, where T, I, and D denote the costs of transposition, insertion, and deletion respectively, their algorithm achieves a time and space complexity of $O(mn)$, where m and n are the lengths of the input strings ($m \leq n$). Since the standard DL distance assigns unit cost to all operations, i.e., $T = D = I = 1$, this condition is satisfied and their algorithm is directly applicable.

Zhao and Sahni reduced the space usage to $O(\sigma m + n)$ [24], where σ is the size of the alphabet, and to $O(n)$ [25] while preserving the $O(mn)$ time complexity.

To reduce computational complexity, Oommen and Loke [20] introduced a restriction whereby no substring may be edited more than once. The DL distance computed under this constraint is referred to as the *restricted DL distance*. For clarity, we refer to the standard DL distance, where repeated edits on the same substring are allowed, as the *unrestricted DL distance*.

To illustrate the effect of this restriction, consider the example from [4]. The unrestricted DL distance between the strings "ba" and "acb" is two, via the sequence: ba → ab → acb, where the first step is a transposition and the second is an insertion. However, in the restricted model, the insertion of "c" modifies a substring that has already been edited by the transposition, which is disallowed. As a result, the shortest valid sequence becomes: ba → aa → ac → acb, requiring three operations. It is worth noting that the restricted DL distance is not a metric, since it does not satisfy the triangle inequality, whereas the unrestricted DL distance is a metric [2]. The restriction limits the effect of transpositions and simplifies the computation to the extent that algorithms for computing the Levenshtein distance can be adapted with only minor modifications.

In many practical applications, the exact value of the DL distance is only of interest when it is below a certain threshold k. For instance, in tasks such as approximate string matching or sequence clustering, strings with a distance exceeding k can often be safely disregarded or treated as dissimilar. Therefore, algorithms that efficiently determine whether the DL distance is at most k, and compute its exact value in that case, are of particular utility.

For the Levenshtein distance, there exist bounded algorithms that run in $O(km)$ time [22] and in $O(n + k^2)$ time [12,16]. These algorithms can also be applied to compute the restricted DL distance with minor modifications [10]. However, for the unrestricted DL distance, efficient algorithms tailored for small k remain largely unexplored. In this work, we address this gap by proposing a new algorithm that efficiently computes the unrestricted DL distance when it is bounded by a small threshold k. Our main results are summarized in Theorem 1.

Definition 1 (Bounded DL distance problem). *Let A and B be two strings of lengths m and n, over an alphabet of size σ, where $m \leq n$. Let k be a distance threshold. Compute the DL distance between the two strings A and B, or report that it exceeds k.*

Theorem 1. *We propose two algorithms that solve the Bounded DL distance problem:*

 - *An algorithm running in $O(\sigma n + k^2)$ time and space. For constant-size alphabets, the complexity becomes $O(n + k^2)$.*
 - *An algorithm running in $O(n + k^2 \log n)$ time and $O(n + k^2)$ space, designed for integer alphabets.*

2 Preliminaries

In this section, we define notations and review the main idea of computing DL distance and the $O(n + k^2)$ algorithm for computing Levenshtein distance.

2.1 Notations

We consider strings over an integer alphabet Σ of size σ (i.e., $|\Sigma| = \sigma$). For a string T, let $T[i]$ be its i-th character, where $1 \leq i \leq |T|$. For two integers $1 \leq i \leq j \leq |T|$, let $T[i..j]$ denote the substring of T that begins at position i and ends at position j. When $i > j$, let $T[i..j]$ be the empty string ε. Let $T[1..i] = T[..i], T[i..|T|] = T[i..]$ and $T[i..j] = T[i..j + 1) = T(i - 1..j]$. $T^R = T[n]T[n - 1] \ldots T[1]$ denotes the reverse string of T, where $n = |T|$. Iverson bracket $[X]$ is equal to 1 if condition X is true and is 0 otherwise.

2.2 Levenshtein Distance and DL Distance

Levenshtein distance between strings A and B can be computed using the following textbook dynamic programming solution, where $L(i, j)$ denotes the Levenshtein distance between $A[..i]$ and $B[..j]$.

$$L(i, j) = \min \begin{cases} 0 & \text{if } i = j = 0 \\ L(i - 1, j) + 1 & \text{if } i > 0 \\ L(i, j - 1) + 1 & \text{if } j > 0 \\ L(i - 1, j - 1) + [A[i] \neq B[j]] & \text{if } i, j > 0 \end{cases} \tag{1}$$

The restricted DL distance can be computed by the following, where $RDL(i,j)$ denotes the restricted DL distance between $A[..i]$ and $B[..j]$.

$$RDL(i,j) = \min \begin{cases} 0 & \text{if } i = j = 0 \\ RDL(i-1,j) + 1 & \text{if } i > 0 \\ RDL(i,j-1) + 1 & \text{if } j > 0 \\ RDL(i-1,j-1) + [A[i] \neq B[j]] & \text{if } i,j > 0 \\ RDL(i-2,j-2) + [A[i] \neq B[j]] & \text{if } i,j > 1 \text{ and} \\ & \qquad A^R[i-1..i] = B[j-1..j] \end{cases}$$

(2)

Removing the restriction of repeated edits to substrings adds significant complexity. As shown by Lowrance and Wagner [23], the unrestricted DL distance can be calculated as the restricted edit distance by adding an edit operation.

Theorem 2 *([23]). Suppose we allow the following edit operations:*

- *unit-cost insertions, deletions, and substitutions of single characters, and*
- *generalized transpositions of the form $aUb \rightarrow bVa$, where $a, b \in \Sigma$ are characters and $U, V \in \Sigma^*$ are (possibly empty) substrings, with cost $|U| + |V| + 1$.*

Then, the minimum total cost of transforming A into B using these operations, under the constraint that no substring is edited more than once, is equal to the unrestricted DL distance between A and B.

Theorem 2 leads to the following recursion, where $DL(i,j)$ denotes the unrestricted DL distance between $A[..i]$ and $B[..j]$.

$$DL(i,j) = \min \begin{cases} 0 & \text{if } i = j = 0 \\ DL(i-1,j) + 1 & \text{if } i > 0 \\ DL(i,j-1) + 1 & \text{if } j > 0 \\ DL(i-1,j-1) + [A[i] \neq B[j]] & \text{if } i,j > 0 \\ \min_{\substack{0 < i' < i, 0 < j' < j \\ A[i]=B[j'], A[i']=B[j]}} DL(i'-1,j'-1) + (i-i') + (j-j') - 1 \end{cases}$$

(3)

2.3 $O(n + k^2)$ Algorithm for Levenshtein Distance

The algorithm of Myers [16] and Landau and Vishkin [12] manages to compute the Levenshtein distance in $O(n + k^2)$ time. They utilize the monotonicity and greedy extension.

Lemma 1 *([22]). For all indices (i,j): $L(i,j) - L(i-1,j-1) \in \{0,1\}$.*

Lemma 1 implies the greedy extension $L(i,j) = L(i-1,j-1)$ when $A[i] = B[j]$. Since $L(i,j) > k$ whenever $|i-j| > k$, we only need to consider the

$2k + 1$ central diagonals. Furthermore, on each of these diagonals, we only need to evaluate the k positions where the value may increase, as characterized by the monotonicity property in Lemma 1. This implies that only $O(k^2)$ entries need to be computed. These values are computed in non-decreasing order with respect to their distance values, rather than in the lexicographic order of the indices.

Definition 2 (*d*-**diagonal**) *For each integer d, we define the d-diagonal as the set of index pairs (i, j) such that $j = i + d$. This corresponds to the diagonal line in the dynamic programming matrix where the column index exceeds the row index by d.*

Let $Lrow^h(d)$ denote the largest row index of a point on the d-diagonal that can be reached with h edit operations. Let $Slide_d(i)$ denote the slide in the d-diagonal starting on row i, which is $Slide_d(i) := \max\{q : A(i..q] = B(i+d..q+d]\}$. Then, the following recurrence holds.

$$Lrow^0(0) = Slide_0(0),$$
$$Lrow^h(d) = Slide_d(\tag{4}$$
$$\max\{Lrow^{h-1}(d+1) + 1, Lrow^{h-1}(d) + 1, Lrow^{h-1}(d-1)\}).$$

The restricted DL distance can be computed by considering $Lrow^{h-1}(d) + 2$ as another candidate for the maximum, whenever the condition $A^R[i..i + 1] = B[j..j + 1]$ holds, where $i = Lrow^{h-1}(d)$ and $j = i + d$.

However, in the case of the unrestricted DL distance, unlike the restricted model, we must consider edit operations with non-unit cost, such as transforming substrings of the form aUb into bVa as was shown in Theorem 2. Since the lengths of U and V may vary, the structure of these operations is inherently more complex. Consequently, a straightforward extension of the recurrence (4) used for the Levenshtein or restricted DL distance is not sufficient.

To address this challenge, we prove an extended version of the greedy extension lemma (see Lemma 4) and change the dynamic programming update strategy from pull-based to push-based.

3 Key Properties

We prove the next lemma for utility.

Lemma 2 *The unrestricted DL distance between strings A and B is equal to that between their reversals, A^R and B^R.*

Proof Let $\mathcal{E}$ be a sequence of edit operations that transforms A into B with minimal cost. We construct a sequence $\mathcal{E}^R$ that transforms A^R into B^R by applying the reverse of each operation in $\mathcal{E}$, in the reverse order. Substitutions, insertions, and deletions are closed under reversal. For generalized transpositions, an operation of the form $aUb \to bVa$ in $\mathcal{E}$ becomes $bU^Ra \to aV^Rb$ in $\mathcal{E}^R$. Therefore, for any edit path from A to B, there exists a corresponding reversed path from A^R to B^R with the same cost, and vice versa. Hence, the distances are equal. $\qquad\square$

Now we are ready to prove monotonicity and greedy extension.

Lemma 3 *For all indices* (i, j): $DL(i, j) - DL(i - 1, j - 1) \in \{0, 1\}$.

Proof By the recurrence (3), one of the candidates for computing $DL(i, j)$ is $DL(i-1, j-1) + [A[i] \neq B[j]]$. Therefore, we have $DL(i, j) \leq DL(i-1, j-1) + 1$. It remains to show the inequality $DL(i - 1, j - 1) \leq DL(i, j)$. We prove this by induction on $i + j$. The optimal edit path to (i, j) must come from the following four cases:

Case 1. From $(i - 1, j - 1)$, which corresponds to match $A[i] = B[j]$ (no edit operations). In this case, $DL(i, j) = DL(i - 1, j - 1)$, so the inequality holds.

Case 2. From $(i - 1, j)$, which corresponds to deletion of $A[i]$. In this case,

$$
\begin{aligned}
DL(i, j) &= DL(i - 1, j) + 1 \\
&\geq DL(i - 2, j - 1) + 1 \qquad \text{from induction hypothesis} \\
&\geq DL(i - 1, j - 1). \qquad\qquad \text{from recurrence(3)}
\end{aligned}
$$

Case 3. From $(i, j - 1)$, which corresponds to insertion of $B[j]$. This is symmetrical to Case 2.

Case 4. From (x, y) such that $x \leq i - 2$, $A[x+1] = B[j]$ and $y \leq j - 2$, $B[y+1] = A[i]$, corresponding to the generalized transposition $A(x..i) \to B(y..j)$. Since the substring $A(x..i - 1]$ can be transformed into $B(y..j - 1]$ using a sequence of standard edit operations (insertions, deletions, and substitutions) with total cost at most $\max\{i - 1 - x, j - 1 - y\}$, we have:

$$
\begin{aligned}
DL(i, j) &= DL(x, y) + (i - x - 2) + (j - y - 2) + 1 \\
&= DL(x, y) + (i - 1 - x) + (j - 1 - y) - 1 \\
&= DL(x, y) + \max\{i - 1 - x, j - 1 - y\} \\
&\quad + \min\{i - 1 - x, j - 1 - y\} - 1 \\
&\geq DL(x, y) + \max\{i - 1 - x, j - 1 - y\} \qquad \text{from } x \leq i - 2, y \leq j - 2 \\
&\geq DL(i - 1, j - 1).
\end{aligned}
$$

Hence, in all cases, the inequality holds, which completes the proof.   □

Lemma 4 *For all indices* $i' < i$ *and* $j' < j$, *if* $A[i' + 1] = B[j' + 1]$, *then the unrestricted DL distance between the substrings* $A(i'..i]$ *and* $B(j'..j]$ *is equal to the unrestricted DL distance between* $A(i' + 1..i]$ *and* $B(j' + 1..j]$.

Proof The unrestircted DL distance between $A(i'..i]$ and $B(j'..j]$ is equal to that between $A^R(i'..i]$ and $B^R(j'..j]$ from Lemma 2, and if $A[i' + 1] = B[j' + 1]$, it is equal to that between $A^R(i' + 1..i]$ and $B^R(j' + 1..j]$ from Lemma 3, and it is equal to that between $A(i' + 1..i]$ and $B(j' + 1..j]$ from Lemma 2.

We next observe that it suffices to consider a restricted form of the generalized transposition $aUb \to bVa$.

Lemma 5 *The set of operations $\{aUb \to bVa\}$ described in Theorem 2 can be restricted to the subset where either $U = \varepsilon$ or $V = \varepsilon$ without affecting the computed distance.*   □

Proof When adjacent character transpositions are not allowed, the transformation $aUb \to bVa$ can alternatively be performed using a sequence of standard edit operations (insertions, deletions, and substitutions) with total cost at most $\max\{|U|, |V|\} + 2$. On the other hand, the cost of the generalized transposition is defined as $|U| + |V| + 1$. When both $|U| \geq 1$ and $|V| \geq 1$, we have

$$\max\{|U|, |V|\} + 2 \leq |U| + |V| + 1.$$

Therefore, in such cases, using standard edits is no more expensive than applying the generalized transposition. Hence, it suffices to consider only those generalized transpositions where at least one of U or V is empty.   □

Lemma 6 *In the sequence of operations transforming A into B, we can assume that all generalized transpositions have the following structure.*
If $A[i] = B[j + 1]$ and $A[i] \neq B[j]$, then the edit operation that transforms the substring $A[i..x]$ into $B[j..j + 1]$, where $i + 1 \leq x$ and $A[x] = B[j]$, can be restricted to the minimal such x.
Similarly, if $A[i + 1] = B[j]$ and $A[i] \neq B[j]$, then the edit operation that transforms $A[i..i + 1]$ into $B[j..y]$, where $j + 1 \leq y$ and $B[y] = A[i]$, can be restricted to the minimal such y.

Proof We prove the first case; the second follows by symmetry. Let $x_{\min}$ be the smallest x such that $i + 1 \leq x$ and $A[x] = B[j]$. Consider any $x' > x_{\min}$ with $A[x'] = B[j]$. The generalized transposition $A[i..x'] \to B[j..j + 1]$ incurs a cost of $x' - i$. Alternatively, we can perform the transposition $A[i..x_{\min}] \to B[j..j+1]$ with cost $x_{\min} - i$, and then delete $A(x_{\min}..x']$ with cost $x' - x_{\min}$, resulting in the same total cost $x' - i$. Therefore, using the minimal x incurs no additional cost, and it suffices to restrict attention to such x.   □

4   Our Algorithm

Since $DL(i, j) > k$ whenever $|i - j| > k$, we only need to consider the d-diagonals with $-k \leq d \leq k$. For each of these $2k+1$ diagonals, there are at most k positions where $DL(i, i + d) < DL(i + 1, i + d + 1) \leq k$, by Lemma 3. From Lemma 4, it suffices to consider edit operations only at these $O(k^2)$ positions, which leads to the push-based update strategy. Furthermore, by Lemma 5 and Lemma 6, the generalized transposition can be restricted to two options. These observations lead to Algorithm 1.

Definition 3 ($DLrow^h(d)$). *Let $DLrow^h(d)$ denote the maximum row index on the d-diagonal that can be reached with cost h.*

Definition 4 ($LCP(i, j)$). *Let $LCP(i, j)$ denote the length of the longest common prefix between suffixes $A(i..]$ and $B(j..]$.*

Algorithm 1. Calculate Bounded DL distance

1: $DLrow^0(0) \leftarrow LCP(0,0)$
2: **for** $h \leftarrow 0$ to $k-1$ **do**
3:    **for** $d \leftarrow -h$ to h **do**
4:        $i \leftarrow DLrow^h(d)$, $j \leftarrow i+d$ ▷ Push from index (i,j)
5:        $DLrow^{h+1}(d) \leftarrow \max\{DLrow^{h+1}(d), i+1+LCP(i+1,j+1)\}$
6:        $DLrow^{h+1}(d-1) \leftarrow \max\{DLrow^{h+1}(d-1), i+1+LCP(i+1,j)\}$
7:        $DLrow^{h+1}(d+1) \leftarrow \max\{DLrow^{h+1}(d+1), i+LCP(i,j+1)\}$
8:        **if** $A[i+1] = B[j+2]$ **then**
9:            $x \leftarrow \min x$ s.t. $x > i+1$ and $A[x] = B[j+1]$ ▷ $A(i..x] \rightarrow B(j..j+2]$
10:           $DLrow^{h+x-i-1}(j-x+2)$
11:               $\leftarrow \max\{DLrow^{h+x-i-1}(j-x+2), x+LCP(x,j+2)\}$
12:       **else if** $A[i+2] = B[j+1]$ **then**
13:           $y \leftarrow \min y$ s.t. $y > j+1$ and $B[y] = A[i+1]$ ▷ $A(i..i+2] \rightarrow B(j..y]$
14:           $DLrow^{h+y-j-1}(y-i-2)$
15:               $\leftarrow \max\{DLrow^{h+y-j-1}(y-i-2), i+2+LCP(i+2,y)\}$
16:       **end if**
17:    **end for**
18: **end for**
19: **return** $\min h$ s.t. $DLrow^h(|B|-|A|) = |A|$, or report that there is no such $h \leq k$

Substitution of $A[i+1]$ with $B[j+1]$ corresponds to line 5 of Algorithm 1. This can be viewed as taking an edge of cost 1 from position (i,j) to $(i+1,j+1)$ in the edit graph, followed by a greedy extension along the longest common prefix. Deletion of $A[i+1]$, corresponding to line 6, can be interpreted as an edge of cost 1 from (i,j) to $(i+1,j)$, followed by greedy extension. Insertion of $B[j+1]$, corresponding to line 7, can likewise be seen as an edge of cost 1 from (i,j) to $(i,j+1)$, followed by greedy extension.

The generalized transposition is restricted to two forms, $aUb \rightarrow ba$ and $ab \rightarrow bVa$, as shown in Lemma 5. Lines 8–11 of Algorithm 1 correspond to the former case, where the substring $A(i..x]$ is transformed into $B(j..j+2]$. The correctness of considering only the minimum x comes from Lemma 6. This can be interpreted as taking an edge of cost $x-i-1$ from (i,j) to $(x,j+2)$ in the edit graph, followed by a greedy extension. Lines 12–15 correspond to the latter form, transforming $A(i..i+2]$ into $B(j..y]$. This is equivalent to taking an edge of cost $y-j-1$ from (i,j) to $(i+2,y)$, followed by a greedy extension.

The longest common prefix can be queried in constant time by precomputation of $O(n)$ time, using the linear time construction of suffix trees for integer alphabets [8], where $n = \max\{|A|, |B|\}$. For the $O(k^2)$ entries of the dynamic programming table $DLrow^h(d)$, most of the push-based update for each entry can be done in constant time. The remaining part is the calculation of x and y for generalized transposition, which corresponds to lines 9 and 13 of Algorithm 1. The calculation can be generalized as the following problem shown in Definition 5.

Algorithm 2. Precompute Next Occur Problem

1: initialize all the entries of *next* with $|T| + 1$
2: **for** $i \leftarrow |T|$ to 1 **do**   ▷ traverse i in reverse order
3:    **for** $c \leftarrow 1$ to σ **do**
4:       $next(i, c) \leftarrow next(i + 1, c)$
5:    **end for**
6:    $next(i, T[i]) \leftarrow i$
7: **end for**
8: **return** *next*

Definition 5 (Next Occur Problem). *Given string T, index i, and character c. Find the minimum index x such that $x \geq i$ and $T[x] = c$.*

We define $next(i, c)$ as the smallest index $x \geq i$ such that $T[x] = c$, or $|T| + 1$ if no such index exists. For convenience, we also define $next(|T| + 1, c) = |T| + 1$ for all characters c. All σn entries can be precomputed in $O(\sigma n)$ time using a simple dynamic programming approach (see Algorithm 2). We can construct such tables for both strings A and B in $O(\sigma n)$ time and space. By querying these precomputed tables, the values of x and y in lines 9 and 13 of Algorithm 1 can be obtained in constant time. Therefore, we obtain an $O(\sigma n + k^2)$ time and space algorithm for solving the bounded DL distance problem, which runs in $O(n + k^2)$ time and space for constant-sized alphabets.

For integer alphabets, however, Algorithm 2 may take $O(n^2)$ time and space in the worst case when $\sigma = O(n)$. In this case, we instead precompute, for each character c, the list of indices $\{i \mid T[i] = c\}$. We precompute such lists for both strings A and B in $O(n)$ time and space. Then, for a given query index i and character c, we can perform a binary search over the list for character c to find the smallest x that is no smaller than i in $O(\log n)$ time. This results in an $O(n + k^2 \log n)$ time and $O(n + k^2)$ space algorithm for the bounded DL distance problem, which leads to Theorem 1.

5 Experiments

We conducted experiments to demonstrate the practical effectiveness of our algorithm. The source code is available at: https://github.com/CoCo-Japan-pan/bounded_Damerau-Levenshtein. To compute longest common prefixes, we used suffix arrays [19] and longest common prefix (LCP) arrays [11] instead of suffix trees. For simplicity of implementation and favorable constant factors, we adopted the sparse table approach for Range Minimum Queries (RMQ), which requires $O(n \log n)$ preprocessing time and supports constant-time queries. Our algorithm is implemented in Rust and compiled using rustc 1.87.0 in release mode. We compared our implementation against the algorithm of Lowrance and Wagner [23], and the linear space algorithm for constant-sized alphabets by Zhao and Sahni [24], for which the authors provide source code. Zhao and Sahni propose several algorithms; here, we use the fastest one, which they call the *Strip*

Damerau-Levenshtein distance algorithm, and set the strip width parameter to 1024 following their instructions. Their implementation, written in C, was compiled using gcc 13.3.0 with the -O3 option. We ran all experiments on Ubuntu 24.04 (WSL 2) with an Intel(R) Core(TM) Ultra 7 155H CPU and 64 GiB of memory.

k	$n = m = 10^4$				$n = m = 10^5$			
	$\sqrt{n}$	$0.1n$	$0.2n$	$0.5n$	$\sqrt{n}$	$0.1n$	$0.2n$	$0.5n$
Ours	0.00287	0.03329	0.13374	0.85088	0.03841	6.11260	32.317	280.23
LW	0.38631				52.595			
ZS	0.17127				16.640			

Fig. 1. The table reports the average computation time (in seconds) for bounded DL distance computation between pairs of substrings randomly selected from DNA sequences of Escherichia coli. We generated 1000 sequence pairs for $n = m = 10^4$, and 100 pairs for $n = m = 10^5$. "LW" refers to the algorithm by Lowrance and Wagner, and "ZS" denotes the algorithm by Zhao and Sahni.

We conducted experiments using real genome sequences obtained from the NCBI (National Center for Biotechnology Information) server [17]. From the complete genome sequence of Escherichia coli K-12 strain MG1655 (approximately 4.6 Mbp), we randomly select pairs of substrings of lengths 10^4 and 10^5. For each pair, we evaluated the performance of our algorithm under thresholds k set to $\sqrt{n}, 0.1n, 0.2n$, and $0.5n$. For comparison, we also measured the average execution time of the k-independent algorithms by Lowrance and Wagner, and by Zhao and Sahni, under the same conditions without specifying k.

The results are shown in Fig. 1. For $k = \sqrt{n}$ and $k = 0.1n$, our algorithm achieved the fastest performance, while for $k = 0.2n$ and $k = 0.5n$, the algorithm by Zhao and Sahni outperformed ours. Our algorithm uses $O(\sigma n + k^2)$ space, whereas Zhao and Sahni's algorithm requires only $O(\sigma n)$. As k increases, the effect of cache misses becomes more pronounced and cannot be neglected.

We also conducted additional experiments to verify the $O(n)$ running time behavior for the case of $k = \sqrt{n}$. Specifically, we randomly selected pairs of substrings of lengths ranging from 10^3 to 10^8 from the Homo sapiens isolate HG04199 chromosome 1 sequence (approximately 247.5 Mbp), and measured the average execution time. For comparison, we also measured the execution time of Zhao and Sahni's algorithm for lengths up to 10^6, as it became prohibitively slow for longer sequences.

The experimental results are shown in the log-log plot in Fig. 2. From the graph, we observe that our algorithm exhibits linear running time with respect to n for $k = \sqrt{n}$, while Zhao and Sahni's algorithm shows quadratic behavior. While existing algorithms become impractical around $n = 10^7$ for comparing substrings, introducing a threshold $k = \sqrt{n}$ allows our algorithm to remain efficient even for sequences as long as $n = 10^8$.

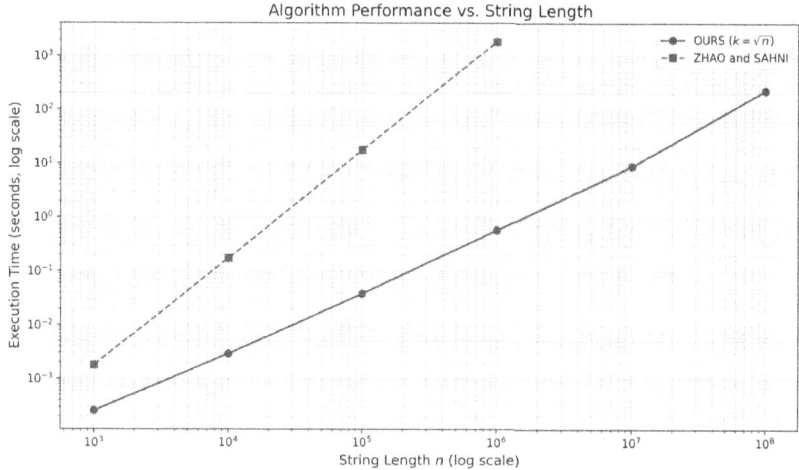

Fig. 2. The log-log plot shows the average computation time (in seconds) for bounded DL distance computation between pairs of substrings randomly selected from DNA sequences of Homo sapiens. For our algorithm, we generated 1000 sequence pairs for $n \leq 10^6$, 100 pairs for $n = 10^7$, and 10 pairs for $n = 10^8$. For Zhao and Sahni's algorithm, we generated 1000 pairs for $n \leq 10^4$, 100 pairs for $n = 10^5$, and 10 pairs for $n = 10^6$.

6   Conclusion

We proposed new algorithms for computing the unrestricted DL distance in $O(n + k^2)$ time for constant-sized alphabets, and in $O(n + k^2 \log n)$ time for integer alphabets. By treating unrestricted transpositions as additional non-unit-cost edit operations, we proved a greedy extension lemma that allows us to restrict attention to only $O(k^2)$ increasing points along the diagonals. Building on this, we employed a push-based update strategy to adapt existing $O(n + k^2)$ time frameworks for the standard Levenshtein distance. Our experimental results demonstrate that the proposed algorithm is practical and significantly faster than previous methods when the threshold k is relatively small.

Acknowledgments. This work was supported by MEXT KAKENHI Grant Numbers 21H05052, 23H03345, and 23K18501.

Disclosure of Interests. The authors have no competing interests to declare that are relevant to the content of this article.

References

1. Apostolico, A., Guerra, C.: The longest common subsequence problem revisited. Algorithmica **2**, 315–336 (1987). https://doi.org/10.1007/BF01840365
2. Bard, G.V.: Spelling-error tolerant, order-independent pass-phrases via the Damerau-Levenshtein string-edit distance metric. Cryptology ePrint Archive (2006). https://eprint.iacr.org/2006/364
3. Biswas, A.K., Gao, J.X.: PR2S2CLUST: patched RNA-SEQ read segments' structure-oriented clustering. J. Bioinform. Comput. Biol. **14**(05), 1650027 (2016). https://doi.org/10.1142/S021972001650027X, pMID: 27455882
4. Boytsov, L.: Indexing methods for approximate dictionary searching: comparative analysis. ACM J. Exp. Algorithmics **16** (2011). https://doi.org/10.1145/1963190.1963191
5. Brill, E., Moore, R.C.: An improved error model for noisy channel spelling correction. In: Proceedings of the 38th Annual Meeting of the Association for Computational Linguistics, pp. 286–293. Association for Computational Linguistics, Hong Kong (2000). https://doi.org/10.3115/1075218.1075255, https://aclanthology.org/P00-1037/
6. Cai, X., Zhang, X.C., Joshi, B., Johnson, R.: Touching from a distance: website fingerprinting attacks and defenses. In: Proceedings of the 2012 ACM Conference on Computer and Communications Security, pp. 605–616. CCS 2012, Association for Computing Machinery, New York, NY, USA (2012). https://doi.org/10.1145/2382196.2382260
7. Damerau, F.J.: A technique for computer detection and correction of spelling errors. Commun. ACM **7**(3), 171–176 (1964). https://doi.org/10.1145/363958.363994
8. Farach-Colton, M., Ferragina, P., Muthukrishnan, S.: On the sorting-complexity of suffix tree construction. J. ACM **47**(6), 987–1011 (2000). https://doi.org/10.1145/355541.355547
9. Gabrys, R., Yaakobi, E., Milenkovic, O.: Codes in the Damerau distance for deletion and adjacent transposition correction. IEEE Trans. Inf. Theory **64**(4), 2550–2570 (2018). https://doi.org/10.1109/TIT.2017.2778143
10. Hyyrö, H.: A bit-vector algorithm for computing Levenshtein and Damerau edit distances. Nordic J. Comput. **10**(1), 29–39 (2003)
11. Kasai, T., Lee, G., Arimura, H., Arikawa, S., Park, K.: Linear-time longest-common-prefix computation in suffix arrays and its applications. In: Amir, A. (ed.) Combinatorial Pattern Matching, pp. 181–192. Springer, Berlin Heidelberg, Berlin, Heidelberg (2001). https://doi.org/10.1007/3-540-48194-X_17
12. Landau, G.M., Vishkin, U.: Fast string matching with k differences. J. Comput. Syst. Sci. **37**(1), 63–78 (1988). https://doi.org/10.1016/0022-0000(88)90045-1
13. Levenshtein, V.I., et al.: Binary codes capable of correcting deletions, insertions, and reversals. In: Soviet Physics Doklady, vol. 10, pp. 707–710. Soviet Union (1966)
14. Li, M., Zhu, M., Zhang, Y., Zhou, M.: Exploring distributional similarity based models for query spelling correction. In: Calzolari, N., Cardie, C., Isabelle, P. (eds.) Proceedings of the 21st International Conference on Computational Linguistics and 44th Annual Meeting of the Association for Computational Linguistics, pp. 1025–1032. Association for Computational Linguistics, Sydney, Australia (2006). https://doi.org/10.3115/1220175.1220304, https://aclanthology.org/P06-1129/
15. Majorek, K.A., et al.: The RNASE h-like superfamily: new members, comparative structural analysis and evolutionary classification. Nucleic Acids Res. **42**(7), 4160–4179 (2014). https://doi.org/10.1093/nar/gkt1414

16. Myers, E.W.: An $O(ND)$ difference algorithm and its variations. Algorithmica **1**(1), 251–266 (1986). https://doi.org/10.1007/BF01840446
17. National Center for Biotechnology Information (NCBI): NCBI Datasets (2025). https://www.ncbi.nlm.nih.gov/datasets/. Accessed 3 June
18. Needleman, S.B., Wunsch, C.D.: A general method applicable to the search for similarities in the amino acid sequence of two proteins. J. Mol. Biol. **48**(3), 443–453 (1970). https://doi.org/10.1016/0022-2836(70)90057-4
19. Nong, G., Zhang, S., Chan, W.H.: Linear suffix array construction by almost pure induced-sorting. In: 2009 Data Compression Conference, pp. 193–202 (2009). https://doi.org/10.1109/DCC.2009.42
20. Oommen, B., Loke, R.: Pattern recognition of strings with substitutions, insertions, deletions and generalized transpositions. Pattern Recogn. **30**(5), 789–800 (1997). https://doi.org/10.1016/S0031-3203(96)00101-X
21. Sankoff, D., Kruskal, J.B. (eds.): Time Warps, String Edits, and Macromolecules: The Theory and Practice of Sequence Comparison. Addison-Wesley, Reading, MA (1983)
22. Ukkonen, E.: Algorithms for approximate string matching. Inf. Control **64**(1), 100–118 (1985). https://doi.org/10.1016/S0019-9958(85)80046-2 international Conference on Foundations of Computation Theory
23. Wagner, R.A., Lowrance, R.: An extension of the string-to-string correction problem. J. ACM **22**(2), 177–183 (1975). https://doi.org/10.1145/321879.321880
24. Zhao, C., Sahni, S.: String correction using the Damerau-Levenshtein distance. BMC Bioinform. **20**, 1–28 (2019). https://doi.org/10.1186/s12859-019-2819-0
25. Zhao, C., Sahni, S.: Linear space string correction algorithm using the Damerau-Levenshtein distance. BMC Bioinform. **21**, 1–21 (2020). https://doi.org/10.1186/s12859-019-3184-8

Author Index

A

Alhadi, Anas   10
Allam, Nour   10
Awofeso, Christine   1

B

Bals, Ben   1
Bannai, Hideo   188, 281
Bansal, Parth   265
Begleiter, Dove   10
Bessière, Chloé   156
Brown, Nathaniel K.   10
Bulteau, Laurent   133

C

Carfagna, Lorenzo   18
Carmona, Gabriel   28

D

De Luca, Alessandro   45
Depuydt, Lore   10
Díaz-Domínguez, Diego   54
Dinklage, Patrick   64
Diseth, Anastasia C.   79
dos Santos, Vinicius   202

F

Fariña, Antonio   95
Fici, Gabriele   45, 133
Fischer, Johannes   64
Fujie, Yuto   109
Fujimaru, Hiroto   124

G

Gabory, Estéban   133
Gagie, Travis   10, 54
Gautheret, Daniel   156
Gawrychowski, Paweł   148

G

Gómez-Brandón, Adrián   95
Gómez-Colomer, Asunción   95
Guerrini, Veronica   54

H

Heljanko, Keijo   79
Hernandez–Courbevoie, Yohan   156

I

Inenaga, Shunsuke   109, 124, 188, 281

J

Jain, Samkith K.   172
Janczewski, Wojciech   148

K

Kai, Kazuki   188
Karpagavalli, Nithin Bharathi Kabilan   10
Khajjayam, Suchith Sridhar   10
Khan, Shabnam   265
Khan, Shahbaz   265
Kishi, Kaisei   188
Kondratovsky, Eitan   248
Köppl, Dominik   281

L

Lachish, Oded   1
Langmead, Ben   10, 54
Limasset, Antoine   156
Lipták, Zsuzsanna   54

M

Manzini, Giovanni   28, 54
Marchet, Camille   156
Masillo, Francesco   54
Mhaskar, Neerja   172
Mieno, Takuya   124, 233
Monteiro, Bruno   202

G. Badkobeh et al. (Eds.): SPIRE 2025, LNCS 16073, pp. 305–306, 2026.
https://doi.org/10.1007/978-3-032-05228-5

N
Nakashima, Yuto   109, 188, 281
Nalbach, Lukas   64
Navarro, Gonzalo   95, 217

P
Pissis, Solon P.   1
Porat, Ely   248
Puglisi, Simon J.   79

R
Romana, Giuseppe   217

S
Salson, Mikaël   156
Sekizaki, Shoma   233
Shalom, B. Riva   248
Shibata, Hiroki   109, 281
Shibuya, Tetsuo   291
Shivakumar, Vikram   54

T
Talera, Saumya   265
Tosoni, Carlo   18

U
Umezaki, Haruki   281
Urbina, Cristian   217

V
Verbeek, Hilde   133

W
Wahed, Hamza   10

X
Xue, Haoliang   156

Y
Yamano, Ryosuke   291

Z
Zakeri, Mohsen   10
Zumbrink, Jan   64

The manufacturer's authorised representative in the EU is Springer
Nature Customer Service Centre GmbH, Europaplatz 3, 69115 Heidelberg,
Germany. If you have any concerns regarding our products, please
contact ProductSafety@springernature.com

Printed and bound by CPI Group (UK) Ltd, Croydon, CR0 4YY

28/04/2026

02098527-0003